Eighth Workshop on Computational Linguistics and Clinical Psychology: Improving Access (CLPsych 2022)

Online
15 July 2022

ISBN: 978-1-7138-5626-9

Printed from e-media with permission by:

Curran Associates, Inc.
57 Morehouse Lane
Red Hook, NY 12571

CURRAN ASSOCIATES INC. proceedings.com

Some format issues inherent in the e-media version may also appear in this print version.

Association for Computational Linguistics
209 N. Eighth Street
Stroudsburg, Pennsylvania 18360

Phone: 1-570-476-8006
Fax: 1-570-476-0860

acl@aclweb.org

Additional copies of this publication are available from:

Curran Associates, Inc.
57 Morehouse Lane
Red Hook, NY 12571 USA
Phone: 845-758-0400
Fax: 845-758-2633
Email: curran@proceedings.com
Web: www.proceedings.com

CLPsych 2022

Eighth Workshop on Computational Linguistics and Clinical Psychology

Proceedings of the Workshop

July 15, 2022

The CLPsych organizers gratefully acknowledge the support from the following sponsors.

Silver

UMIACS

University of Maryland Institute for Advanced Computer Studies

Bronze

Receptiviti

LANGUAGE PSYCHOLOGY API

Copper

REBECCA RESNIK & ASSOCIATES PSYCHOLOGICAL CARE

AMERICAN ASSOCIATION OF SUICIDOLOGY

Order copies of this and other ACL proceedings from:

Association for Computational Linguistics (ACL)
209 N. Eighth Street
Stroudsburg, PA 18360
USA
Tel: +1-570-476-8006
Fax: +1-570-476-0860
acl@aclweb.org

Introduction

Mental health is a pressing concern. Worldwide, mental health conditions are among the leading causes of disability [,], and the global economic cost of mental health issues between 2011 and 2030, including neurological and substance use disorders, is projected to be more than $16 trillion [ ]. In the U.S. in 2020, suicide was in the top nine leading causes of death for people ages 10-64, and the second leading cause of death for people ages 10-14 and 25-34 [ ]. Over the past several years, COVID-19 has created additional challenges to mental health. For instance, Sheridan et al. [ ] found that suicide attempts in young children 10-12 have increased more than five-fold between 2010 and 2020. Furthermore, U.S. Surgeon General Vivek Murthy in 2021 called for a nationwide response to the mental health crisis that youth especially are facing during the pandemic [ ].

For the Eighth Workshop on Computational Linguistics and Clinical Psychology (CLPsych), we adopt the theme "mental health in the face of change". This includes the kind of aspects natural language processing technologies need to address to deliver explainable and fair solutions that can be integrated in the clinical setting. Additionally, how these solutions can capture changes in mood over longitudinal and temporal data, which has been the focus of this year's shared task.

CLPsych was a hybrid workshop that accommodated both in-person and remote participation. It was collocated with NAACL'22, which took place in Seattle, Washington, USA on July 15th, 2022.

Since 2014, CLPsych has been successful in bringing together people from different backgrounds (e.g. mental health experts, clinicians, and computational linguists), to share and discuss their work and results. Its central goal is to build bridges so that these different disciplines can integrate to improve our understanding of mental health issues, and to deliver better mental health treatments and diagnoses to everybody.

CLPsych'22 included a shared task that focused on using longitudinal data to understand mood changes and relate them to risk assessment for suicidality. The shared task was organized by Adam Tsakalidis, Federico Nanni, and Maria Liakata. The overview of the shared task in this volume [ ] discusses the tasks, team approaches and results, and lessons learned.

Our program committee included mental health and technological experts, in order to provide all the papers with more informative feedback that address both aspects. CLPsych'22 received a total of 23 papers for the main workshop, of which 15 were accepted; all 9 submitted shared task papers were also accepted. The organizing committee, with the help of the program committee scores, and feedback chose seven main workshop papers and two shared task papers as oral presentations, and the rest were presented in the poster session.

CLPsych'22 also hosted excellent invited speakers and panelists. Our keynote speakers were Finale Doshi-Velez (Harvard University), Shri Narayanan (University of Southern California), and Elizabeth Shriberg (Ellipsis Health and Johns Hopkins University). The talks were followed by a discussion moderated by April Foreman (Department of Veterans Affairs). Additionally, we hosted invited talks by David Crepaz (Mental Health Foundation in UK), Munmun De Choudhury (Georgia Tech), Mark Dredze (Johns Hopkins University), and Zac Imel (University of Utah). This was followed by a panel moderated by Paul Middlebrooks (creator and host of the Brain Inspired podcast).

The CLPsych organizing committee would like to extend special thanks to all the people that helped make the workshop a success. This includes and is not limited to our authors, shared task participants and organizers, program committee members, and the NORC team that helped in setting up the secure system for the shared task teams. We also would like to thank the North American chapter of the Association for Computational Linguistics for making this workshop possible. Philip Resnik assisted with acquisition of sponsors, shared task data, and general advice. Special thanks to our generous sponsors: University of Maryland Institute for Advanced Computer Studies (silver sponsor), Receptiviti (bronze sponsor), Rebecca Resnik & Associates (copper sponsor), and the American Association of Suicidology (copper sponsor). Their funds helped to support the workshop and its program, and provided support for attendees from underrepresented minorities and/or people with financial difficulties by covering their registration costs.

Ayah Zirikly, Dana Atzil-Slonim, Maria Liakata, Steven Bedrick, Bart Desmet, Molly Ireland, Andrew Lee, Sean MacAvaney, Matthew Purver, Rebecca Resnik, and Andrew Yates

References

[1] David E Bloom, Elizabeth Cafiero, Eva Jané-Llopis, Shafika Abrahams-Gessel, Lakshmi Reddy Bloom, Sana Fathima, Andrea B Feigl, Tom Gaziano, Ali Hamandi, Mona Mowafi, et al. 2012. The global economic burden of noncommunicable diseases. Technical report, Program on the Global Demography of Aging.

[2] Daniel C Ehlman. 2022. Changes in suicide rates—united states, 2019 and 2020. *MMWR. Morbidity and Mortality Weekly Report*, 71.

[3] Christopher JL Murray, Alan D Lopez, World Health Organization, et al. 1996. *The global burden of disease: a comprehensive assessment of mortality and disability from diseases, injuries, and risk factors in 1990 and projected to 2020: summary*. World Health Organization.

[4] Office of the Surgeon General et al. 2021. US surgeon general issues advisory on youth mental health crisis further exposed by COVID-19 pandemic. *HHS.gov*. Retrieved June 7, 2022.

[5] David C Sheridan, Sara Grusing, Rebecca Marshall, Amber Lin, Adrienne R Hughes, Robert G Hendrickson, and B Zane Horowitz. 2022. Changes in suicidal ingestion among preadolescent children from 2000 to 2020. *JAMA pediatrics*.

[6] Adam Tsakalidis, Jenny Chim, Iman Munire Bilal, Ayah Zirikly, Dana Atzil-Slonim, Federico Nanni, Philip Resnik, Manas Gaur, Kaushik Roy, Becky Inkster, Jeff Leintz, and Maria Liakata. 2022. Overview of the CLPsych 2022 shared task: Capturing moments of change in longitudinal user posts. In *Proceedings of the Eighth Workshop on Computational Linguistics and Clinical Psychology: Mental Health in the Face of Change*, Seattle, Washington. Association for Computational Linguistics.

[7] Ma-Li Wong and Julio Licinio. 2001. Research and treatment approaches to depression. *Nature Reviews Neuroscience*, 2(5):343–351.

Organizing Committee

Workshop Co-chairs

Ayah Zirikly, Johns Hopkins University
Dana Atzil-Slonim, Bar-Ilan University
Maria Liakata, QMUL & The Alan Turing Institute

Organizing Committee

Steven Bedrick, Oregon Health & Science University
Bart Desmet, National Institutes of Health
Molly Ireland, Receptiviti
Andrew Lee, University of Michigan
Sean MacAvaney, University of Glasgow
Matthew Purver, QMUL
Rebecca Resnik, Rebecca Resnik and Associates, LLC
Andrew Yates, University of Amsterdam

Shared Task Organizers

Adam Tsakalidis, QMUL & The Alan Turing Institute
Federico Nanni, The Alan Turing Institute
Maria Liakata, QMUL & The Alan Turing Institute

Program Committee

Keynote Speakers

Finale Doshi-Velez, Harvard University
Shri Narayana, University of Southern California
Elizabeth Shriberg, Ellipsis Health and Johns Hopkins University
April Foreman, Department of Veterans Affairs (Moderator)

Invited Speakers and Panelists

David Crepaz, Mental Health Foundation in UK
Munmun De Choudhury, Georgia Tech
Mark Dredze, Johns Hopkins University
Zac Imel, University of Utah
Paul Middlebrooks, Brain Inspired (Moderator)

Program Committee

Carlos Aguirre, Johns Hopkins University
Kfir Bar, Basis Technology
Laura Biester, University of Michigan
Jenny Chim, QMUL
Trevor Cohen, University of Washington
Shauna Concannon, University of Cambridge
Kim De-Jong, University of Leiden
April Foreman, Department of Veterans Affairs
Manas Gaur, University of South Carolina
Keith Harrigian, Johns Hopkins University
Zac Imel, University of Utah/ Lyssn
Loring Ingraham, George Washington University
Lorenzo Lorenzo-Luaces, Indiana University
Sean MacAvaney, University of Glasgow
Adam Miner, Stanford Psychiatry
Sarah Morgan, University of Cambridge
Yaakov Ophir, Technion (Israel)
Rob Procter, University of Warwick
Emily Tucker Prud'hommeaux, Boston College
Brian Roark, Google
Philip Resnik, University of Maryland
Julian Rubel, Giessen University
Frank Rudzicz, University of Toronto
Jonathan Schler, Bar-Ilan University
Andrew Schwartz, Stony Brook University & University of Pennsylvania
Richard Sproat, Google
Rob Stewart, King's College London
Michael Tanana, University of Utah
Adam Tsakalidis, QMUL
Bo Wang, Massachusetts General Hospital

Cody Weston, Johns Hopkins University
Maria Wolters, University of Edinburgh
Elad Yom-Tov, Microsoft Research

Table of Contents

DEPAC: a Corpus for Depression and Anxiety Detection from Speech

Mashrura Tasnim, Malikeh Ehghaghi, Brian Diep, Jekaterina Novikova
{mashrura, malikeh, brian, jekaterina}@winterlightlabs.com
Winterlight Labs
Toronto, Canada

Abstract

Mental distress like depression and anxiety contribute to the largest proportion of the global burden of diseases. Automated diagnosis system of such disorders, empowered by recent innovations in Artificial Intelligence, can pave the way to reduce the sufferings of the affected individuals. Development of such systems requires information-rich and balanced corpora. In this work, we introduce a novel mental distress analysis audio dataset DEPAC, labelled based on established thresholds on depression and anxiety standard screening tools. This large dataset comprises multiple speech tasks per individual, as well as relevant demographic information. Alongside, we present a feature set consisting of hand-curated acoustic and linguistic features, which were found effective in identifying signs of mental illnesses in human speech. Finally, we justify the quality and effectiveness of our proposed audio corpus and feature set in predicting depression severity by comparing the performance of baseline machine learning models built on this dataset with baseline models trained on other well-known depression corpora.

1 Introduction

Effective treatment for psychiatric diseases requires characterizing disease profiles with high accuracy. The traditional schema for diagnosis is based on clustering of non-specific physical and behavioral symptoms, which makes the diagnostic process challenging. For example, in major depressive disorder (MDD), high disease heterogeneity and lack of agreed-upon assessment standards necessitate a high degree of clinical experience and training to make an accurate diagnosis. Both clinician-administered and self-rated clinical assessments for MDD, such as the Hamilton Depression Scale (HAM-D) (Hamilton and Guy, 1976), Montgomery Asberg Depression Scale (MADRS) (Montgomery and Åsberg, 1979), Beck Depression Inventory (BDI) (Beck et al., 1988), and Patient Health Questionnaire (PHQ-9) (Löwe et al., 2004) are suboptimal in many ways. Each assess the illness through different symptom domains, have low construct validity, lack specific behavioral references, and are subjective (Berman et al., 1985; Nemeroff, 2007; Wakefield, 2013). Moreover, participants are often reluctant to fill-out the self rated assessment in regular intervals. These issues can lead to misdiagnosis, which impacts treatment timelines and can lead to poor clinical outcomes.

In contrast, language can be an effective alternative to objectively characterize psychiatric illness. For example, emotion and cognition are both affected in MDD. As a result, depressed patients demonstrate negative emotional bias in memory, attention, and event-interpretation (Mathews and MacLeod, 2005), as well as more general impairment in attention, memory, and decision-making (Cohen et al., 1982; Blanco et al., 2013). These effects are manifested in patients' language in a variety of ways, for example, slowed rate of speech, volume, prosody, as well as increased use of first-person pronouns, negatively valenced speech content, and use of absolute words (Flint et al., 1992; Fineberg et al., 2016). Therefore, automated computational analysis of speech represents an excellent data source to develop digital biomarkers for mental illness. This kind of automated assessment takes only a few minutes of audio recording, therefore is less time-consuming, and would reduce burden on the individuals. However, such model development requires access to datasets of sufficient quality and size.

The recent development of speech-based computational models for measuring depression prevalence and severity has been accelerated by the introduction of Audio-Visual Emotion Recognition Challenge (AVEC) in 2013. A subset of the audio-visual depressive language corpus (AViD-Corpus) was introduced as challenge corpus for 2013 (Val-

Proceedings of the Eighth Workshop on Computational Linguistics and Clinical Psychology, pages 1 - 16
July 15, 2022

star et al., 2013) and 2014 (Valstar et al., 2014) Depression Recognition Sub-Challenge (DSC) of the event. This dataset comprises 150 recordings in German language, divided equally into training, development and test partitions. Predicting depression severity on BDI-II scale was the challenge specified task.

Another popular dataset in this area is the Distress Analysis Interview Corpus (DAIC) (Gratch et al., 2014). It contains semi-structured clinical interviews in English language formulated to support diagnosis of psychological conditions such as anxiety, depression, and post-traumatic stress disorder. Different subsets of this dataset were used as the challenge corpus of AVEC 2016, 2017 and 2019 (Valstar et al., 2016; Ringeval et al., 2017, 2019) where challenge participants reported PHQ-8 scores predicted by their respective regression models.

However, the depression corpora used in previous research suffer from two vital limitations. Firstly, the small sample size in the existing depression datasets increases the risk of overfitting in the machine learning models. For example, the number of recordings in the AVEC challenges available for model training range from 50 to 189, which is far from sufficient. Secondly, the datasets in the previous works lack in linguistic variety, as they only contain a small number (only one or two) of samples per subject. To mitigate these challenges, in this work we introduce the **DEP**ression and **A**nxiety **C**rowdsourced corpus (DEPAC) as a novel dataset that is rich in the diversity of speech tasks and subjects and is tailored to capture the signs of anxiety and depression to make accurate prediction on subjects' psychological state. We also present a set of acoustic and linguistic features extracted from the corpus which incorporates domain knowledge of clinical and machine learning experts. Finally, we benchmark our dataset with several baseline machine learning models that use this set of features, to show that this novel dataset is well-suited for the machine learning-based methods with the goal of generating speech biomarkers for depression.

2 DEPAC Corpus

The DEPAC corpus introduced in this work was collected with the goal of gathering a large training dataset to identify candidate speech and language features that are specific to a given psychiatric disease. Data collection for the corpus was approved by the Institutional Review Board (IRB). This is a proprietary dataset, collected via crowdsourcing and consists of a variety of self-administered speech tasks. The participants completed these tasks using Amazon Mechanical Turk[1] (mTurk), a platform where individuals are paid to complete short tasks online (Paolacci et al., 2010). The speech samples were then manually transcribed and compiled along with participant demographic information into the final corpus.

2.1 Platform and Instrumentation

Once recruited for this study via mTurk, participants were able to remotely complete a range of tasks including surveys and responding to audio prompts. Participants were required to have:

1. A desktop or laptop computer
2. A working microphone
3. Chrome or Mozilla Firefox browser

Amazon facilitated payment between the experimenter and the participant.

2.2 Recruitment and Screening

Participation in the study was voluntary. Participant eligibility was configured to only permit individuals located in Canada and the United States. Amazon verified the location of participants by confirming their address and associated credit card. Locations were used to assess eligibility only.

The platform also restricted participation to individuals with an mTurk approval rating of at least 95%. This preliminary criterion attempted to ensure that participation was restricted to those who had historically consistently followed task instructions.

During the study, participants saw a short description of the task, the approximate length of the task (5 to 8 minutes, depending on the condition they were randomly placed into), and the per-minute payment for their time. Participants were compensated at a rate of $0.16 per minute. This is well above the average payment rate for mTurk tasks and above the recommended rate of $0.10/minute (Chandler and Shapiro, 2016).

As part of our exclusion criteria, individuals with a history of chronic alcohol or drug dependency within the past 5 years, as well as participants with clinically significant vision or hearing impairment, were excluded from the study.

[1] https://www.mturk.com

2.3 Transcription and Quality Assurance

Each audio sample gathered from the mTurk platform was assigned to a trained transcriptionist to follow the protocols and annotation formats detailed in the CHAT manual (MacWhinney, 2000) that was used to transcribe TalkBank, which is the largest open repository of spoken data (MacWhinney, 2007). The transcriptionists annotated via an internally developed tool where they had access to the recording and a platform for transcribing the content of the audio file, separating the audio file into utterances, and performing quality assurance. Samples with minor audio issues not impacting the transcriptionist's ability to produce an accurate transcript were processed as normal. Samples that could not be properly transcribed due to excessive background noise, poor audio quality, or other external issues such as the presence of multiple speakers in the file were tagged as unusable and were omitted from the corpus. In total, 91 samples out of 2765 collected samples were tagged as such and omitted.

2.4 Demographic Data Collection

Upon consenting, participants were asked to indicate whether they are native English speakers (i.e., whether they learned the English language before the age of 5 years old). They were also asked to indicate their age, gender, and education level.

2.5 Speech tasks

During each recording session, the subjects completed the following standard tasks, selected to elicit speech patterns that can be analyzed for acoustic and linguistic features that correlate to psychiatric state:

Figure 1: 'Family in the kitchen' image used in the picture description task.

Criteria	AVEC2013, 2014	DAIC-WoZ	DEPAC (our)
Language	German	English	English
# of speech tasks	2	1	5
# of samples total / per subj.	150 / 2	189 / 1	2674 / 5
Depression scale	BDI-II	PHQ-8	PHQ-9
Anxiety scale	-	-	GAD-7
Avg. depression score	15.34(± 12.13)	6.65 (± 6.11)	6.56 (± 5.56)
Depression score range in the corpus	0-45	0-23	0-27

Table 1: Description of our DEPAC dataset and its comparison to existing depression/anxiety corpora.

- **Phoneme Task:** Participants were asked to sustain a phoneme sound (e.g., /ā/) for as long as they could, up to one minute. They could cease making the sound whenever they choose. Due to difficulty in finding voiced parts in continuous speech, sustaining vowels would be optimal for measuring source and respiration features (e.g., shimmer) (Low et al., 2020).
- **Phonemic fluency:** Phonemic verbal fluency was evaluated using the FAS ('F', 'A', 'S') (Borkowski et al., 1967) task (letter "F"). This assessment has been used widely in a variety of populations, including individuals with Alzheimer's Disease (AD). The average duration of this speech task was 22.13 seconds in DEPAC dataset.
- **Picture description**: A static image depicting an event was presented to the subject, and they were asked to describe what is happening in their own words. The average length of picture-based narratives was 46.60 seconds. Tasks of this type have been shown to be good proxies for spontaneous discourse (Giles et al., 1996). Picture description was found to be an effective speech task in evoking situations that required more cognitive effort and caused noticeable changes in speech for detecting depression (Jiang et al., 2017). In this study, a proprietary image 'Family in the kitchen' (Figure 1) was used, which was designed to match the 'Cookie theft' picture (Goodglass et al., 2001) in style and content units. The picture was a line drawing of an everyday scene, containing three to four characters, two salient action items (e.g., broken bottle, or steaming pot), and a similar number of object units (20-25), action items (9-10) and locations (2) (Forbes-McKay and Venneri, 2005). Our core design guidelines to develop this picture are

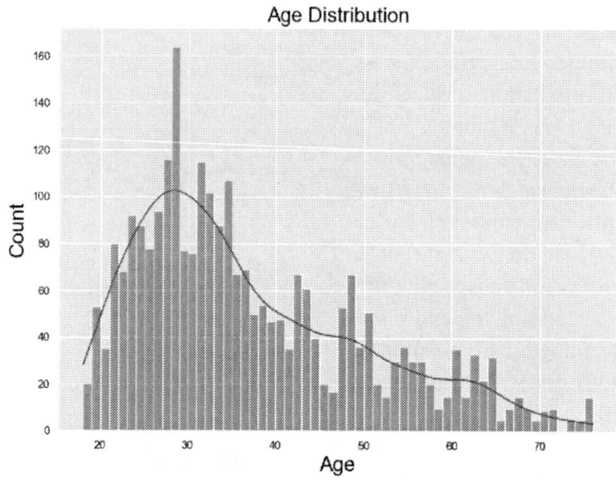

(a) Age distribution

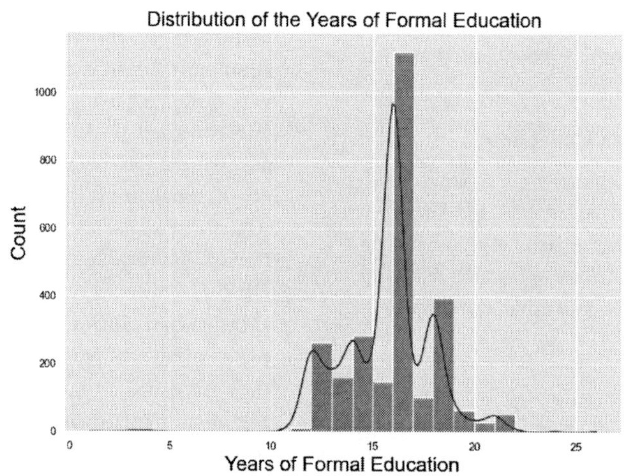

(b) Distribution of the formal years of education

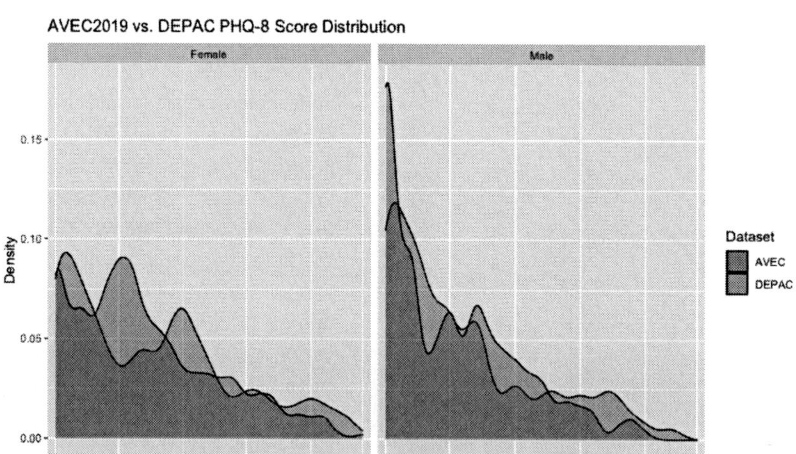

(c) Distribution of PHQ-8 scores in AVEC 2019 and DEPAC dataset by gender

Figure 2: Distribution of the participants' demographics in mTurk Study.

listed in A.1.

- **Semantic fluency:** Participants were asked to list as many positive future experiences as they can within one minute. They were given time parameters to guide them, such as future events predicted to happen within three weeks, within one month, within one year, and so on. They were allowed to describe as little or as much as they choose. Performance on verbal fluency tasks are found to correlate with executive deficits caused by depression (Fossati et al., 2003). The length of speech in this task was 43.76 seconds on average.
- **Prompted narrative:** Participants were asked to describe an event, interest, or hobby based on a single prompt, e.g., "Describe your day", "Describe a travel experience" and "Describe a hobby you have". Participants were allowed to describe as much or as little as they choose. Narrative speech provides an opportunity to elicit speech containing the linguistic structures and acoustic information that is known to contain indicators of depression (Trifu et al., 2017). The average duration of the prompted speech in the collected dataset was 45.34 seconds.

2.6 Clinical Assessments

The following two mental health assessment questionnaires were completed by the participants after the recording session:

Patient Health Questionnaire (PHQ-9): The PHQ-9 is a well established 3-point self-rated measure for **depressive symptoms** that has been validated against clinician rated measures (Kroenke et al., 2001). It contains 9 questions which correspond to the core criteria of the Diagnostic And Statistical Manual of Mental Disorders (DSM) for depression. Scores on this scale range from 0 to 27 with diagnostic cut-off thresholds for depression severity. Scores less than 5 represent the individuals with no depression; individuals with a mild or subthreshold depressive disorder are reflected by scores from 5 to 9; scores between 10 and 14 indicate moderate severity level of depression, and scores 15 or higher signify major depressive disorder in the participants (Kroenke et al., 2001).

Generalized Anxiety Disorder - 7 (GAD-7): The GAD-7 is a popular self-rated measure of general **anxiety symptoms** that is scored from 0 to 21 (Spitzer et al., 2006). It has been validated against clinical diagnosis and has been shown to be robust as a screening tool and a continuous measure of symptom severity. Scores of 10 or above indicate a reasonable threshold for detecting individuals with generalized anxiety disorder. Similar to the levels of depressive disorder in PHQ-9, 5, 10, and 15 are the cut points on the GAD-7 scale to classify anxiety severity level into minimal, mild, moderate and severe groups (Spitzer et al., 2006).

2.7 Corpus Composition

The dataset consists of 2,674 audio samples collected from 571 subjects (Table 1). 54.67% of the study subjects are female and 45.33% are male. The age of the subjects ranges between 18 and 76, and they received 1 to 26 years of formal education.

Figure 2 illustrates the demographic distribution of the mTurk study. The age distribution is shifted toward the left around its average value, which is equal with 36.85, indicating that most of the dataset is made up of young or middle-aged adults (Figure 2(a)). Moreover, it is witnessed in the education level distribution plot that the most of the participants received higher education, with on average around 15 years of formal education (Figure 2(b)).

Figure 3 (Appendix A.2) demonstrates that the distribution of both GAD-7 and PHQ-9 scores are skewed-right, representing that the majority of the dataset is composed of either no or subthreshold level of the disorders. In addition, the number of samples with moderate to severe level of both disorders are higher among women compared with men.

3 Feature Sets

In this section, we introduce a set of hand-crafted features extracted from the DEPAC audio records and the associated transcripts. The set of features comprises various linguistic and acoustic features that have been found in previous psychiatric literature to be effective in detection of depression and anxiety from spoken language (Low et al., 2020; Smirnova et al., 2018).

3.1 Acoustic Features:

We extracted 220 acoustic features from each audio sample. The feature set includes:

Generic Linguistic Features

Feature Category	**Description**
Discourse mapping (18)	**Utterance distances** and **speech-graph** (Mota et al., 2012) features extracted from the graph representation of the transcripts.
Local coherence (15)	Average, maximum, and minimum similarity between Word2Vec (Mikolov et al., 2013) representations of the successive utterances.
Lexical complexity and richness (103)	**Vocabulary richness**: Such as Brunet's index (Brunet et al., 1978) and Honore's statistic (Honoré et al., 1979). **Psycholinguistics norms**: Average norms across all words, nouns only and verbs only for imageability, age of acquisition, familiarity (Stadthagen-Gonzalez and Davis, 2006) and frequency (commonness) (Brysbaert and New, 2009). **Grammatical constituents**: The constituents comprising the parse tree in a set of Context-Free Grammar (CFG) features.
Syntactic complexity (143)	**Constituency-parsing based features**: Scores based on the parse tree (Chae and Nenkova, 2009) (e.g., the height of the tree, the statistical functions of Yngve depth (a measure of embeddedness) (Yngve, 1960), and the frequencies of various production rules(Chae and Nenkova, 2009)). **Lu's syntactic complexity features**: Metrics of syntactic complexity suggested by (Lu, 2010) such as the length of sentences, T-units, and clauses, etc. **Utterance length**: Average, maximum and minimum utterance length.
Utterance cohesion (1)	Number of switches in verb tense across utterances divided by total number of utterances.
Sentiment (9)	Variables such as valence, arousal, and dominance scores (Warriner et al., 2013) for all words and word types describing the sentiment of the words used.
Word finding difficulty (11)	**Pauses and fillers**: Variables like speech rate, hesitation, duration of words and number of filled (e.g., um, uh) and unfilled pauses as signs of word finding difficulty, which result in less fluid or fluent speech (Pope et al., 1970). **Invalid words**: Not in Dictionary (NID) indicating proportion of words not in the English dictionary.

Task-Specific Linguistic Features

Speech Task	**Description**
Phonemic Fluency (2)	Includes the raw number of words starting with the correct letter with/without explicit filtering out of proper nouns by their Part of Speech (POS) tags.
Picture Description (25)	**Global coherence**: Average, minimum and maximum cosine distance between GloVe (Pennington et al., 2014) word vector representation of each utterance and its closest content unit centroid utterances. **Information units**: The number of objects, subjects, locations and actions used to measure the number of items correctly named in the picture description task.
Semantic Fluency (1)	Includes the raw number of words of the correct category.

Table 2: List of linguistic features in our conventional feature set. The number of features in each subtype is shown in the parentheses.

- **Spectral features:** Intensity (auditory model based), MFCC 0-12, Zero-Crossing Rate (ZCR)
- **Voicing-related features:** Fundamental frequency (F_0), Harmonic-to-Noise Ratio (HNR), shimmer and jitter, durational features, pauses and fillers, phonation rate

Statistical functionals including minimum, maximum, average, and variance were computed on the low-level descriptors. Additionally, skewness and kurtosis were calculated on MFCCs, first and second order derivatives of MFCCs, and Zero Crossing Rate (ZCR) (Low et al., 2020) (Table 7 in appendix elaborates on detailed descriptions of these features as well as previous literature motivating their selection as the indicators of psychiatric conditions).

A Python implementation of Praat phonetic analysis toolkit (Boersma and Van Heuven, 2001) has been used to extract the majority of these features. The MFCC features and their functionals were computed using `python_speech_features`[2] library.

3.2 Linguistic Features:

We also applied standard natural language processing libraries (e.g., spaCy[3] and Stanford Parser[4]) to extract 300 generic and 28 task-specific linguistic features from the associated transcripts of the audio files (Table 2). For simplification, we classified the generic features into the categories including discourse mapping, local coherence, lexical complexity and richness, syntactic complexity, utterance cohesion, sentiment, and word finding difficulty (the selection motivations of our linguistic features are explained in Appendix A.3, Table 6).

4 Intended Usage

The study aimed to collect a high quality training dataset with the intention of developing a speech-based digital biomarker for the psychiatric diseases of depression and anxiety. The dataset is well-suited for exploratory analysis involving statistical and machine learning methods to generate potential speech biomarkers and test their validity. In Section 5, we present the baseline models to predict depression severity using this dataset, that can be used as benchmarks for the future research.

[2]https://pypi.org/project/python_speech_features/

[3]https://spacy.io/

[4]https://nlp.stanford.edu/software/lex-parser.html

Range of scores	AVEC PHQ-9	DEPAC PHQ-9	DEPAC GAD-7
[0 - 5)	77	240	261
[5 - 10)	36	178	152
[10 - 15)	26	84	87
[15 - 20)	17	40	45
[20 - 27]	7	10	7

Table 3: Counts for the PHQ-8/GAD-7 scores in AVEC and DEPAC datasets

5 Baseline Models for Depression Analysis

5.1 Data Preprocessing

Standardization: Once the acoustic and linguistic features were extracted from the data records, we standardized them using z-scores, i.e., subtracting the mean and dividing by standard deviation. The standard score of a sample x of feature f_i is calculated as:

$$y = \frac{x - \mu}{\sigma} \tag{1}$$

here μ and σ are the mean and standard deviation of the values of f_i in all training samples.

5.2 Model Training

To compare the efficacy of different modalities in predicting depression, we trained a combination of linear and non-linear Machine Learning (ML) models: Support Vector Regressors (SVR), Linear Regression (LR), and Random Forest Regressor (RF) separately on the following feature categories:

1. Demographic features (i.e., age, gender, and education)
2. Acoustic features
3. Linguistic features

We further investigated the effectiveness of each speech task for predicting depression severity on the PHQ-8 scale. The main reason for excluding the last question in PHQ-9 questionnaire was to make the results comparable to the performances with AVEC 2016 (Valstar et al., 2016) and AVEC 2019 (Ringeval et al., 2019) baselines, which are reported on PHQ-8 scale. The audio samples in AVEC challenges are subsets of Distress Analysis Interview Corpus (DAIC-WoZ) (Gratch et al.,

2014), which includes interviews of the participants conducted by a virtual agent. The length of the speech samples of the DAIC-WoZ dataset range from 5 to 25 minutes, including both participants' and interviewer's speech.

Figure 2(c) compares how the PHQ-8 scores are distributed in male versus female participants in AVEC 2019 and DEPAC datasets. Higher PHQ scores indicates the higher depression severity in the subjects. The distributions are skewed-right both for the male and female participants, representing that the majority of both datasets is composed of either no or mild level of depression. The number of samples in each level of depression in each of the two datasets is summarized in Table 3.

To validate the comparison of our models' performance with the ones trained on the AVEC datasets, we performed independent t-test on the PHQ-8 score distribution of the DEPAC dataset and AVEC 2019 corpus. The outcome of the test showed that the two datasets do not exhibit significant differences ($t = 0.65, p > 0.05$)and as such, these two datasets are similar enough to compare the performance of the baseline ML models.

Compared with previous datasets, our dataset is enriched with a greater variety of speech tasks. Thus, in addition to an analysis using data from all the included tasks, we evaluate models trained on task-subsets of the corpus and report their performance in predicting depressive disorder. Each model is evaluated with regard to the Mean Absolute Error (MAE) and Root Mean Square Error (RMSE) scales, following the baseline set by AVEC challenge (Valstar et al., 2016), (Ringeval et al., 2017). The performance metrics are described in Appendix A.4.

We trained an SVR model on the combination of acoustic and linguistic features extracted from all five speech tasks (See Section 2.5), and also separately on each of the speech (See Table 5).

For all the experiments, all model hyperparameters were set to their default values as on the Scikit-learn implementation (Pedregosa et al., 2011). Models were trained using grouped 10-fold cross validation, where samples from the same individual do not appear in both the training folds and test fold. All results are reported as the mean MAE/RMSE scores across the 10 folds.

Features	Algorithm	RMSE	MAE
Demographic	LR	6.94	5.18
	RF	6.34	4.93
	SVR	**5.20**	**4.06**
Acoustic	LR	7.51	5.86
	RF	5.41	4.41
	SVR	5.48	4.40
	AVEC 2016 baseline (Valstar et al., 2016)	7.78	5.72
	AVEC 2019 baseline (Ringeval et al., 2019)	8.19	-
Linguistic	LR	5.72	4.60
	RF	5.40	4.37
	SVR	5.37	4.24

Table 4: Regression results of the models predicting PHQ-8 score on different categories of features. Bold indicates the best performance.

5.3 Baseline Model Result and Discussion

We present and discuss the results of baseline model training across different modalities of input features, i.e. demographic, acoustic and linguistic, as well as across five different speech tasks, using DEPAC speech data.

Model Performance across Modalities: Among the three modalities, SVR model trained on demographic features performs the best, achieving the lowest MAE and RMSE, followed by the SVR model trained on linguistic features. Both acoustic and linguistic baseline models attain less than 20% MAE in the range of scores (0 to 24). Marginal deviation of both MAE and RMSE between acoustic and linguistic models suggests that these two modalities are effective for the task of recognizing signs of depression from speech. It is noteworthy that, the audio files did not undergo any preprocessing or enhancement before extracting the acoustic features. Yet, models trained on acoustic features exhibit competitive performance with the linguistic model, indicating that the quality of the recordings is sufficient and is a valuable foundation for future research.

In terms of predicting PHQ-8 scores, our baseline models perform substantially better than the baseline models specified by challenge organizers

Speech task	RMSE	MAE
Phoneme Task	5.49	4.32
Phonemic fluency	5.44	4.31
Picture description	5.36	4.25
Positive fluency	**5.19**	**4.11**
Prompted narrative	5.30	4.20
All tasks	5.38	4.27

Table 5: Regression results of SVR models predicting PHQ-8 score on different speech tasks. Bold indicates the best performance.

of AVEC 2016 (Valstar et al., 2016) and AVEC 2019 (Ringeval et al., 2019) (Table 4), despite the shorter length of samples than the AVEC corpus, which justify the robustness of the hand-curated acoustic features introduced in this work, as well as the quality of the dataset.

Surprisingly, the SVR model using only demographic features outperforms both acoustic and linguistic models (Table 4). This demographic information was previously found to be highly correlated to one's level of depression in literature (Akhtar-Danesh and Landeen, 2007). However, in real-world application, the demographic model may not be completely reliable due to ambiguity of these features.

Model Performance across Speech Tasks: In our task-specific analysis, comparatively lower RMSE and MAE are scored by models trained on picture description, positive fluency and prompted narrative than the phoneme task, phonemic fluency and all tasks combined. The possible reason behind this observation is that the picture description, positive fluency and prompted narrative tasks produce longer audio samples, resulting in more informative acoustic and linguistic features, leading to more accurate models. This observation shows that long recordings of narrative tasks can be rich sources of markers to predict depressive disorder from speech.

6 Conclusion

In this work, we introduce DEPAC, a rich audio dataset for mental health research which is labelled with scores on standard scales of two highly prevalent mental disorders: PHQ-9 scores for depression and GAD-7 scores for anxiety assessment. The dataset offers a remarkably larger sample size in comparison to other publicly available corpora. One other source of novelty of the presented corpus is its richness in the diversity of speech tasks and participants with various degrees of education, genders, and age groups. We also introduce a hand-curated set of acoustic and linguistic features incorporating domain knowledge of clinical and ML experts, which are used as the predictors of models for quantifying depression severity. We present the performance of baseline models in prediction of depression severity level, that can be applied by future researchers as a benchmark. Our baseline models achieve competitive performance when compared to the AVEC 2016 and AVEC 2019 baseline models and demonstrate the quality of the DEPAC dataset and effectiveness of our proposed feature set in measuring depression severity.

References

Noori Akhtar-Danesh and Janet Landeen. 2007. Relation between depression and sociodemographic factors. *International journal of mental health systems*, 1(1):1–9.

Murray Alpert, Enrique R Pouget, and Raul R Silva. 2001. Reflections of depression in acoustic measures of the patient's speech. *Journal of affective disorders*, 66(1):59–69.

RG Bachu, S Kopparthi, B Adapa, and BD Barkana. 2008. Separation of voiced and unvoiced using zero crossing rate and energy of the speech signal. In *American Society for Engineering Education (ASEE) zone conference proceedings*, pages 1–7. American Society for Engineering Education.

Megan S Barker, Breanne Young, and Gail A Robinson. 2017. Cohesive and coherent connected speech deficits in mild stroke. *Brain and language*, 168:23–36.

Aaron T Beck, Robert A Steer, and Margery G Carbin. 1988. Psychometric properties of the beck depression inventory: Twenty-five years of evaluation. *Clinical psychology review*, 8(1):77–100.

Jeffrey S Berman, R Christopher Miller, and Paul J Massman. 1985. Cognitive therapy versus systematic desensitization: Is one treatment superior? *Psychological bulletin*, 97(3):451.

Nathaniel J Blanco, A Ross Otto, W Todd Maddox, Christopher G Beevers, and Bradley C Love. 2013. The influence of depression symptoms on exploratory decision-making. *Cognition*, 129(3):563–568.

Paul Boersma and Vincent Van Heuven. 2001. Speak and unspeak with praat. *Glot International*, 5(9/10):341–347.

John G Borkowski, Arthur L Benton, and Otfried Spreen. 1967. Word fluency and brain damage. *Neuropsychologia*, 5(2):135–140.

Étienne Brunet et al. 1978. *Le vocabulaire de Jean Giraudoux structure et évolution*. Slatkine.

Marc Brysbaert and Boris New. 2009. Moving beyond kučera and francis: A critical evaluation of current word frequency norms and the introduction of a new and improved word frequency measure for american english. *Behavior research methods*, 41(4):977–990.

Jieun Chae and Ani Nenkova. 2009. Predicting the fluency of text with shallow structural features: Case studies of machine tanslation and human-written text.

Jesse Chandler and Danielle Shapiro. 2016. Conducting clinical research using crowdsourced convenience samples. *Annual review of clinical psychology*, 12:53–81.

Robert M Cohen, Herbert Weingartner, Sheila A Smallberg, David Pickar, and Dennis L Murphy. 1982. Effort and cognition in depression. *Archives of general psychiatry*, 39(5):593–597.

Sarah Kathryn Fineberg, J Leavitt, Sasha Deutsch-Link, Samson Dealy, Christopher D Landry, Kevin Pirruccio, Samantha Shea, Savannah Trent, Guillermo Cecchi, and Philip R Corlett. 2016. Self-reference in psychosis and depression: a language marker of illness. *Psychological medicine*, 46(12):2605–2615.

Alastair J Flint, Sandra E Black, Irene Campbell-Taylor, Gillian F Gailey, and Carey Levinton. 1992. Acoustic analysis in the differentiation of parkinson's disease and major depression. *Journal of Psycholinguistic Research*, 21(5):383–399.

Katrina E Forbes-McKay and Annalena Venneri. 2005. Detecting subtle spontaneous language decline in early alzheimer's disease with a picture description task. *Neurological sciences*, 26(4):243–254.

Philippe Fossati, Anne-Marie Ergis, Jean-François Allilaire, et al. 2003. Qualitative analysis of verbal fluency in depression. *Psychiatry research*, 117(1):17–24.

Kathleen C Fraser, Jed A Meltzer, and Frank Rudzicz. 2016. Linguistic features identify alzheimer's disease in narrative speech. *Journal of Alzheimer's Disease*, 49(2):407–422.

Lior Galili, Ofer Amir, and Eva Gilboa-Schechtman. 2013. Acoustic properties of dominance and request utterances in social anxiety. *Journal of social and clinical psychology*, 32(6):651–673.

Eva Gilboa-Schechtman, Lior Galili, Yair Sahar, and Ofer Amir. 2014. Being "in" or "out" of the game: subjective and acoustic reactions to exclusion and popularity in social anxiety. *Frontiers in human neuroscience*, 8:147.

Elaine Giles, Karalyn Patterson, and John R Hodges. 1996. Performance on the boston cookie theft picture description task in patients with early dementia of the alzheimer's type: missing information. *Aphasiology*, 10(4):395–408.

Harold Goodglass, Edith Kaplan, and Sandra Weintraub. 2001. *BDAE: The Boston Diagnostic Aphasia Examination*. Lippincott Williams & Wilkins Philadelphia, PA.

Jonathan Gratch, Ron Artstein, Gale Lucas, Giota Stratou, Stefan Scherer, Angela Nazarian, Rachel Wood, Jill Boberg, David DeVault, Stacy Marsella, et al. 2014. The distress analysis interview corpus of human and computer interviews. In *Proceedings of the Ninth International Conference on Language Resources and Evaluation (LREC'14)*, pages 3123–3128.

M Hamilton and W Guy. 1976. Hamilton depression scale. *Group*, 1:4.

Antony Honoré et al. 1979. Some simple measures of richness of vocabulary. *Association for literary and linguistic computing bulletin*, 7(2):172–177.

Christine Howes, Matthew Purver, and Rose McCabe. 2014. Linguistic indicators of severity and progress in online text-based therapy for depression. In *Proceedings of the workshop on computational linguistics and clinical psychology: from linguistic signal to clinical reality*. Association for Computational Linguistics.

Haihua Jiang, Bin Hu, Zhenyu Liu, Lihua Yan, Tianyang Wang, Fei Liu, Huanyu Kang, and Xiaoyu Li. 2017. Investigation of different speech types and emotions for detecting depression using different classifiers. *Speech Communication*, 90:39–46.

Emil Kraepelin. 1921. Manic depressive insanity and paranoia. *The Journal of Nervous and Mental Disease*, 53(4):350.

Kurt Kroenke, Robert L Spitzer, and Janet BW Williams. 2001. The phq-9: validity of a brief depression severity measure. *Journal of general internal medicine*, 16(9):606–613.

Kwok-Keung Leung, Tatia MC Lee, Paul Yip, Leonard SW Li, and Michael MC Wong. 2009. Selective attention biases of people with depression: Positive and negative priming of depression-related information. *Psychiatry research*, 165(3):241–251.

Daniel M Low, Kate H Bentley, and Satrajit S Ghosh. 2020. Automated assessment of psychiatric disorders using speech: A systematic review. *Laryngoscope Investigative Otolaryngology*, 5(1):96–116.

Bernd Löwe, Jürgen Unützer, Christopher M Callahan, Anthony J Perkins, and Kurt Kroenke. 2004. Monitoring depression treatment outcomes with the patient health questionnaire-9. *Medical care*, pages 1194–1201.

Xiaofei Lu. 2010. Automatic analysis of syntactic complexity in second language writing. *International journal of corpus linguistics*, 15(4):474–496.

Brian MacWhinney. 2000. The childes project: Tools for analyzing talk: Volume i: Transcription format and programs, volume ii: The database.

Brian MacWhinney. 2007. The talkbank project. In *Creating and digitizing language corpora*, pages 163–180. Springer.

Andrew Mathews and Colin MacLeod. 2005. Cognitive vulnerability to emotional disorders. *Annu. Rev. Clin. Psychol.*, 1:167–195.

Tomas Mikolov, Ilya Sutskever, Kai Chen, Greg S Corrado, and Jeff Dean. 2013. Distributed representations of words and phrases and their compositionality. In *Advances in neural information processing systems*, pages 3111–3119.

Stuart A Montgomery and MARIE Åsberg. 1979. A new depression scale designed to be sensitive to change. *The British journal of psychiatry*, 134(4):382–389.

Natalia B Mota, Nivaldo AP Vasconcelos, Nathalia Lemos, Ana C Pieretti, Osame Kinouchi, Guillermo A Cecchi, Mauro Copelli, and Sidarta Ribeiro. 2012. Speech graphs provide a quantitative measure of thought disorder in psychosis. *PloS one*, 7(4):e34928.

James C Mundt, Peter J Snyder, Michael S Cannizzaro, Kara Chappie, and Dayna S Geralts. 2007. Voice acoustic measures of depression severity and treatment response collected via interactive voice response (ivr) technology. *Journal of neurolinguistics*, 20(1):50–64.

Charles B Nemeroff. 2007. Prevalence and management of treatment-resistant depression. *Journal of Clinical Psychiatry*, 68(8):17.

Gabriele Paolacci, Jesse Chandler, and Panagiotis G Ipeirotis. 2010. Running experiments on amazon mechanical turk. *Judgment and Decision making*, 5(5):411–419.

Rupal Patel and Kathryn Connaghan. 2014. Park play: A picture description task for assessing childhood motor speech disorders. *International Journal of Speech-Language Pathology*, 16(4):337–343.

F. Pedregosa, G. Varoquaux, A. Gramfort, V. Michel, B. Thirion, O. Grisel, M. Blondel, P. Prettenhofer, R. Weiss, V. Dubourg, J. Vanderplas, A. Passos, D. Cournapeau, M. Brucher, M. Perrot, and E. Duchesnay. 2011. Scikit-learn: Machine learning in Python. *Journal of Machine Learning Research*, 12:2825–2830.

Jeffrey Pennington, Richard Socher, and Christopher D Manning. 2014. Glove: Global vectors for word representation. In *Proceedings of the 2014 conference on empirical methods in natural language processing (EMNLP)*, pages 1532–1543.

Benjamin Pope, Thomas Blass, Aron W Siegman, and Jack Raher. 1970. Anxiety and depression in speech. *Journal of Consulting and Clinical Psychology*, 35(1p1):128.

Thomas F Quatieri and Nicolas Malyska. 2012. Vocal-source biomarkers for depression: A link to psychomotor activity. In *Thirteenth annual conference of the international speech communication association*.

Anupama Ray, Siddharth Kumar, Rutvik Reddy, Prerana Mukherjee, and Ritu Garg. 2019. Multi-level attention network using text, audio and video for depression prediction. In *Proceedings of the 9th international on audio/visual emotion challenge and workshop*, pages 81–88.

Emna Rejaibi, Ali Komaty, Fabrice Meriaudeau, Said Agrebi, and Alice Othmani. 2022. Mfcc-based recurrent neural network for automatic clinical depression recognition and assessment from speech. *Biomedical Signal Processing and Control*, 71:103107.

Fabien Ringeval, Björn Schuller, Michel Valstar, Nicholas Cummins, Roddy Cowie, Leili Tavabi, Maximilian Schmitt, Sina Alisamir, Shahin Amiriparian, Eva-Maria Messner, et al. 2019. Avec 2019 workshop and challenge: state-of-mind, detecting depression with ai, and cross-cultural affect recognition. In *Proceedings of the 9th International on Audio/Visual Emotion Challenge and Workshop*, pages 3–12.

Fabien Ringeval, Björn Schuller, Michel Valstar, Jonathan Gratch, Roddy Cowie, Stefan Scherer, Sharon Mozgai, Nicholas Cummins, Maximilian Schmitt, and Maja Pantic. 2017. Avec 2017: Real-life depression, and affect recognition workshop and challenge. In *Proceedings of the 7th Annual Workshop on Audio/Visual Emotion Challenge*, pages 3–9.

Daun Shin, Won Ik Cho, C Hyung Keun Park, Sang Jin Rhee, Min Ji Kim, Hyunju Lee, Nam Soo Kim, and Yong Min Ahn. 2021. Detection of minor and major depression through voice as a biomarker using machine learning. *Journal of clinical medicine*, 10(14):3046.

Daria Smirnova, Paul Cumming, Elena Sloeva, Natalia Kuvshinova, Dmitry Romanov, and Gennadii Nosachev. 2018. Language patterns discriminate mild depression from normal sadness and euthymic state. *Frontiers in psychiatry*, 9:105.

Hannah R Snyder, Roselinde H Kaiser, Mark A Whisman, Amy EJ Turner, Ryan M Guild, and Yuko Munakata. 2014. Opposite effects of anxiety and depressive symptoms on executive function: The case of selecting among competing options. *Cognition & emotion*, 28(5):893–902.

Robert L Spitzer, Kurt Kroenke, Janet BW Williams, and Bernd Löwe. 2006. A brief measure for assessing generalized anxiety disorder: the gad-7. *Archives of internal medicine*, 166(10):1092–1097.

Hans Stadthagen-Gonzalez and Colin J Davis. 2006. The bristol norms for age of acquisition, imageability, and familiarity. *Behavior research methods*, 38(4):598–605.

Raluca Nicoleta Trifu, Bogdan NEMEŞ, Carolina Bodea-Haţegan, and Doina Cozman. 2017. Linguistic indicators of language in major depressive disorder (mdd). an evidence based research. *Journal of Evidence-Based Psychotherapies*, 17(1).

Michel Valstar, Jonathan Gratch, Björn Schuller, Fabien Ringeval, Denis Lalanne, Mercedes Torres Torres, Stefan Scherer, Giota Stratou, Roddy Cowie, and Maja Pantic. 2016. Avec 2016: Depression, mood, and emotion recognition workshop and challenge. In *Proceedings of the 6th international workshop on audio/visual emotion challenge*, pages 3–10.

Michel Valstar, Björn Schuller, Kirsty Smith, Timur Almaev, Florian Eyben, Jarek Krajewski, Roddy Cowie, and Maja Pantic. 2014. Avec 2014: 3d dimensional affect and depression recognition challenge. In *Proceedings of the 4th international workshop on audio/visual emotion challenge*, pages 3–10.

Michel Valstar, Björn Schuller, Kirsty Smith, Florian Eyben, Bihan Jiang, Sanjay Bilakhia, Sebastian Schnieder, Roddy Cowie, and Maja Pantic. 2013. Avec 2013: the continuous audio/visual emotion and depression recognition challenge. In *Proceedings of the 3rd ACM international workshop on Audio/visual emotion challenge*, pages 3–10.

Jerome C Wakefield. 2013. The dsm-5 debate over the bereavement exclusion: Psychiatric diagnosis and the future of empirically supported treatment. *Clinical psychology review*, 33(7):825–845.

Amy Beth Warriner, Victor Kuperman, and Marc Brysbaert. 2013. Norms of valence, arousal, and dominance for 13,915 english lemmas. *Behavior research methods*, 45(4):1191–1207.

Victor H Yngve. 1960. A model and an hypothesis for language structure. *Proceedings of the American philosophical society*, 104(5):444–466.

Jörg Zinken, Katarzyna Zinken, J Clare Wilson, Lisa Butler, and Timothy Skinner. 2010. Analysis of syntax and word use to predict successful participation in guided self-help for anxiety and depression. *Psychiatry research*, 179(2):181–186.

Chiara Zucco, Barbara Calabrese, and Mario Cannataro. 2017. Sentiment analysis and affective computing for depression monitoring. In *2017 IEEE international conference on bioinformatics and biomedicine (BIBM)*, pages 1988–1995. IEEE.

A Appendix

A.1 Picture Design Guidelines

To develop the 'Family in the kitchen' image (Figure 1) for our picture description task, we used the core design principles (Patel and Connaghan, 2014) described below:

1. Image content breakdown should contain:
   (a) **2 scenes/locations** (e.g., kitchen, or living room)
   (b) **20 to 25 objects** (e.g., knife, pan, or cupboard)
   (c) **9 to 10 actions** (e.g., chop, cook, steam, or fall)
   (d) **3 to 4 people/subjects** (e.g., dad, dog, mom, or daughter)
   (e) **2 "dangerous" elements** (e.g., broken bottle, or steaming pot)
2. Images should display **relationships between components** in a scene.
3. Images should depict **familiar themes**, but they must be accessible to adults with diverse cultural backgrounds, sexual orientations, and various socioeconomic strata.
4. Images should be designed appropriately for **older adults** with varied levels of visual impairment.
5. Images should provoke spontaneous discourse useful in **diagnosis and assessment** of mental health conditions. It should:
   (a) Elicit tokens whose labels **span the phonetic range** useful in diagnosing motor speech difficulty.
   (b) Elicit tokens whose labels **span lexical norms** (varying age of acquisition (AoA), familiarity, and imageability). Representing a varied range of lexical norms allows for using the same image to test speakers with varying degrees of cognitive and language impairment.
   (c) **Contain sub-scenarios** (Patel and Connaghan, 2014) which would be useful generally for generating longer speech samples, and specifically in assessing discourse structure (e.g., coherence, repetition, trajectory (what order are the sub-scenarios described in), content units (which sub-scenarios are mentioned and which left out), reasoning/inferences (e.g., interconnections and causation between the sub-scenarios)).

The goal of these guidelines was to keep the content generalizable across diverse cultures and to control the similarity with the 'Cookie theft' (Goodglass et al., 2001) image in lexico-syntactic complexity and the amount of information content units.

A.2 Distribution of Assessment Scores

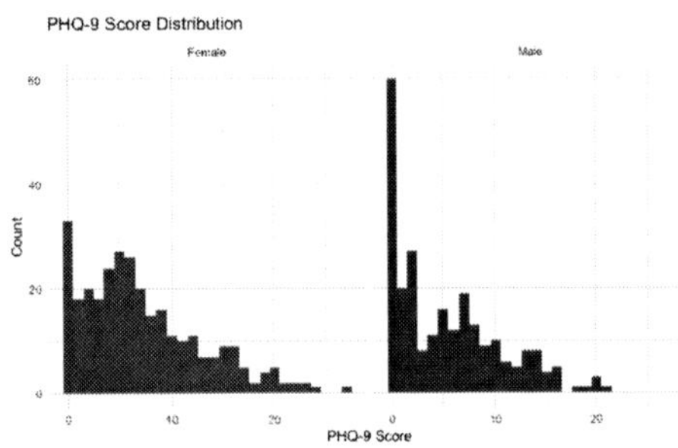

(a) Distribution of PHQ-9 scores per gender

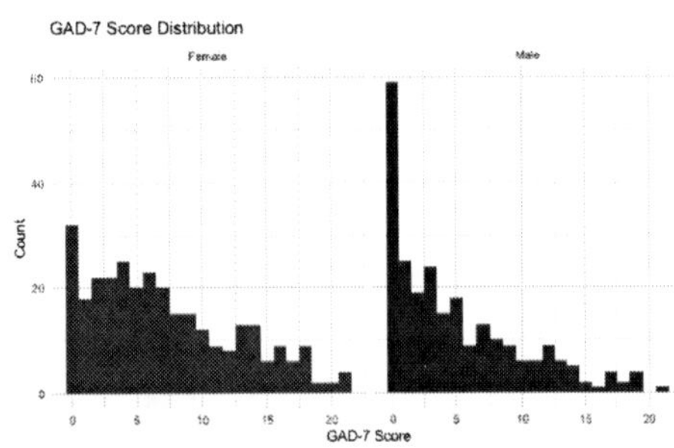

(b) Distribution of GAD-7 scores per gender

Figure 3: Distribution of the participants' PHQ-9 and GAD-7 scores in mTurk Study.

A.3 Feature Selection Motivations

The prior studies supporting the choice of our conventional feature set are described in Table 6 and 7. Table 7 displays the selection motivations of our acoustic features derived from the audio files, including spectral and energy related as well as voicing related features. In addition, Table 6 represents the motivations behind the choice of the generic and task-specific linguistic features extracted from the associated transcripts.

Generic Linguistic Features

Feature Category	Motivations
Discourse mapping	Techniques to formally quantify utterance similarity and disordered speech via distance metrics or graph-based representations have been used to differentiate speech from those suffering from various other mental health issues that are known to affect speech production (Mota et al., 2012; Fraser et al., 2016).
Local coherence	Coherence and cohesion in speech is associated with the ability to sustain attention and executive functions (Barker et al., 2017). Depression and anxiety are both known to impair such cognitive processes (Leung et al., 2009; Snyder et al., 2014).
Lexical complexity and richness	Language pattern changes in particular related to the irregular usage patterns of words of certain grammatical categories such as pronouns or verb tenses have been found to differentiate depression from normal fluctuations in mood from healthy individuals (Smirnova et al., 2018).
Syntactic complexity	Previous literature suggests that syntactic complexity of utterances, can be used to predict symptoms of depression (Smirnova et al., 2018), including utterances elicited in self-administered contexts (Zinken et al., 2010).
Utterance cohesion	Rates of verb tense use (in particular the past-tense) is known to be changed in individuals with depression. (Smirnova et al., 2018).
Sentiment	Emotional state and speech are connected, and sentiment scores in speech have been used to predict depression and anxiety levels in past research (Howes et al., 2014; Zucco et al., 2017).
Word finding difficulty	Previous work has found relationships between speech disturbance, filled, and unfilled speech of individuals with anxiety and depression (Pope et al., 1970).

Task-Specific Linguistic Features

Speech Task	Motivations
Phonemic Fluency	Measures of individual performance at the phonemic fluency task (Borkowski et al., 1967).
Picture Description	Measures of individual performance at picture description task as defined in (Giles et al., 1996; Jiang et al., 2017).
Semantic Fluency	Measures of individual performance at the semantic fluency task (Fossati et al., 2003).

Table 6: Support literature motivating the selection of the linguistic features in our conventional feature set.

Spectral and Energy Related Features

Feature	Motivations
Intensity (auditory model based)	Perceived loudness in dB relative to normative human auditory threshold. In 1921, Emil Kraepelin recognized lower sound intensity in the voices of depressed patients (Kraepelin, 1921).
MFCC 0-12	MFCC 0-12 and energy, their first and second order derivatives are calculated on every 16 ms window and step size of 8 ms, and then, averaged over the entire sample. MFCCs and their derivatives were included as baseline features in AVEC since 2013 (Valstar et al., 2013), (Valstar et al., 2016), (Ringeval et al., 2019) and found to be effective in predicting depression severity in the literature (Ray et al., 2019), (Rejaibi et al., 2022).
Zero-crossing rate (ZCR)	Zero crossing rate across all the voiced frames showing how intensely the voice was uttered. It was used as a speech biomarker of depression in previous studies (Bachu et al., 2008; Shin et al., 2021).
Voicing Related Features	
F_0	Fundamental frequency in Hz. A drop in F_0 and F_0 range indicates monotonous speech, which is common in depression (Low et al., 2020). In addition, many studies have discovered a considerable rise in mean F_0 in people suffering from social anxiety disorder (Gilboa-Schechtman et al., 2014; Galili et al., 2013).
Harmonics-to-noise-ratio (HNR)	Degree of acoustic periodicity in dB using both auto-correlation and cross-correlation method. Decreasing HNR ratio has been found to correlate with increasing severity of depression (Quatieri and Malyska, 2012).
Jitter and shimmer	Jitter is the period perturbation quotient and shimmer is the amplitude perturbation quotient representing the variations in the fundamental frequency. In previous studies, anxious patients indicated substantially higher shimmer and jitter. In addition, rise in jitter and shimmer variability was observed in subjects with major depressive disorder (Low et al., 2020).
Durational features	Total audio and speech duration in the sample. In prior studies, depression severity increased the total duration of speech because of longer pauses resulting in lower speech to pause ratio (Alpert et al., 2001; Mundt et al., 2007).
Pauses and fillers	Number and duration of short ($< 1s$), medium ($1 - 2s$) and long ($> 2s$) pauses, mean pause duration, and pause-to-speech ratio. Depression and anxiety are known to affect the rate of pauses/speech in individuals (Pope et al., 1970).
Phonation rate	Number of voiced time windows over the total number of time windows in a sample.

Table 7: Support literature motivating the selection of the acoustic features in our conventional feature set.

A.4 Performance Metrics

Root Mean Square Error (RMSE) and Mean Absolute Error (MAE) are calculated using the formulas shown below.

$$RMSE = \sqrt{\frac{\sum_{i=1}^{N}(x_i - y_i)^2}{N}} \tag{2}$$

$$MAE = \frac{\sum_{i=1}^{N}|x_i - y_i|}{N} \tag{3}$$

In the above, x_i and y_i are the true and predicted scores respectively.

The ethical role of computational linguistics in digital psychological formulation and suicide prevention

Martin P. Orr
Auckland University of Technology
martinorr521@gmail.com

Kirsten van Kessel
Auckland University of Technology
kirsten.vankessel@aut.ac.nz

David Parry
Murdoch University
david.parry@murdoch.edu.au

Abstract

Formulation is central to clinical practice. Formulation has a factor weighing, pattern recognition and explanatory hypothesis modelling focus. Formulation attempts to make sense of why a person presents in a certain state at a certain time and context, and how that state may be best managed to enhance mental health, safety and optimal change. Inherent to the clinical need for formulation is an appreciation of the complexities, uncertainty and limits of applying theoretical concepts and symptom, diagnostic and risk categories to human experience; or attaching meaning or weight to any particular factor in an individual's history or mental state without considering the broader biopsychosocial and cultural context. With specific reference to suicide prevention, this paper considers the need and potential for the computational linguistics community to be both cognisant of and ethically contribute to the clinical formulation process.

1 Introduction

Modelling is central to mental healthcare. Deficits in modelling, or failure to understand and manage those deficits, can lead to deficits in care.

Risk prediction, the diagnostic process, and key phenomena identification and monitoring such as mood symptoms are valid targets for the application of computational linguistics to suicide prevention. However, from a clinical perspective each of these targets and the research and categorical conceptual modelling that underlie them has major limitations, complexity, and contention (Chakraborty, 2020; Franklin et al., 2017; Fried, 2015; Large, 2018; Turner et al., 2021; Waszczuk et al., 2017).

Many aspects of mental health clinical practice are based on limited theoretical models, limited data and limited resources and involve varying presentations, preferences and levels of understanding, strengths, insight, and engagement. Formulation is the key clinical process for attempting to integrate these multiple interacting limited models and factors, to create an overall working model on which to base future action and interventions (de Beer, 2017; Carey and Pilgrim, 2010; Challoner and Papayianni, 2018).

Clinical formulation has a pattern recognition, factor weighing and explanatory hypothesis modelling focus. Formulation attempts to make sense of why a person presents in a certain state at a certain time and context and how given the known vulnerabilities, strengths, preferences and available resources that state may be best changed in a safe and effective way (Critchfield et al., 2022; Fernando et al., 2012; Johnstone and Dallos, 2013; Mace and Binyon, 2005; Manjunatha, 2019).

In keeping with the evolution and variation of mental health practice, formulation has historically taken varying forms and had varying drivers, and had questions raised about its validity and utility. However formulation retains a central role in care delivery, is considered as requiring the highest level of clinical expertise and is a key component of examination for specialist qualification (de Beer, 2017; Challoner and Papayianni, 2018; Sullivan et al., 2020).

Inherent to the clinical need for formulation is an appreciation of the complexities, uncertainty and limits of applying categories and theoretical concepts to human experience or attempting to attach meaning or weight to any particular factor in an individual's history or mental state without considering the broader context.

This paper considers the opportunities and challenges for computational linguistics in emulating and augmenting the clinical formulation process and contributing to broader related digital mental health developments. Highlighted is the need to appreciate the ethical and clinical safety risks, particularly if developments in the computational linguistics field are misperceived or exaggerated in terms of their certainty and capacity for suicide

Proceedings of the Eighth Workshop on Computational Linguistics and Clinical Psychology, pages 17 - 29

prediction and reduction.

The paper discusses the clinical assessment and planning process and the phenomenological psychopathology analysis, nosological diagnostic classification, individual psychodynamics and risk prediction complexities that drive the need for formulation. The concepts of mood, affect and emotion are discussed to illustrate some of the issues around the standardised interpretation of human experience and classification into diagnoses. The ethics and difficulties of attempting to predict or modify the risk of low base rate complex emergent events such as suicide is highlighted (Woodford et al., 2019; World Health Organization, 2014). A structure for formulation is provided to highlight the key components and where computational linguistics may be of assistance.

The central arguments will be that the data gathering, pattern recognition, factor weighing, and modelling of clinical formulation are areas in which computational linguistics could and should assist. Pattern recognition and modelling around words and language in context is central to mental health clinical practice and computational linguistics. Mental health and computational linguistics specialists can synergically use language as a method to gain insight and formulate a model of another's consciousness, intent and experience. This can contribute to risk, diagnostic and psychodynamic formulation. However, appreciation of the limitations of modelling and prediction particularly in application to suicide prevention will remain central. The Artificial Intelligence (AI) ethical principles of autonomy, justice, beneficence non-maleficence and explicability will remain a challenge and a duty for the CLPsych community (Floridi and Cowls, 2019). Appreciating the rationale for the utilisation of formulation in clinical practice and seeking to place the ethos and process of formulation at the heart of computational linguistics practice to enhance explicability will assist in addressing that duty. Machine learning and computational linguistics may play a role in more accurately identifying the contextual and contingent factors and the level of certainty or uncertainty inherent in the formulation modelling and explanatory hypothesis.

The primary purpose of this work is to provoke thought and facilitate further conceptual and operational ethical co-design of digital formulation. The aim is to help build a shared understanding of the rationale, structure and process of clinical formulation and call upon the CLPsych community to consider what contributions they could make to digitally enable and improve it particularly within a suicide prevention context. It is recognized there is a concept-reality gap between what clinicians might ideally desire and what the computational linguistic field is currently able to offer (Orr and Sankaran, 2007). However, the CLPsych community could play an important role in clarifying and developing the conceptual vision for digital formulation, and the required technological and methodological steps to get there.

The paper is intentionally largely technology and data source agnostic and focused on the clinical need and related medicolegal and ethical principles. The aim is to stimulate rather than limit thought or argue for a particular technological or methodological direction. The paper will touch on the initial steps to cross the concept-reality gap the authors are taking. This includes a focus on ethics, digital transformation, sleep and suicide, social media data and integrated thematic analysis and topic modelling.

2 The role and place of formulation in clinical practice

This next section aims to briefly set out some key concepts on which to build a shared understanding of the need for and place of formulation in clinical practice particularly in suicide prevention. These concepts are complex and contentious with differing definitions and scopes and varying degrees of clinical understanding and application in practice. Highlighted are the roles and limitations of language, phenomenology, nosology and risk prediction.

3 Language as a window into mental and brain state

There is limited understanding of the nature of consciousness or the mind and how this relates to brain function (Frith, 2021; Graziano, 2021). However, there is a general understanding that integrated biological, psychological and sociological factors impact on brain function and impact on the integrated experience and expression of thoughts, emotion, and behaviour. Machine learning affords the capacity to dynamically identify and analyse multiple signals indicative of an individual's mental state and intent. These signals may be neurophysiological, behavioural and of increasing interest to suicide prevention natural language, including that occurring in social media (Resnik et al., 2020; Chancellor and De Choudhury, 2020; Coppersmith et al., 2018; Fonseka et al., 2019).

To gain a greater timely understanding of the lived experience and meaning of suicidal thoughts and behaviour we need a greater appreciation of the dynamic cognitions and emotion and contexts that colour an individual's thoughts and drive them to action (Harris and Barraclough, 1997; Liu et al., 2020; Marsh, 2018). Social media data may provide an additional window and insights into this experience and an opportunity to intervene in a timely way.

Clinically language is a key tool for assessing and communicating thoughts, emotion and behaviour. Language is central to the assessment of mental state and from this potential brain state. Language assists in making hypotheses about electrochemical and cognitive processes in a section or circuit of the brain at a particular point in time that hence drive physical, biological, psychopharmacological and psychosocial interventions.

Language is a significant window into human experience but may not always provide an accurately drawn picture of reality. The image may be skewed and distorted by faulty mental models, cognitive biases, and misinterpretations by both the experiencer and the observer. Computational linguistics as the study of language using computational methods and theoretical models, similarly to clinical practice, has an inherent interest in ensuring any model deficits or conflicts are understood, minimised and managed.

4 Phenomenological psychopathology and nosology

Phenomenological psychopathological analysis is the process that underlies the clinical perception and interpretation of the experience and behaviour of others (Aftab and Ryznar, 2021; Chakraborty, 2020; Nelson et al., 2021).

Nosology is the classification of medical diseases. Nosological modelling can occur at three levels: aetiological (disease cause is known) pathogenetic (disease process is known) and symptom (only reported or interpreted experience is known). Mental disorder diagnoses are typically at the symptom modelling syndrome level (Kendler, 2009; Aftab and Ryznar, 2021).

Human experience and behaviour are characterised by a dimensional nature and multifactorial temporal contextual determinants. Complexity, and ambiguity is inherent. There is only limited knowledge of the causes and mechanisms by which mental disorders and perceived aberrant experience and behaviour arise. Accordingly, there are only theoretical models of varying fidelity and evidence base and agreement around the nature and classification of mental disorder, and how experience and behaviour should be interpreted and determined to be pathological. Similarly, the selection and mechanism of action of interventions, their benefits and harms, and predictions and determinants of prognosis all require the interpretation and weighing of various population research models as to what may be best and available for a specific patient in a specific mental state, in a specific time and context.

Risk categories, and diagnostic categories based on the identification and interpretation of phenomena and syndromes have significant reliability, validity and intervention, prognostic and safety limitations (Michelini et al., 2021; Nelson et al., 2021).

Although the terms affect, emotion and mood are often used interchangeably, they have a broad historical range of interrelated but separate specific meanings, definitions and perceived implications arising from variations (Berrios, 1985).

An emotion can be understood as the subjective personal experience and interpretation of a feeling state. Affect refers to an assessor's interpretation of the emotional experience of another, and typically includes not just reference to the type, but also the range and stability and appropriateness of expressed emotion within a specific context.

Emotions may be of short duration and fluctuate and represent the subjective interpretation of chemically induced physiological experience. The interpretation of this physiological emotional experience may be influenced by the longer standing and more prominent mood state, which may have a complex biopsychosocial basis.

A report of a mood symptom such as depression may have significant differing impact, relevance and meaning depending on the pattern intensity, duration, associations and context of occurrence. It may be a sign of a brief adjustment to a stressor, an indication of emotional dysregulation in someone with a personality disorder, form part of various levels and presentations of a major depressive disorder, be associated with medical and neurological disorders from dementia to Parkinson's disease, be associated with or secondary to drug use prescribed and illicit, form part of a broader bipolar disorder, or be an early presentation or association with schizophrenia. Weight may be given to one diagnosis over another if there is a clear family or personal history or pattern of a particular disorder and other known risk, symptom, sign and contextual factors are present.

Diagnoses can be of use in care planning, funding, research and making predictions about the future. However, they have significant limitations, not least if it is forgotten they are syndromal level models, that tell little of the personal story and context of the individual. The symptom and sign and temporal components of the diagnostic model may be subject to deficits or non-standardisation in interpretation and report. Race, culture, gender, age, language, education, intellectual and sensory impairment and economic status and societal marginalisation may all have an impact on the expression and interpretation of experience and behaviour. These factors may contribute to significant inter-rater variability as to what diagnosis or diagnoses are ascribed to an individual. Individuals that receive a specific mental disorder diagnosis, may have significant variation in terms of what criteria they meet, their individual experience, and underlying causal and mechanism of development factors. For example, in the DSM classification system 227 combinations of criteria can lead to a diagnosis of major depression, including 64 combinations which don't require a report of depressed mood. Some combinations may be more common and more meaningful from a clinical priority and potential to intervene perspective and computational linguistics could assist with identifying these (Zimmerman et al., 2015).

5 Suicide risk prevention

There are major challenges, limitations and clinical and ethical risks in trying to predict complex multi-factorial emergent low base rate events such as suicide that have a high magnitude of adverse consequence if that prediction is wrong (Pridmore, 2015; Nock et al., 2019; Large et al., 2017). The majority of those classified as being at high risk of suicide, do not commit suicide, and the majority of suicides will emerge from those classified as low risk, or who have not been assessed for suicide risk, have not expressed suicidal ideation or whom are not engaged in services (McHugh et al., 2019; Kessler et al., 2020; Durie, 2017; Large, 2018). The expression and actioning of suicidal intent can vary in intensity and fluctuate rapidly and be influenced by ambivalence, mood and emotional state, change in perceived circumstances, level of trust, wish to protect others, shame, denial, rationalisation, coping patterns, cognitive impairment, gain, and impulsivity (Yaseen et al., 2019; Galynker et al., 2017; Deisenhammer et al., 2009; Freedenthal, 2007).

There is a need to appreciate that even if increase the specificity and sensitivity of a technology capable of screening for a particular disorder, behaviour or risk, the positive predictive and negative predictive value will vary as a factor of prevalence in the targeted community. Suicide is a low base rate event making prediction complex and making the capacity for undue harm and intrusive unnecessary interventions higher. Even if the sensitivity and specificity of a test for suicide is significantly improved the positive predictive value may still be relatively low. This is not an argument to stop researching computational linguistics' capacity to improve suicide prediction but is a call to be cognisant of the limitations and to take a broader view on how machine learning and computational linguistics may contribute to suicide risk management.

Different people will have different pathways, processes, contexts, and timelines that take them to suicide. Some may have a more linear escalating suicide risk chain they follow; others will display a more complex emergence pattern where multiple factors came together at a certain point in time and a chain is only apparent with retrospective coherence (Kurtz and Snowden, 2003).

While making suicide predictions has inherent complexity and limitation, enhancing the capacity for machine learning to detect risk signals and offer support to reach out and seek help, would provide a chance to positively change that pathway and context (Tielman et al., 2019; Ryan, 2015). Machine learning may be able to assist in identifying what key potential contextual risk factors are for an individual or community; assist in the triage and prioritisation of attention for that individual or community; and do this at a speed and scale over multiple sources that exceeds human capabilities (Resnik et al., 2020; Shing et al., 2020).

Clinical risk including suicide risk needs to be considered in relation to a specific population and in relation to the individual's own baseline or typ-

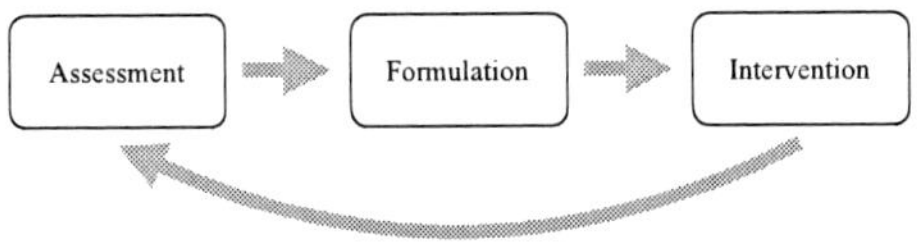

Figure 1: Central Role of Formulation

ical risk level or pattern. Timing and context is important including the presence or absence of key precipitants or protective factors. Ascertaining what factors are able to be managed, minimised or developed, and which resources are ideally and in reality, available are also important concerns. Rather than just provide a simple risk category, formulation aims to provide an integrated weighted contextualised view of all factors that may have a role in intervention planning and risk management (Pisani et al., 2016; Wyder et al., 2021; Fitzpatrick, 2018; Kessler et al., 2020).

6 Formulation at the centre of clinical practice

The clinical assessment and intervention process typically involves the key dynamic, integrated, iterative stages of history taking, mental state examination, formulation, diagnosis and care planning. Formulation is at the centre, prioritising and integrating key aspects of the assessment as a foundation for the personalised intervention planning.

Formulation can be perceived as a form of clinical storytelling. Clinical formulation includes the recurrent patterns, key themes, plot points and relationships, and cultural and contextual factors that characterise and help draw a mental model of an individual and their world.

Many individuals even if they have never had a formal mental health diagnosis before, may be found on assessment to have had recurrent patterns suggestive of previous episodes or prodrome or vulnerabilities. Those that have an established recurrent relapsing disorder, may have patterns of risk behaviours and contexts and early warning signs, that the client has varying and fluctuating levels of insight into, but that may be well recognised by families and supports.

Storytelling in written and spoken language has traditionally been a way to transmit knowledge and understanding to a group and through generations. The narrative structure of storytelling may assist in human recall and motivational understanding. Clinical formulation is a structured way to make sense and meaning of another's consciousness and experience and convey the story of their life, with a view to positively influencing the next stage of that story. Multiple factors and models are weighed and weaved using a structured process of analysis and reporting. Computational linguistics with its strengths in factor identification, weighing and integrated pattern recognition across multiple sources and contexts could assist this process.

Unlike describing the precise formulation or composition of a chemical compound or drug, the specific elements or processes by which a human experience arises is unknown. There is limited knowledge of the aetiology or pathogenesis of mental disorders, nor the nature of consciousness or emotions or experience. However, there are a range of biological, psychological and sociological research-based models that may guide the formulation process. The quality of the formulation is dependent on the knowledge and skill of the clinician in history taking and mental state examination and being able to identify key patterns, vulnerabilities, strengths, relationships and structures and integrate the findings with appropriate theoretical models. The quality of the formulation can be iteratively improved by the availability of additional data sources and the input of the multiple stakeholders, family, supports and caregivers that may play a role in an individual's life (Ford et al., 2019; Geach et al., 2018; Johnstone, 2018).

7 Clinical formulation structure

The clinical formulation may be approached in a structured manner, with data and key findings captured under a series of interrelated headings, each capturing a different but interrelated descriptive, theoretical or explanatory perspective (Chang and Lundahl, 2019; Weerasekera, 1993). Machine learning and computational linguistics operating under various levels of autonomy, could assist in augmenting this clinical pattern recognition and modelling process.

Some variant of a series of "P" headings such as problem, predisposing, precipitating, perpetuating, protective factors and prognosis headings is common clinical practice. Patterns, preferences, and priorities have been added for this work to emphasise these key attributes of the formulation process and where machine learning and computational linguistics could play a key role in capturing and weighing and providing decision support.

Problem	What are the key findings from the presenting complaint and history of presenting complaint, and the mental state examination, that characterize the problem or disorder?
Predisposing	What biopsychosocial and cultural contextual factors may have predisposed the individual to the disorder?
Precipitating	What biopsychosocial and cultural contextual factors may have precipitated the problem or exacerbation of the disorder?
Perpetuating	What biopsychosocial and cultural contextual factors may perpetuate or exacerbate the problem or provide a barrier to recovery?
Protective	What are the biopsychosocial and cultural contextual factors that may offer protection, assist in recovery or prevent further harm or adverse outcomes?
Prognosis	What is the expected response and outcome for this individual given what is known from population research and their specific history and level of insight, impairment, vulnerabilities, strengths and resources? How might interventions work or not work or cause harm and in what context and time?
Patterns	What patterns may be evident in the history and how may these relate to known psychological models?
Preferences	How does the individual prefer to understand or model their problem(s) for themselves and what resources and interventions do they prefer to utilize and how may these preferences be impacted on by insight and judgement?
Priorities	What are the priorities for the intervention plan given the knowledge about the individual, their past response and preferences, available resources and logistics, local and professional best practice guidelines, and relevant science?

Table 1: Clinical Formulation Structure

By considering biological, psychological and social facets and individual and systemic contextual components of each heading, the key formulation issues are often captured in a biopsychosocial grid structure, before being converted into an integrated coherent written form (Weerasekera, 1993). Some structures consider culture as inherent to the biopsychosocial analysis; others draw it out as a separate heading or separate cultural formulation process to ensure this important factor is focused on and not neglected.

Conceptual models created from international data may have transcultural limitations. This may be a significant issue when conceptual models are being utilised in formulation and particularly when involving the interpretation of experience and behaviour. DSM5 diagnoses are essentially conceptual models built on international data that contribute to care by providing a framework to describe an individual's perceived experience. However, the framework may negatively impact on care if the transcultural limitations are not adequately addressed or understood (Bredström, 2019; Rangihuna et al., 2018; La Roche et al., 2015).

Similarly, computational linguistics research and application in the mental health and suicide prevention domains needs to be designed and interpreted with a sociocultural contextual awareness as is emphasised by the formulation ethos (Durie, 2017; Hatcher et al., 2017; Lawson-Te Aho Dr, 2017; McClintock and McClintock, 2017).

8 Ethical and regulatory issues

The following section outlines a range of ethical and regulatory issues that are important considerations when developing and deploying digital mental health interventions particularly in the area of suicide prevention. Health interventions should be evidence based, and subject to academic, clinical governance, regulatory and ethical review. There is a need to be ethically cognisant of the risk-benefit profile, the relative utility and costs and the numbers of people that may additionally benefit or be harmed by an intervention within a specific context. The relevance, meaning, sensitivity and specificity of screening and diagnostic tests must be described with reference to a stipulated time period, prevalence and clinical context (Andrade, 2015). Digital mental health interventions including computational linguistic based suicide prevention interventions need to be subject to similar standards.

Primum non nocere or "first, do no harm" is a fundamental principle of bioethics. Failure to understand the complexity and limitations of suicide risk prediction has significant capacity to cause harm. Simplistic, generalised or static risk categorisation can lead to unintended harm and there is a need for dynamic formulation based assessment that recognises the importance of context for an individual's strengths and vulnerabilities.

The analytic power, reach, personalisation, timeliness and vigilance of AI based digital care affords major potential benefit. However, AI can be intrusive, discriminative, unwanted, and wrong. In suicide prevention resources could be allocated to the wrong groups, to the wrong individual, or be of the wrong type or quality and quantity. Some individuals may have unnecessary protections or intrusions placed on their lives which are damaging or disabling (McKernan et al., 2018). AI algorithms are subject to bias, misuse, undue trust, and unintended consequences and require continuous ethically based and sociocultural aware research, co-design, and governance (Yu, 2020; Floridi et al., 2020; Stein and Reed, 2019; Challen et al., 2019).

Continually striving to improve AI based risk prediction and management at an individual to societal level and getting to zero people dying by suicide is a morally worthy goal. However, there is a need to consider how the nature and current status of attaining that goal may be societally interpreted or misinterpreted. Stigma can have an impact on suicide bereavement and is an important consideration for suicide postvention. Bereaved family and caregivers can experience significant stigma, shame and blame and societal judgement based on a belief that they should have seen the signals of pending suicide and predicted and prevented the death (Evans and Abrahamson, 2020). In the reporting of improvements in suicide risk prediction, it is important that the CLPsych community highlight the ongoing complexities and limitations in identifying, seeing, analysing and acting on the signals and do not unintentionally contribute to exacerbating suicide bereavement and stigma.

If a digital system claims a clinical or therapeutic intervention function, then the system can be expected to be held to a high ethical and regulatory standard. This includes requiring a high level of mandated understanding of how the system integrates into broader clinical care processes, medicolegal responsibility and governance frameworks and whether it potentially requires software as a medical device type certification. There are varying developing regulatory standards and definitions for medical device type software. An AI based system providing triage and treatment advice where an individual may be at risk of suicide would likely present some of the highest ethical and clinical risks for development and deployment in a healthcare context and attract the highest regulatory categorisation and governance requirements (NEAC, 2019; Fernandes and Chaltikyan, 2020; Keutzer and Simonsson, 2020).

There is increasing interest in the use of social media and AI in suicide prevention. The international literature on social media research highlights various contentions including defining public vs private data, consent and anonymity and minimising bias and algorithmic harm (Townsend and Wallace, 2016; Chiauzzi and Wicks, 2019). There is increasing recognition of a need for social media-based research to have ethical overview to ensure that quality research is being proposed that understands the limitations and context of the data analysis and is protective and respectful of potentially vulnerable communities (British Psychological Society, 2017; Townsend and Wallace, 2016; Pagoto and Nebeker, 2019; Chiauzzi and Wicks, 2019; Benton et al., 2017).

Tutelary law and ethics, relates to those aspects of the legal and ethical system that have a focus on guardianship and protection (Unsworth, 1991). Mental healthcare services have had a long and difficult history with care and protection and guardian roles. The legal system is aware that good protective intents do not always result in good or optimal outcomes and there is always a need to consider who will guard the guardians. Clinical decisions and opinions about risk that impact on an individual's civil liberties, are often subject to review by tutelary mental health courts and tribunals; decisions influenced by AI based categorisation or predictions should similarly be expected to be reviewed by the tutelary system (Szmukler, 2014).

Floridi and Cowls (2019) have argued AI ethics can be reduced to five core principles. Four of these are the traditional bioethical principles of autonomy, justice, beneficence and non-maleficence to which they have added explicability. Explicability aims to capture the concepts of intelligibility and accountability. Building suicide prevention interventions and research on faulty, limited or poorly understood or described models affords significant ethical and clinical risk. Clinicians and researchers need to take a lead in ethically shaping and governing the emergent capacity for greater levels of social media and AI based suicide prevention research and development(Hom et al., 2017; Hunter et al., 2018; Pagoto and Nebeker, 2019). Before the deployment of AI in a mental health setting, stakeholders should have an adequate understanding of how it was co-designed and works and who is accountable and liable for how it works (Floridi and Cowls, 2019; Price et al., 2019). The formulation process could improve explicability in that there should be a clearer, intelligible and accountable process as to why intervention decisions were made. This should include having an understanding of the mental health theoretical models on which or for what, the machine learning algorithms were built (de Andrade et al., 2018).

9 Discussion and conclusion

In clinical practice there are significant standardisation, ethical, safety and effectiveness issues when classifying an individual as in or out of some binary diagnostic or risk category. This is particularly so when the constructs or models that underlie each criterion are limited in their scientific basis and are not operationally defined and there is significant variation in training, interpretation and application and perceived clinical utility.

Similarly, when computational linguistics developments aim to assist in symptom, diagnostic and suicide risk prediction categorisation there may be significant theoretical, ethical, utility and clinical safety concerns and limitations. There is a need to move beyond risk categorisation to risk formulation as part of the broader clinical formulation and intervention context. Any risk prediction categorisation produced needs to be treated like the output from any screening or diagnostic test; that is as another datapoint for the formulation, that is to be iteratively weighed, integrated and interpreted within the broader dynamic clinical context and not considered definitive or static.

Human experience is often time and context dependent, dynamic and multidimensional and occurs along a spectrum rather than within discrete categories. Formulation is the key focus of natural clinical intelligence and ought to be a key focus for artificial intelligence.

Computational linguistics could help in the development and assessment of a broader contextual understanding of an individual's history and mental state. A diagnostic and risk formulation process affords the opportunity to present a richer personalised explanatory model that links all the factors and highlights the complexity, uncertainty and importance of context and dynamic change.

The clinical formulation process of iterative factor identification and weighing, pattern recognition and modelling, is in keeping with the strengths of machine learning and the computational linguistics process. There are opportunities for significant synergy. Computational linguistics can operate at a speed and scale of factor identification and analysis across multiple sources beyond human capability. The machine learning process may be refined on previous clinical assessments, with emphasis given to mental state examinations and formulations. Clinician in the loop training and curation processes may assist with explicable and reflexive algorithmic improvement and production of meaningful safe ethical outputs. Though such formal clinical data may be difficult to access for current researchers, this can be expected to improve as machine learning is integrated and normalised as part of care delivery.

Machine learning and computational linguistics could improve the explicable quality of the acquisition, analysis and description of formulation data. Machine learning could also improve the quality of the theoretical models applied by improving the quality of research that underlies those models. There may be different and changing reasons and typologies for suicide and AI enabled research may be able to better timely categorise, trend and define these at an individual and community level (Clapperton et al., 2020; Martin et al., 2020).

In formulation every current and emergent finding needs to be iteratively analysed in context, and with knowledge of the strengths and limitations of the related clinical theoretical models. Particularly in suicide risk management there is a need to be highly cognisant of the difficulty and contention of predicting complex low base rate events and the harm that can result from both false negatives and false positives.

Digital transformation and co-design in AI empowered suicide prevention requires the working together of clinicians, communities, consumers, and digital media companies. The leverage, reach and analysis of AI empowered digital media make taking a co-design and societal perspective more meaningful and achievable and anything more limited, less ethically justifiable.

Looking to the future computational linguistics could assist in the creation of a self-constructing and updating digital formulation drawing on multiple sources from social media to email to clinical notes and assessments to conducting autonomous interviews in oral and written format. These services could be delivered in the form of customisable digital guardians, coaches or clinicians that address, with varying levels of expertise, medicolegal responsibility and autonomy, the assessment, formulation and intervention process.

In terms of an example of potential next research steps the authors are currently integrating qualitative thematic analysis with machine learning based topic modelling to study sleep related concerns in a large social media based suicidality dataset. Sleep disturbances from insomnia to nightmares to sleep disordered breathing are associated with an increased risk of suicidal behaviour and night-time is a high-risk period for suicide (Braun and Clarke, 2006, 2019; Blei et al., 2003; Blei, 2012; Fast et al., 2016; Shing et al., 2018, 2020; Zirikly et al., 2019; Porras-Segovia et al., 2019; Tubbs et al., 2019). The research is exploring whether this integrated thematic analysis and topic modelling approach can contribute to the development of an explicable conceptual linguistic sleep signal model for AI empowered clinical formulation, prioritisation, treatment category recommendation, and psychoeducation in the area of suicide prevention. Identifying key topics and themes and a related lexicon are central to these clinical processes. Suicide is complex and multifactorial. By focusing on one potential signal (sleep), one machine learning technique (topic modelling) and one dataset the aim is a greater conceptual understanding of the opportunities and challenges that could be presented by an multi-signal, multi-source, explicable AI and formulation based suicide prevention system. The current focus is on social media data, but a range of biopsychosocial data sources might be integrated into a future system. Formulation and broader data contained in clinical assessments and discharge summaries could be a future key target for both analysis and enrichment (Adnan et al., 2013).

Developments in digital formulation from a clinician augmentation or decision support role to a more autonomous social media focused digital guardian, coach or clinician role will have major clinical and societal impact. Suicide is a complex time and context dependent phenomenon. There is increasing recognition of the need to broaden the clinical service focus of suicide prevention to a more societal level focus that has more timely vigilance and leverages a greater range of resources.

Clinic-based services accessing social media data, and social media based services accessing formal clinical data, and the integration of such services raises significant medicolegal, security and ethical issues (Williams et al., 2017; Price et al., 2019; Bhatia-Lin et al., 2019). However, machine learning assisting in the expansion from a clinical service to societal focus allows for more protective layers and opportunities for integrated formulation and intervention at more time points. Social media machine learning based interventions have the advantage that even if they have only a relatively small effect size on reducing suicidal behaviour they can be deployed at such scale and minimal marginal cost that they may have a significant impact at a population societal level (Torok et al., 2020).

Socialising emergent concepts, among research and practice leaders, is an important stage in the innovation diffusion and health practice change process (Beausoleil, 2018; Taherdoost, 2018; Rahimi et al., 2018). This can lead to critical analysis of relative advantage, adoption challenges and health impact , and feed through to strategic research, implementation and governance plans(Renken and Heeks, 2019). This paper has aimed to socialise the concept of digital psychological formulation with the goal of making a positive health impact on suicide prevention by promoting adoption and development of the concept by the CLPsych community.

There is a significant concept reality gap between current developments and getting to a stage of digital formulation being utilised by human and digital clinicians as part of standard mental health and suicide prevention practice (Heeks, 2006; Orr and Sankaran, 2007). However, it is a concept reality gap that is potentially fast narrowing and that the CLPsych community has both the developing expertise and ethical duty to take a leadership role in crossing, in a safe and clinically effective manner.

References

Mehnaz Adnan, Jim Warren, and Martin Orr. 2013. Semlink—dynamic generation of hyperlinks to enhance patient readability of discharge summaries. In *Proceedings of the 26th IEEE International Symposium on Computer-Based Medical Systems*, pages 35–40. IEEE.

Awais Aftab and Elizabeth Ryznar. 2021. Conceptual and historical evolution of psychiatric nosology. *International Review of Psychiatry*, 33(5):486–499.

Chittaranjan Andrade. 2015. The numbers needed to treat and harm (nnt, nnh) statistics: what they tell us and what they do not. *The Journal of clinical psychiatry*, 76(3):12971.

Angele Marie Beausoleil. 2018. Revisiting rogers: the diffusion of his innovation development process as a normative framework for innovation managers, students and scholars. *Journal of Innovation Management*, 6(4):73–97.

Adrian Benton, Glen Coppersmith, and Mark Dredze. 2017. Ethical research protocols for social media health research. In *Proceedings of the First ACL Workshop on Ethics in Natural Language Processing*, pages 94–102.

German E Berrios. 1985. The psychopathology of affectivity: conceptual and historical aspects. *Psychological medicine*, 15(4):745–758.

Ananya Bhatia-Lin, Alexandra Boon-Dooley, Michelle K Roberts, Caroline Pronai, Dylan Fisher, Lea Parker, Allison Engstrom, Leah Ingraham, and Doyanne Darnell. 2019. Ethical and regulatory considerations for using social media platforms to locate and track research participants. *The American Journal of Bioethics*, 19(6):47–61.

David M Blei. 2012. Probabilistic topic models. *Communications of the ACM*, 55(4):77–84.

David M Blei, Andrew Y Ng, and Michael I Jordan. 2003. Latent dirichlet allocation. *the Journal of machine Learning research*, 3:993–1022.

Virginia Braun and Victoria Clarke. 2006. Using thematic analysis in psychology. *Qualitative research in psychology*, 3(2):77–101.

Virginia Braun and Victoria Clarke. 2019. Reflecting on reflexive thematic analysis. *Qualitative Research in Sport, Exercise and Health*, 11(4):589–597.

Anna Bredström. 2019. Culture and context in mental health diagnosing: Scrutinizing the dsm-5 revision. *Journal of Medical Humanities*, 40(3):347–363.

British Psychological Society. 2017. Ethics guidelines for internet-mediated research. *Leicester, UK: British Psychological Society.*

Timothy A Carey and David Pilgrim. 2010. Diagnosis and formulation: What should we tell the students? *Clinical Psychology & Psychotherapy*, 17(6):447–454.

Nandini Chakraborty. 2020. The importance of embedding psychopathology and phenomenology in clinical practice and training in psychiatry. *BJPsych Advances*, 26(5):287–295.

Robert Challen, Joshua Denny, Martin Pitt, Luke Gompels, Tom Edwards, and Krasimira Tsaneva-Atanasova. 2019. Artificial intelligence, bias and clinical safety. *BMJ Quality & Safety*, 28(3):231–237.

Harriet Challoner and Fani Papayianni. 2018. Evaluating the role of formulation in counselling psychology: A systematic literature review. *The European Journal of Counselling Psychology*, 7(1).

S. Chancellor and M. De Choudhury. 2020. Methods in predictive techniques for mental health status on social media: a critical review. *NPJ Digit Med*, 3(1):43.

Anita Kumar Chang and Leslie H Lundahl. 2019. Which problem are we addressing today? the utility of a multifaceted formulation approach to a complex case. *American Journal of Psychotherapy*, 72(1):29–33.

E. Chiauzzi and P. Wicks. 2019. Digital trespass: Ethical and terms-of-use violations by researchers accessing data from an online patient community. *J Med Internet Res*, 21(2):e11985.

A. Clapperton, L. Bugeja, S. Newstead, and J. Pirkis. 2020. Identifying typologies of persons who died by suicide: Characterizing suicide in victoria, australia. *Arch Suicide Res*, 24(1):18–33.

Glen Coppersmith, Ryan Leary, Patrick Crutchley, and Alex Fine. 2018. Natural language processing of social media as screening for suicide risk. *Biomedical informatics insights*, 10:1178222618792860.

Kenneth L Critchfield, Francesco Gazzillo, and Ueli Kramer. 2022. Case formulation of interpersonal patterns and its impact on the therapeutic process: Introduction to the issue. *Journal of Clinical Psychology*.

Norberto Nuno Gomes de Andrade, Dave Pawson, Dan Muriello, Lizzy Donahue, and Jennifer Guadagno. 2018. Ethics and artificial intelligence: suicide prevention on facebook. *Philosophy & Technology*, 31(4):669–684.

Wayne A de Beer. 2017. Original opinion: the use of bloom's taxonomy to teach and assess the skill of the psychiatric formulation during vocational training. *Australasian Psychiatry*, 25(5):514–519.

Eberhard A Deisenhammer, Chy-Meng Ing, Robert Strauss, Georg Kemmler, Hartmann Hinterhuber, and

Elisabeth M Weiss. 2009. The duration of the suicidal process: how much time is left for intervention between consideration and accomplishment of a suicide attempt? *Journal of Clinical Psychiatry*, 70(1):19.

Mason Durie. 2017. Indigenous suicide: the turamarama declaration. *Journal of Indigenous Wellbeing*, 2(2):59–67.

Amy Evans and Kathleen Abrahamson. 2020. The influence of stigma on suicide bereavement: A systematic review. *Journal of Psychosocial Nursing and Mental Health Services*, 58(4):21–27.

Ethan Fast, Binbin Chen, and Michael S Bernstein. 2016. Empath: Understanding topic signals in large-scale text. In *Proceedings of the 2016 CHI Conference on Human Factors in Computing Systems*, pages 4647–4657.

Fara Aninha Fernandes and Georgi V Chaltikyan. 2020. Analysis of legal and regulatory frameworks in digital health: A comparison of guidelines and approaches in the european union and united states. *Journal of the International Society for Telemedicine and eHealth*, 8:e11 (1–13).

Irosh Fernando, Martin Cohen, and Frans Henskens. 2012. Pattern-based formulation: a methodology for psychiatric case formulation. *Australasian Psychiatry*, 20(2):121–126.

Scott J Fitzpatrick. 2018. Reshaping the ethics of suicide prevention: responsibility, inequality and action on the social determinants of suicide. *Public Health Ethics*, 11(2):179–190.

Luciano Floridi and Josh Cowls. 2019. A unified framework of five principles for ai in society. *Harvard Data Science Review*.

Luciano Floridi, Josh Cowls, Thomas C King, and Mariarosaria Taddeo. 2020. How to design ai for social good: Seven essential factors. *Science and Engineering Ethics*, 26(3):1771–1796.

Trehani M Fonseka, Venkat Bhat, and Sidney H Kennedy. 2019. The utility of artificial intelligence in suicide risk prediction and the management of suicidal behaviors. *Australian & New Zealand Journal of Psychiatry*, 53(10):954–964.

E. Ford, K. Curlewis, A. Wongkoblap, and V. Curcin. 2019. Public opinions on using social media content to identify users with depression and target mental health care advertising: Mixed methods survey. *JMIR Ment Health*, 6(11):e12942.

Joseph C Franklin, Jessica D Ribeiro, Kathryn R Fox, Kate H Bentley, Evan M Kleiman, Xieyining Huang, Katherine M Musacchio, Adam C Jaroszewski, Bernard P Chang, and Matthew K Nock. 2017. Risk factors for suicidal thoughts and behaviors: a meta-analysis of 50 years of research. *Psychological bulletin*, 143(2):187.

Stacey Freedenthal. 2007. Challenges in assessing intent to die: can suicide attempters be trusted? *OMEGA-Journal of death and dying*, 55(1):57–70.

Eiko I Fried. 2015. Problematic assumptions have slowed down depression research: why symptoms, not syndromes are the way forward. *Frontiers in psychology*, 6:309.

Chris D Frith. 2021. The neural basis of consciousness. *Psychological medicine*, 51(4):550–562.

Igor Galynker, Zimri S Yaseen, Abigail Cohen, Ori Benhamou, Mariah Hawes, and Jessica Briggs. 2017. Prediction of suicidal behavior in high risk psychiatric patients using an assessment of acute suicidal state: the suicide crisis inventory. *Depression and anxiety*, 34(2):147–158.

Nicole Geach, Nima G Moghaddam, and Danielle De Boos. 2018. A systematic review of team formulation in clinical psychology practice: definition, implementation, and outcomes. *Psychology and Psychotherapy: Theory, Research and Practice*, 91(2):186–215.

Michael SA Graziano. 2021. Understanding consciousness. *Brain*, 144(5):1281–1283.

E Clare Harris and Brian Barraclough. 1997. Suicide as an outcome for mental disorders. a meta-analysis. *British journal of psychiatry*, 170(3):205–228.

Simon Hatcher, Allison Crawford, and Nicole Coupe. 2017. Preventing suicide in indigenous communities. *Current opinion in psychiatry*, 30(1):21–25.

Richard Heeks. 2006. Health information systems: Failure, success and improvisation. *International journal of medical informatics*, 75(2):125–137.

M. A. Hom, M. C. Podlogar, I. H. Stanley, and T. E. Joiner. 2017. Ethical issues and practical challenges in suicide research. *Crisis*, 38(2):107–114.

R. F. Hunter, A. Gough, N. O'Kane, G. McKeown, A. Fitzpatrick, T. Walker, M. McKinley, M. Lee, and F. Kee. 2018. Ethical issues in social media research for public health. *Am J Public Health*, 108(3):343–348.

Lucy Johnstone. 2018. Psychological formulation as an alternative to psychiatric diagnosis. *Journal of Humanistic Psychology*, 58(1):30–46.

Lucy Johnstone and Rudi Dallos. 2013. *Introduction to formulation*, pages 21–37. Routledge.

Kenneth S Kendler. 2009. An historical framework for psychiatric nosology. *Psychological medicine*, 39(12):1935–1941.

R. C. Kessler, R. M. Bossarte, A. Luedtke, A. M. Zaslavsky, and J. R. Zubizarreta. 2020. Suicide prediction models: a critical review of recent research with recommendations for the way forward. *Mol Psychiatry*, 25(1):168–179.

Lina Keutzer and Ulrika SH Simonsson. 2020. Medical device apps: an introduction to regulatory affairs for developers. *JMIR mHealth and uHealth*, 8(6):e17567.

Cynthia F Kurtz and David J Snowden. 2003. The new dynamics of strategy: Sense-making in a complex and complicated world. *IBM systems journal*, 42(3):462–483.

Martin J La Roche, Milton A Fuentes, and Devon Hinton. 2015. A cultural examination of the dsm-5: Research and clinical implications for cultural minorities. *Professional Psychology: Research and Practice*, 46(3):183.

M. M. Large. 2018. The role of prediction in suicide prevention. *Dialogues Clin Neurosci*, 20(3):197–205.

Matthew Michael Large, Daniel Thomas Chung, Michael Davidson, Mark Weiser, and Christopher James Ryan. 2017. In-patient suicide: selection of people at risk, failure of protection and the possibility of causation. *BJPsych open*, 3(3):102–105.

Keri Rose Lawson-Te Aho Dr. 2017. The case for re-framing māori suicide prevention research in aotearoa/new zealand: Applying lessons from indigenous suicide prevention research. *Journal of Indigenous Research*, 6(2017):1.

Richard T Liu, Alexandra H Bettis, and Taylor A Burke. 2020. Characterizing the phenomenology of passive suicidal ideation: a systematic review and meta-analysis of its prevalence, psychiatric comorbidity, correlates, and comparisons with active suicidal ideation. *Psychological medicine*, 50(3):367–383.

Chris Mace and Sharon Binyon. 2005. Teaching psychodynamic formulation to psychiatric trainees: Part 1: Basics of formulation. *Advances in Psychiatric Treatment*, 11(6):416–423.

Narayana Manjunatha. 2019. Case presentation in academic psychiatry: The clinical applications, purposes, and structure of formulation and summary. *Indian Journal of Psychiatry*, 61(6):644.

Ian Marsh. 2018. *Historical phenomenology: Understanding experiences of suicide and suicidality across time*, pages 1–12. Springer.

Jeffery Martin, Jessica M LaCroix, Laura A Novak, and Marjan Ghahramanlou-Holloway. 2020. Typologies of suicide: A critical literature review. *Archives of suicide research*, 24(sup1):25–40.

Kahu McClintock and Rachel McClintock. 2017. Hoea te waka: Indigenous suicide prevention outcomes framework and evaluation processes-part 1. *Journal of Indigenous Wellbeing*, 2(2):68–76.

Catherine M McHugh, Amy Corderoy, Christopher James Ryan, Ian B Hickie, and Matthew Michael Large. 2019. Association between suicidal ideation and suicide: meta-analyses of odds ratios, sensitivity, specificity and positive predictive value. *BJPsych open*, 5(2).

Lindsey C McKernan, Ellen W Clayton, and Colin G Walsh. 2018. Protecting life while preserving liberty: ethical recommendations for suicide prevention with artificial intelligence. *Frontiers in psychiatry*, 9:650.

Giorgia Michelini, Isabella M Palumbo, Colin G DeYoung, Robert D Latzman, and Roman Kotov. 2021. Linking rdoc and hitop: A new interface for advancing psychiatric nosology and neuroscience. *Clinical psychology review*, 86:102025.

NEAC. 2019. National ethical standards for health and disability research and quality improvement. *Wellington: Ministry of Health.*

Barnaby Nelson, Patrick D McGorry, and Anthony V Fernandez. 2021. Integrating clinical staging and phenomenological psychopathology to add depth, nuance, and utility to clinical phenotyping: a heuristic challenge. *The Lancet Psychiatry*, 8(2):162–168.

Matthew K Nock, Franchesca Ramirez, and Osiris Rankin. 2019. Advancing our understanding of the who, when, and why of suicide risk. *JAMA psychiatry*, 76(1):11–12.

Martin Orr and Shankar Sankaran. 2007. Mutual empathy, ambiguity, and the implementation of electronic knowledge management within the complex health system. *Emergence: Complexity & Organization*, 9.

Sherry Pagoto and Camille Nebeker. 2019. How scientists can take the lead in establishing ethical practices for social media research. *Journal of the American Medical Informatics Association*, 26(4):311–313.

A. R. Pisani, D. C. Murrie, and M. M. Silverman. 2016. Reformulating suicide risk formulation: From prediction to prevention. *Acad Psychiatry*, 40(4):623–9.

Alejandro Porras-Segovia, Maria M Perez-Rodriguez, Pilar López-Esteban, Philippe Courtet, Jorge López-Castromán, Jorge A Cervilla, Enrique Baca-García, et al. 2019. Contribution of sleep deprivation to suicidal behaviour: a systematic review. *Sleep medicine reviews*, 44:37–47.

W Nicholson Price, Sara Gerke, and I Glenn Cohen. 2019. Potential liability for physicians using artificial intelligence. *Jama*, 322(18):1765–1766.

S. Pridmore. 2015. Mental disorder and suicide: a faulty connection. *Aust N Z J Psychiatry*, 49(1):18–20.

Bahlol Rahimi, Hamed Nadri, Hadi Lotfnezhad Afshar, and Toomas Timpka. 2018. A systematic review of the technology acceptance model in health informatics. *Applied clinical informatics*, 9(03):604–634.

Diana Rangihuna, Mark Kopua, and David Tipene-Leach. 2018. Mahi a atua: A pathway forward for māori mental health. *New Zealand Medical Journal*, 131(1471):79–83.

Jaco Renken and Richard Heeks. 2019. Champions of is innovations. *Communications of the Association for Information Systems*, 44(1):38.

Philip Resnik, April Foreman, Michelle Kuchuk, Katherine Musacchio Schafer, and Beau Pinkham. 2020. Naturally occurring language as a source of evidence in suicide prevention. *Suicide and Life-Threatening Behavior*.

C. J. Ryan. 2015. Suicide explained! *Aust N Z J Psychiatry*, 49(1):83–4.

Han-Chin Shing, Suraj Nair, Ayah Zirikly, Meir Friedenberg, Hal Daumé III, and Philip Resnik. 2018. Expert, crowdsourced, and machine assessment of suicide risk via online postings. In *Proceedings of the Fifth Workshop on Computational Linguistics and Clinical Psychology: From Keyboard to Clinic*, pages 25–36.

Han-Chin Shing, Philip Resnik, and Douglas W Oard. 2020. A prioritization model for suicidality risk assessment. In *Proceedings of the 58th Annual Meeting of the Association for Computational Linguistics*, pages 8124–8137.

Dan J Stein and Geoffrey M Reed. 2019. Global mental health and psychiatric nosology: Dsm-5, icd-11, and rdoc. *Brazilian Journal of Psychiatry*, 41(1):3–4.

Mark Sullivan, Michael F Walton, Elizabeth L Auchincloss, and Julie B Penzner. 2020. Teaching psychiatric formulation to residents and faculty. *Academic Psychiatry*, pages 1–4.

George Szmukler. 2014. Fifty years of mental health legislation: Paternalism, bound and unbound. *Psychiatry: Past, present, and prospect*, pages 133–153.

Hamed Taherdoost. 2018. A review of technology acceptance and adoption models and theories. *Procedia manufacturing*, 22:960–967.

Myrthe L Tielman, Mark A Neerincx, Claudia Pagliari, Albert Rizzo, and Willem-Paul Brinkman. 2019. Considering patient safety in autonomous e-mental health systems–detecting risk situations and referring patients back to human care. *BMC medical informatics and decision making*, 19(1):47.

Michelle Torok, Jin Han, Simon Baker, Aliza Werner-Seidler, Iana Wong, Mark E Larsen, and Helen Christensen. 2020. Suicide prevention using self-guided digital interventions: a systematic review and meta-analysis of randomised controlled trials. *The Lancet Digital Health*, 2(1):e25–e36.

Leanne Townsend and Claire Wallace. 2016. Social media research: A guide to ethics. *University of Aberdeen*, pages 1–16.

Andrew S Tubbs, Michael L Perlis, and Michael A Grandner. 2019. Surviving the long night: the potential of sleep health for suicide prevention. *Sleep medicine reviews*, 44:83.

Kathryn Turner, Nicolas JC Stapelberg, Jerneja Sveticic, and Anthony R Pisani. 2021. Suicide risk classifications do not identify those at risk: where to from here? *Australasian psychiatry*, page 10398562211032233.

Clive Unsworth. 1991. Mental disorder and the tutelary relationship: from pre-to post-carceral legal order. *Journal of Law and Society*, 18(2):254–278.

Monika A Waszczuk, Mark Zimmerman, Camilo Ruggero, Kaiqiao Li, Annmarie MacNamara, Anna Weinberg, Greg Hajcak, David Watson, and Roman Kotov. 2017. What do clinicians treat: Diagnoses or symptoms? the incremental validity of a symptom-based, dimensional characterization of emotional disorders in predicting medication prescription patterns. *Comprehensive Psychiatry*, 79:80–88.

Priyanthy Weerasekera. 1993. Formulation: A multiperspective model. *The Canadian Journal of Psychiatry*, 38(5):351–358.

Matthew L Williams, Pete Burnap, Luke Sloan, Curtis Jessop, and Hayley Lepps. 2017. *Users' views of ethics in social media research: Informed consent, anonymity, and harm*. Emerald Publishing Limited.

R. Woodford, M. J. Spittal, A. Milner, K. McGill, N. Kapur, J. Pirkis, A. Mitchell, and G. Carter. 2019. Accuracy of clinician predictions of future self-harm: A systematic review and meta-analysis of predictive studies. *Suicide Life Threat Behav*, 49(1):23–40.

World Health Organization. 2014. *Preventing suicide: A global imperative*. World Health Organization.

Marianne Wyder, Manaan Kar Ray, Samara Russell, Kieran Kinsella, David Crompton, and Jeremy van den Akker. 2021. Suicide risk assessment in a large public mental health service: do suicide risk classifications identify those at risk? *Australasian Psychiatry*, 29(3):322–325.

Zimri S Yaseen, Mariah Hawes, Shira Barzilay, and Igor Galynker. 2019. Predictive validity of proposed diagnostic criteria for the suicide crisis syndrome: an acute presuicidal state. *Suicide and Life-Threatening Behavior*, 49(4):1124–1135.

Peter K Yu. 2020. The algorithmic divide and equality in the age of artificial intelligence. *Florida Law Review*, 72:19–44.

Mark Zimmerman, William Ellison, Diane Young, Iwona Chelminski, and Kristy Dalrymple. 2015. How many different ways do patients meet the diagnostic criteria for major depressive disorder? *Comprehensive psychiatry*, 56:29–34.

Ayah Zirikly, Philip Resnik, Ozlem Uzuner, and Kristy Hollingshead. 2019. Clpsych 2019 shared task: Predicting the degree of suicide risk in reddit posts. In *Proceedings of the Sixth Workshop on Computational Linguistics and Clinical Psychology*, pages 24–33.

Explaining Models of Mental Health via Clinically Grounded Auxiliary Tasks

Ayah Zirikly
Johns Hopkins University
aziriklyθjhu.edu

Mark Dredze
Johns Hopkins University
mdredzeθcs.jhu.edu

Abstract

Models of mental health based on natural language processing can uncover latent signals of mental health from language. Models that indicate whether an individual is depressed, or has other mental health conditions, can aid in diagnosis and treatment. A critical aspect of integration of these models into the clinical setting relies on explaining their behavior to domain experts. In the case of mental health diagnosis, clinicians already rely on an assessment framework to make these decisions; that framework can help a model generate meaningful explanations.

In this work we propose to use PHQ-9 categories as an auxiliary task to explaining a social media based model of depression. We develop a multi-task learning framework that predicts both depression and PHQ-9 categories as auxiliary tasks. We compare the quality of explanations generated based on the depression task only, versus those that use the predicted PHQ-9 categories. We find that by relying on clinically meaningful auxiliary tasks, we produce more meaningful explanations.

1 Introduction

Mental illness has a huge impact on the health and well-being of the United States and world populations. In the US, 25% of the population suffered at some point from mental illness [1]. The urgency to address the mental health crisis became even more critical with the COVID-19 pandemic and its negative impact on mental health, burdening kids and seniors especially (Loades et al., 2020). Depression is among the most prevalent mental disorders. In the United States alone, 21 million adults had at least one major depressive episode [2].

Computational linguistics and natural language processing (NLP) research on mental health has received increased attention in the last decade, with work on suicide risk assessment (Zirikly et al., 2019; Shing et al., 2018; De Choudhury et al., 2016; Coppersmith et al., 2018), anxiety prediction and classification (Osadchiy et al., 2020), and depression prediction and classification (Coley et al., 2021; De Choudhury et al., 2013), among many other tasks. Although clinical data was used for some models (Penfold et al., 2021), prior work also utilized other sources of data, such as social media to overcome challenges in data access and to better understand what influences mental health on a daily basis. The majority of the NLP research on depression classification is focused on improving performance to achieve state-of-the-art models. Such models typically act like a black box, and predictions are therefore not explainable. This results in poor integration of these models into clinical settings, given that clinicians need to understand why a patient is identified as depressed, so that they can make informed decisions in regards to diagnosis and evidence-based treatment (Zhou et al., 2015). Additionally, it has been shown that black-box models are not generalizable across different data genres or domains (Harrigian et al., 2020). This accentuates the need for explainable models, as they could help to troubleshoot and understand the transfer between datasets – e.g. within social media or from social media to electronic health records (EHR).

Recently, and with the proliferation of deep learning in particular, explainable AI (XAI) has attracted significant attention, with the field publishing multiple techniques that provide explanations for machine learning models. Techniques like LIME (Ribeiro et al., 2016) and SHAP (Lundberg and Lee, 2017) have been widely adopted and proven to work in different domains, mental health being one of them (Hu and Sokolova, 2021; Spruit et al., 2022; Uddin et al., 2022).

Clinicians rely on ongoing assessment of pa-

[1] https://www.nimh.nih.gov/health/statistics/mental-illness

[2] https://www.nimh.nih.gov/health/statistics/major-depression

Proceedings of the Eighth Workshop on Computational Linguistics and Clinical Psychology, pages 30 - 39
July 15, 2022

tient progress and well-being for therapeutic decisions. Many assessment instruments exist, including questionnaires such as the Patient Health Questionnaire (PHQ-9) for depression and the General Anxiety Disorder (GAD-7) screener for anxiety. PHQ-9 (Kroenke et al., 2001) is one of the most commonly used and validated depression assessment tools that mental health clinicians and primary care physicians use. The questionnaire addresses the presence and severity of nine symptoms or categories such as problems with sleep, eating, and self-harm to assess and monitor a patient's depression severity.

In this work, we leverage the availability of PHQ-9, a clinically accepted and interpretable tool to measure depression severity, and integrate its items into depression classification models as auxiliary classification tasks. We claim and prove that LIME explanations generated for models that use such clinically grounded auxiliary tasks are better and more informative than explanations on other *black-box* models that do not use these auxiliary models in the decision process.

We summarize our contributions as follows:

- We created a manually labeled dataset that highlights the most prominent terms in a tweet as the explanation for depression,
- designed a multi-task learning framework that uses PHQ-9 categories for depression classification, and
- showed that using auxiliary models (PHQ-9) improves the explainability of depression detection models, regardless of the complexity of the underlying model.

2 Related work

Depression classification has been an important area of focus in mental health NLP in social media data and electronic health records (EHR). To overcome the challenges of data access and to create community-based datasets, many initiatives started using Twitter and Reddit platforms to create depression annotated datasets. These datasets were collected using self-reported terms and regular expressions such as *I was diagnosed with depression* (Coppersmith et al., 2015), or in the case of Reddit, using mental health related subreddits (e.g. r/ADHD) as a proxy to retrieve relevant posts (Pirina and Çöltekin, 2018; Cohan et al., 2018; Yates et al., 2017). Many common techniques related to linguistic features are used to perform the classification task such as using LIWC in social media (Morales et al., 2017; Loveys et al., 2018) and EHR (Bittar et al., 2021). Researchers used a variety of machine learning techniques that range from conventional methods such as SVM (Tadesse et al., 2019; Yazdavar et al., 2017) and LR (Yazdavar et al., 2017; Karmen et al., 2015), to deep learning techniques such as feed-forward networks (Geraci et al., 2017), CNN and LSTM (Mumtaz and Qayyum, 2019; Kour and Gupta, 2022). Many recent work also explored the use of recent pre-trained language models to improve the depression classification task performance, such as BERT-CNN in (Rodrigues Makiuchi et al., 2019) and ALBERT (Owen et al., 2020). There has been a line of research that focused on predicting the symptoms (PHQ categories). (Delahunty et al., 2019) introduced a deep neural network model to predict PHQ-4 scores in Reddit depression dataset (Losada and Crestani, 2016) and DAIC-WOZ transcribed clinical interviews (Gratch et al., 2014). (Yadav et al., 2020) proposed identifying the presence of the depressive symptoms using the auxiliary task of figurative usage detection.

In the area of explainable AI (XAI), most the work that has been done focused on using explainable techniques to highlight the most important features in depression prediction. (Nemesure et al., 2021) used SHAP values (Lundberg and Lee, 2017) to highlight which features were most salient in the depression classification model. (Choi et al., 2020) used LIME to understand which features weighed the most in identifying college students at high risk of depressive disorder. In a recent work by (Nguyen et al., 2022), the authors showed the positive impact of using depression classifiers that are constrained by PHQ-9 symptoms, on their generalizability across different datasets.

3 Data

In this work, we focus on social media data because public access to clinical datasets is limited. The publicly available social media datasets that address depression classification only contain labels for depression (Coppersmith et al., 2015; Cohan et al., 2018), and it is challenging to find publicly available data that has annotations for both depression and PHQ-9 categories.

For our experiments, we use the **D**epression to (**2**) **S**ymptoms (D2S) dataset (Yadav et al., 2020). It is a collection of English only tweets that was crawled using depression-related terms that can be categorized into one of the PHQ-9 categories (symptoms): (S1) lack of interest, (S2) feeling down or depressed, (S3) trouble with sleeping, (S4) lack of energy, (S5) eating disorder, (S6) low self-esteem, (S7) concentration problems, (S8) hyper/lower activity, and (S9) self-harm.

The dataset contains the list of annotated tweet IDs, and a total of 3738 tweets labeled as depressive and 8417 as not depressive (control). The depressive tweets are further annotated with symptoms, where a label of 1 is assigned for $S9$ if the tweet has mentions of self-harm thoughts, 0 otherwise, and so forth for all 9 categories, where multiple categories can receive a 1 annotation. It is worth mentioning that the data, unlike PHQ-9 questionnaire, does not have scores for each category, but only a binary label. Additionally, the original dataset has annotations for sarcasm and metaphor labels for the depressive tweets, since Yadav et al. (2020) focused on the task of understanding how to classify PHQ-9 categories using the sarcasm and metaphor language labels. However, in our work we focus on the depressive and PHQ-9 symptoms annotations, with all annotations scoped at the tweet level, not at the user level as is the case in some other datasets.

We collected the tweets corresponding to the tweet IDs described in D2S using the Twitter API. Some tweets had become unavailable since the publication of D2S, resulting in a reduced dataset with 2132 depressive tweets and 5698 control tweets. Notwithstanding the change in dataset size, we adopt the train, dev, and test splits of D2S to maintain consistency. Table 1 shows the characteristics of the dataset splits, and the distribution of PHQ-9 annotations.

4 Depression classification models

Understanding the domain and the task should be the foundation in designing an NLP model, as opposed to simply applying NLP state-of-the-art models that are hard to interpret. This is especially true for clinical and mental health NLP, where a lack of explainability would result in poor integration in clinical settings. In our work, we aim to build models that mimic a clinical setting, where the clinician uses the scores from PHQ-9 questionnaires to screen if a patient is suffering from depression and to assess its severity.

In this section we discuss the approaches we used to build models that predict if a tweet is depressive or not. We propose three models; the first two, similarly to previous literature, focus on the depression classification task as a standalone problem, without considering how symptoms, in our case the PHQ-9 categories, can help interpret and influence the model's performance. The last model aims to study how predicting symptoms can help in classifying depressive tweets.

In the following subsections, we will describe our models and the balancing techniques we used to address the skewed distributions for the depressive and symptom labels.

4.1 Single task classification models

The task formulation for these models is as follows: given tweet t, classify if t is depressive (dep) or has any of the symptoms (PHQ-9 categories) enabled. For this task, we propose two simple models: logistic regression (LR) and multilayer perceptron (MLP). Both of these models take as input the pre-processed tweet. The preprocessing steps we have used include: lowercasing, tokenizing the tweet, normalizing the numbers (e.g. 123 -> 000), and removing tokens that occur less than 3 times in the training set. The preprocessed tweet text was then vectorized using a term frequency-inverse document frequency (TF-IDF) vectorizer with L2 regularization. We are aware of more sophisticated methods to build representations of the input text, such as applying and fine-tuning BERT contextual embeddings (Devlin et al., 2018; Brown et al., 2020), that could improve results. However, the focus of this paper is not to provide the best performance, but rather to show how using auxiliary models can help in providing better explanations for the depression classification models. Additionally, we believe simpler input forms can make the explainability process cleaner.

For each of the single task approaches, we build 10 different models that can address the classification tasks separately (dep + 9 symptoms).

Logistic Regression In this model we use logistic regression with a maximum of 50 iterations and L2 regularizer. For balancing the data, we apply a higher weight class for the *1:enabled* class for each of the depressive and symptom classes.

Table 2 shows the results of this model on the test data, where the symptoms models are trained

split	control	dep	S1	S2	S3	S4	S5	S6	S7	S8	S9
train	3989	1615	237	235	97	140	173	426	69	51	468
dev	570	140	16	19	5	5	26	53	6	4	28
test	1139	377	32	110	35	29	51	89	6	6	113
all	5698	2132	285	364	137	174	250	568	81	61	609

Table 1: Data statistics

	dep	S1	S2	S3	S4	S5	S6	S7	S8	S9
Precision	0.725	0.214	0.353	0.923	0.8	0.677	0.324	0	0	0.513
Recall	0.629	0.188	0.109	0.343	0.414	0.412	0.528	0	0	0.513
F1	0.673	0.2	0.167	0.5	0.546	0.512	0.402	0	0	0.513

Table 2: Logistic regression results for the single classification task

	dep	S1	S2	S3	S4	S5	S6	S7	S8	S9
Precision	0.249	0.093	0.081	0.308	0.186	0.16	0.14	0.008	0.007	0.25
Recall	1	0.625	0.518	0.571	0.552	0.726	0.652	0.167	0.167	0.558
F1	0.398	0.161	0.139	0.4	0.278	0.262	0.231	0.015	0.013	0.345

Table 3: MLP results for the single-task classification

	dep	S1	S2	S3	S4	S5	S6	S7	S8	S9
Precision	0.765	0.761	0.764	0.768	0.765	0.767	0.766	0.77	0.767	0.767
Recall	0.508	0.489	0.487	0.503	0.487	0.517	0.512	0.49	0.502	0.503
F1	0.611	0.595	0.595	0.608	0.595	0.618	0.614	0.595	0.607	0.608

Table 4: MTL results for the multitask classification

on the depressive only tweets and tested on all the test data (dep + control). We do not report accuracy given how skewed the dataset is.

Multilayer Perceptron Our multilayer perceptron model (MLP) is a three-layer fully connected feedforward network with a hidden layer of size 256. The best parameters obtained for this model on the dev data are: learning rate of 1e-3, a batch size of 32, dropout probability of 0.5, Adam optimizer (Kingma and Ba, 2014), and cross-entropy as the loss function. We minimize the impact of imbalanced data by balancing each batch separately for the 0/1 classes. Similarly to our LR model, the symptoms classifiers use only the depressive tweets for the training and development sets to minimize the imbalance, and because the non-depressive tweets are automatically given label 0 for each of the symptoms. However, for testing we use both depressive and control tweets to mimic the real-life scenario where we don't know the depression status of a patient. The results of our MLP model are depicted in table 3.

4.2 Multitask classification model

Our research question is based on studying the impact of using auxiliary models (symptoms) to generate better explanations for the depression classification model. Given that, we adopt a multitask learning (MTL) framework that classifies each tweet as depressive or not, in addition to each of the 9 PHQ-9 categories (symptoms), simultaneously. For comparability with our MLP model, we adopt the same neural network design choices. the MTL framework consists of multiple MLP networks, one for each of the tasks, with the same parameters in terms of dropout, learning rate, number of hidden layers, optimizer, and loss function. Table 4 shows the results of our MTL proposed model. Similarly to LR and MLP, the symptoms classification task uses only depressive tweets from the training and development sets.

5 Depression model explanations

It has been argued that depression classification models that use machine learning, and deep learning techniques in particular, have been hard to integrate into clinical settings due to the difficulty of interpreting and explaining their results (Sendak et al., 2019). In the literature, there are many initiatives to generate explanations for blackbox models such as LIME (Ribeiro et al., 2016) and SHAP (Lundberg and Lee, 2017), which are highly adopted and used. In our work, to test our hypothesis, we compare the explanations generated by LIME for each of the models listed in section 4 with our in-house gold annotated explanations dataset.

5.1 Explanations dataset

We randomly sampled 105 tweets from the test dataset that are depressive (*D2S-explain*), and manually annotated them. We had one annotator that is experienced in mental health research and its intersection with computational linguistics that read the tweets, and for each tweet identified the tokens that signal depression or that are most relevant to it. To evaluate the quality of the annotation, 25 randomly selected tweets from *D2S-explain* were checked by another annotator that is also an expert in mental health research with a degree in psychology. The first annotator is not a native English speaker, but has full professional proficiency in English, while the second annotator is a native English speaker. The second annotator had three options: accept, modify, or reject an explanation. This process was repeated until we reached 85% agreement for *accept* on the 25 tweets, after which the rest of *D2S-explain* was re-annotated by the first annotator.

Figure 1 shows an example tweet and its corresponding manual annotations [3]. Table 5 shows the details of the number of tweets that have any of the 9 symptoms enabled. Upon acceptance we plan to make the dataset publicly available under a DUA as discussed in 7.

```
I feel that existence is
pointless and everything is
hurting me to a point that I
can't sleep anymore.  My stomach
hurts every time I eat and I
feel that I need to throw up.

existence is pointless |
everything is hurting me
```

Figure 1: Example of manually labeling explanation terms in a tweet

[3] All example tweets are paraphrased for privacy.

5.2 Explanations evaluation

For each of the three models we developed, we employ LIME to generate explanations for the D2S-explain dataset. LIME is able to generate explanations by creating an interpretable model that is an approximation of the original model for each data point (tweet) from the dataset. The LIME explanations look like probability scores for all inputs (in our case, tokens) that indicate how much they are expected to have contributed to the output classification. By looking at the highest-probability tokens of a tweet, we can get a sense of what information the model has used to make its prediction for that tweet.

We identify the following three scenarios for generating and evaluating the explanations:

- **(D)** We generate explanations for each of the three models (LR, MLP, and MTL) for the **depressive** classification task. We rank the explanations (tokens) generated by LIME based on their top relevance probabilities and use the first ten tokens.
- **(S/S-comb)** In this scenario we generate the explanations for the **symptoms** prediction task for only the tweets that were predicted to have the corresponding symptom (S) enabled and that are correctly predicted by the model as depressive. Additionally, we combine all the explanations from the 9 symptoms models and rank the relevance/contributing probabilities of the tokens then pick the top ten tokens (S-comb).
- **(D+S)** In this scenario, we combine the explanations from the 9 symptoms models – same criteria as scenario S, with the explanations from scenario (D). Similarly to S-comb, we rank the relevance probabilities for all the explanations and pick the top ten.

The reason behind structuring the scenarios as proposed is to reflect the research question we formulated earlier and study the impact of using the explanations generated from the auxiliary tasks to help explain and interpret the depression models' outputs, as opposed to using the depression classification models alone.

For evaluation, we use the recall metric since we are mainly interested whether the models were able to generate explanation tokens that match the ones in the gold explanations. A prediction is considered

S1(1)	S2	S3	S4	S5	S6	S7	S8	S9
9	31	8	1	6	16	1	1	46

Table 5: Annotated test data sample stats

a true positive if the predicted explanation is fully or partially in the gold explanation. For instance, if the generated explanation is *lost hope* and the gold explanation is *lost hope in life*, the explanation is considered to be correct and the number of true positives increases by 1. However, this partial matching strategy only applies if the generated explanation contains more than only function words, stop words or pronouns; no credit is given for partial matches of that type. The reason behind the choice for a partial match evaluation is that it is sufficient for a clinician or mental health expert to see part of the term highlighted to understand why a model signaled depression.

D	Recall
LR	0.61
MLP	0.267
MTL	0.524

Table 6: Recall explanation results for scenario (D)

Tables 6, 7, and 8 show the recall performance for each of the scenarios listed above, which will be discussed in the next section.

6 Discussion

When we look at the results of the depression and symptoms classification task in tables 2, 3, and 4, we note that MLP yields the worst results across almost all the labels (dep and symptoms), whereas LR provides the best results for dep. However, its performance on the symptoms is poor, especially for concentration (S7) and activity (S8), where the number of positive instances is very limited. The MTL model, meanwhile, performs slightly worse than LR for the dep class, but is able to perform much better for all the symptoms and is not susceptible to the imbalanced nature of the data. For instance, the F1-scores for S7 and S8 in the MTL setting improve drastically. This observation supports the claim that using the symptoms with the depression labels can provide more reliable performance where we can think of the symptoms predictions as the first layer of explanations we can provide to the clinicians.

After applying LIME on each of the models, we note that the recall of the LR explanations is the highest among the three models at 0.61 (table 6). This is expected, given that LR performance on the dep class is the highest. When we qualitatively examined the explanations' output and compare the results between LR, MLP and MTL, we note that both LR and MTL explanations contain more relevant terms. Additionally, the fact that MLP performs much worse on correctly predicting the dep label affects the performance of the explanations recall. Figure 2 shows a paraphrased tweet example with the explanations generated from the three models.

```
At certain times and without any
trigger, I think I am probably
not even mentally ill, but
rather just an attention seeking
sh**
mentally ill | attention seeking
sh** [gold annotation]
even, mentally, ..., seeking,
ill, attention [LR]
even, time, trigger, ...,
mentally, ill [MLP]
mentally, seeking, sh**,
trigger, mentally,..., ill,
attention [MTL]
```

Figure 2: Example of the explanations from LR and MTL

The main research question we aimed to address is: does augmenting the explanations for depression models with those for PHQ-9 models provide more meaningful explanation to clinicians than those for depression models alone? To answer this, we need to check the recall performance for each of the three models for the D+S scenario in table 8. For the LR model, although the performance was poor for symptoms, the explanation recall increased 1.9% when augmenting with the symptoms' explanations. For MLP and MTL, the increase in recall performance is smaller with almost 1%. In MTL, we reason the smaller increase is caused by the fact that the MTL model already utilizes the symptoms to optimize the depression classification performance in its network design, thus MTL explanations produced for (D) reflects, to some extent,

S/S-avg	S1	S2	S3	S4	S5	S6	S7	S8	S9	S-comb
LR	0.057	0	0.01	0	0.019	0.076	0	0	0.152	0.152
MLP	0.267	0.267	0.267	0.267	0.267	0.267	0.267	0.267	0.267	0.267
MTL	0.533	0.533	0.495	0.524	0.533	0.524	0.514	0.533	0.533	0.533

Table 7: Recall explanation results for scenario (S/S-comb)

D+S	dep
LR	0.629
MLP	0.276
MTL	0.533

Table 8: Recall explanation results for scenario (D+S)

augmenting with the symptoms. We note that in the case of LR and MPL, the symptom models are independent from the depression model, so LIME explanations generated for those symptoms cannot technically be interpreted as having explained the depression model outputs. However, our results show that augmenting with explanations from these disjoint models improves recall of input tokens that would aid a clinician in evaluating tweets that get flagged as depressive, by focusing their attention on clinically relevant information.

To further support our claim and to make sure that ensembling multiple models will not also produce better results than D, we implement bagging techniques. We create 9 random samples to mimic the 9 symptoms sample size. For instance, sample 1 will randomly select 237 depressive and 1378 control tweets to mimic the size of the S1 dataset; the same technique would apply for each of the samples. We report the F1-score, in table 9, for the worst and best model based on which sample it has used. The results show a variance in performance which made us further investigate the recall performance of the explanations if we combined the explanations from (D) with the random 9 models explanations. The results are depicted in table 10 and show that (9samples+D) generates worse results than (D) and (D+S).

Limitations We understand that our work and results are limited in a number of ways. First, the D2S dataset is a Twitter dataset, which by itself can raise some questions about its reliability, however, we justify our decision due to the lack of clinical data access and this can be a proxy to prove our hypothesis using the symptoms models. We are also aware that the dataset is small and its distributions are skewed. In future work, we hope that we or other researchers can generate a large scale dataset for depression with PHQ-9 score annotations. Additionally, describing symptoms in tweets can be challenging due to the short text that cannot provide enough information about symptoms and/or depression. Another limitation is that the PHQ-9 annotations in D2S are binary, unlike the 4-point scale that is used in the PHQ-9 questionnaire, which allows to capture severity of symptoms. Choosing between 0 and 1 can be difficult in gray area cases, and degrades annotation quality. Finally, the manually annotated explanations in *D2S-explain* are only a proxy for what a clinician might find most informative in assessing tweets that are automatically flagged as depressive. Evaluating the informativeness of explanations in a true clinical setting would shed more light on this, but is beyond the scope of this paper.

7 Ethics statement

Although Tweets are publicly available, given the sensitivity of the task, we took the following extra measures, in light of what has been previously published by (Benton et al., 2017) and (Šuster et al., 2017).

- We obtained access to the D2S dataset after signing a data use agreement (DUA), and we followed all the agreements and instructions stated in the DUA. The dataset is stored on a secure server and not published with other researchers but those mentioned in the DUA and got approval.
- We did not obtain institutional review board (IRB) approval, since the dataset falls under *exempt determination* and not IRB approval, as stated in the code of federal regulations CFR 46.101(b)(4)[4] published by the United States Department of Health and Human Services (HHS).
- Our publicly available annotated explanations dataset will enforce a DUA that will respect

[4] https://www.hhs.gov/ohrp/sites/default/files/ohrp/policy/ohrpregulations.pdf

	worst model	best model	reported model
LR	0.645	0.684	0.692
MLP	0.384	0.42	0.398
MTL	0.668	0.692	0.611

Table 9: F1 performance results with bagging

	original model	9samples	9samples+D
LR	0.629	0.5	0.6
MLP	0.276	0.21	0.24
MTL	0.533	0.472	0.51

Table 10: Explanations recall performance with bagging

all the requirements of the D2S dataset DUA, in addition to any extra needed regulations and instructions.

8 Conclusion

Providing models that are explainable and adopt clinically grounded questionnaires is critical in building NLP solutions that can be integrated in clinical settings. In this work, we show that using auxiliary models, namely for PHQ-9 categories/symptoms, in combination with the depression classification models, allows us to generate explanations that are more meaningful and have higher recall when evaluated against a gold standard dataset of manually annotated explanations. This implies that we need to conduct more studies that can benefit from clinical practices and measures, and integrate it into the modeling design choices. To the best of our knowledge, we are the first to produce gold annotations of explanations for the depression classification task, and conduct a thorough analysis on how augmenting with the symptoms can improve the quality of the explanations.

References

Adrian Benton, Glen Coppersmith, and Mark Dredze. 2017. Ethical research protocols for social media health research. In *Proceedings of the First ACL Workshop on Ethics in Natural Language Processing*, pages 94–102, Valencia, Spain. Association for Computational Linguistics.

André Bittar, Sumithra Velupillai, Angus Roberts, Rina Dutta, et al. 2021. Using general-purpose sentiment lexicons for suicide risk assessment in electronic health records: corpus-based analysis. *JMIR medical informatics*, 9(4):e22397.

Tom Brown, Benjamin Mann, Nick Ryder, Melanie Subbiah, Jared D Kaplan, Prafulla Dhariwal, Arvind Neelakantan, Pranav Shyam, Girish Sastry, Amanda Askell, et al. 2020. Language models are few-shot learners. *Advances in neural information processing systems*, 33:1877–1901.

Bongjae Choi, Geumsook Shim, Bumseok Jeong, and Sungho Jo. 2020. Data-driven analysis using multiple self-report questionnaires to identify college students at high risk of depressive disorder. *Scientific reports*, 10(1):1–13.

Arman Cohan, Bart Desmet, Andrew Yates, Luca Soldaini, Sean MacAvaney, and Nazli Goharian. 2018. SMHD: a large-scale resource for exploring online language usage for multiple mental health conditions. In *Proceedings of the 27th International Conference on Computational Linguistics*, pages 1485–1497, Santa Fe, New Mexico, USA. Association for Computational Linguistics.

R Yates Coley, Jennifer M Boggs, Arne Beck, and Gregory E Simon. 2021. Predicting outcomes of psychotherapy for depression with electronic health record data. *Journal of affective disorders reports*, 6:100198.

Glen Coppersmith, Mark Dredze, Craig Harman, Kristy Hollingshead, and Margaret Mitchell. 2015. Clpsych 2015 shared task: Depression and ptsd on twitter. In *Proceedings of the 2nd workshop on computational linguistics and clinical psychology: from linguistic signal to clinical reality*, pages 31–39.

Glen Coppersmith, Ryan Leary, Patrick Crutchley, and Alex Fine. 2018. Natural language processing of social media as screening for suicide risk. *Biomedical informatics insights*, 10:1178222618792860.

Munmun De Choudhury, Michael Gamon, Scott Counts, and Eric Horvitz. 2013. Predicting depression via social media. In *Seventh international AAAI conference on weblogs and social media*.

Munmun De Choudhury, Emre Kiciman, Mark Dredze, Glen Coppersmith, and Mrinal Kumar. 2016. Discovering shifts to suicidal ideation from mental health content in social media. In *Proceedings of the 2016 CHI conference on human factors in computing systems*, pages 2098–2110.

Fionn Delahunty, Robert Johansson, and Mihael Arcan. 2019. Passive diagnosis incorporating the phq-4 for depression and anxiety. In *Proceedings of the Fourth Social Media Mining for Health Applications (# SMM4H) Workshop & Shared Task*, pages 40–46.

Jacob Devlin, Ming-Wei Chang, Kenton Lee, and Kristina Toutanova. 2018. Bert: Pre-training of deep bidirectional transformers for language understanding. *arXiv preprint arXiv:1810.04805*.

Joseph Geraci, Pamela Wilansky, Vincenzo de Luca, Anvesh Roy, James L Kennedy, and John Strauss. 2017. Applying deep neural networks to unstructured text notes in electronic medical records for phenotyping youth depression. *Evidence-based mental health*, 20(3):83–87.

Jonathan Gratch, Ron Artstein, Gale Lucas, Giota Stratou, Stefan Scherer, Angela Nazarian, Rachel Wood, Jill Boberg, David DeVault, Stacy Marsella, et al. 2014. The distress analysis interview corpus of human and computer interviews. In *Proceedings of the Ninth International Conference on Language Resources and Evaluation (LREC'14)*, pages 3123–3128.

Keith Harrigian, Carlos Aguirre, and Mark Dredze. 2020. Do models of mental health based on social media data generalize? In *Findings of the Association for Computational Linguistics: EMNLP 2020*, pages 3774–3788, Online. Association for Computational Linguistics.

YuanZheng Hu and Marina Sokolova. 2021. Explainable multi-class classification of the camh covid-19 mental health data. *arXiv preprint arXiv:2105.13430*.

Christian Karmen, Robert C Hsiung, and Thomas Wetter. 2015. Screening internet forum participants for depression symptoms by assembling and enhancing multiple nlp methods. *Computer methods and programs in biomedicine*, 120(1):27–36.

Diederik P Kingma and Jimmy Ba. 2014. Adam: A method for stochastic optimization. *arXiv preprint arXiv:1412.6980*.

Harnain Kour and Manoj K Gupta. 2022. An hybrid deep learning approach for depression prediction from user tweets using feature-rich cnn and bidirectional lstm. *Multimedia Tools and Applications*, pages 1–37.

Kurt Kroenke, Robert L Spitzer, and Janet BW Williams. 2001. The phq-9: validity of a brief depression severity measure. *Journal of general internal medicine*, 16(9):606–613.

Maria Elizabeth Loades, Eleanor Chatburn, Nina Higson-Sweeney, Shirley Reynolds, Roz Shafran, Amberly Brigden, Catherine Linney, Megan Niamh McManus, Catherine Borwick, and Esther Crawley. 2020. Rapid systematic review: the impact of social isolation and loneliness on the mental health of children and adolescents in the context of covid-19. *Journal of the American Academy of Child & Adolescent Psychiatry*, 59(11):1218–1239.

David E Losada and Fabio Crestani. 2016. A test collection for research on depression and language use. In *International Conference of the Cross-Language Evaluation Forum for European Languages*, pages 28–39. Springer.

Kate Loveys, Jonathan Torrez, Alex Fine, Glen Moriarty, and Glen Coppersmith. 2018. Cross-cultural differences in language markers of depression online. In *Proceedings of the fifth workshop on computational linguistics and clinical psychology: from keyboard to clinic*, pages 78–87.

Scott M Lundberg and Su-In Lee. 2017. A unified approach to interpreting model predictions. *Advances in neural information processing systems*, 30.

Michelle Morales, Stefan Scherer, and Rivka Levitan. 2017. A cross-modal review of indicators for depression detection systems. In *Proceedings of the Fourth Workshop on Computational Linguistics and Clinical Psychology — From Linguistic Signal to Clinical Reality*, pages 1–12, Vancouver, BC. Association for Computational Linguistics.

Wajid Mumtaz and Abdul Qayyum. 2019. A deep learning framework for automatic diagnosis of unipolar depression. *International journal of medical informatics*, 132:103983.

Matthew D Nemesure, Michael V Heinz, Raphael Huang, and Nicholas C Jacobson. 2021. Predictive modeling of depression and anxiety using electronic health records and a novel machine learning approach with artificial intelligence. *Scientific reports*, 11(1):1–9.

Thong Nguyen, Andrew Yates, Ayah Zirikly, Bart Desmet, and Arman Cohan. 2022. Improving the generalizability of depression detection by leveraging clinical questionnaires. *arXiv preprint arXiv:2204.10432*.

Vadim Osadchiy, Jesse Nelson Mills, Sriram Venkata Eleswarapu, et al. 2020. Understanding patient anxieties in the social media era: qualitative analysis and natural language processing of an online male infertility community. *Journal of Medical Internet Research*, 22(3):e16728.

David Owen, Jose Camacho Collados, and Luis Espinosa-Anke. 2020. Towards preemptive detection of depression and anxiety in twitter. *arXiv preprint arXiv:2011.05249*.

Robert B Penfold, Eric Johnson, Susan M Shortreed, Rebecca A Ziebell, Frances L Lynch, Greg N Clarke, Karen J Coleman, Beth E Waitzfelder, Arne L Beck, Rebecca C Rossom, et al. 2021. Predicting suicide attempts and suicide deaths among adolescents following outpatient visits. *Journal of affective disorders*, 294:39–47.

Inna Pirina and Çağrı Çöltekin. 2018. Identifying depression on reddit: The effect of training data. In *Proceedings of the 2018 EMNLP Workshop SMM4H: The 3rd Social Media Mining for Health Applications Workshop & Shared Task*, pages 9–12.

Marco Tulio Ribeiro, Sameer Singh, and Carlos Guestrin. 2016. " why should i trust you?" explaining the predictions of any classifier. In *Proceedings of the 22nd ACM SIGKDD international conference on knowledge discovery and data mining*, pages 1135–1144.

Mariana Rodrigues Makiuchi, Tifani Warnita, Kuniaki Uto, and Koichi Shinoda. 2019. Multimodal fusion of bert-cnn and gated cnn representations for depression detection. In *Proceedings of the 9th International on Audio/Visual Emotion Challenge and Workshop*, pages 55–63.

Mark Sendak, Michael Gao, Marshall Nichols, Anthony Lin, and Suresh Balu. 2019. Machine learning in health care: a critical appraisal of challenges and opportunities. *EGEMs*, 7(1).

Han-Chin Shing, Suraj Nair, Ayah Zirikly, Meir Friedenberg, Hal Daumé III, and Philip Resnik. 2018. Expert, crowdsourced, and machine assessment of suicide risk via online postings. In *Proceedings of the Fifth Workshop on Computational Linguistics and Clinical Psychology: From Keyboard to Clinic*, pages 25–36, New Orleans, LA. Association for Computational Linguistics.

Marco Spruit, Stephanie Verkleij, Kees de Schepper, and Floortje Scheepers. 2022. Exploring language markers of mental health in psychiatric stories. *Applied Sciences*, 12(4):2179.

Simon Šuster, Stéphan Tulkens, and Walter Daelemans. 2017. A short review of ethical challenges in clinical natural language processing. In *Proceedings of the First ACL Workshop on Ethics in Natural Language Processing*, pages 80–87, Valencia, Spain. Association for Computational Linguistics.

Michael M Tadesse, Hongfei Lin, Bo Xu, and Liang Yang. 2019. Detection of depression-related posts in reddit social media forum. *IEEE Access*, 7:44883–44893.

Md Zia Uddin, Kim Kristoffer Dysthe, Asbjørn Følstad, and Petter Bae Brandtzaeg. 2022. Deep learning for prediction of depressive symptoms in a large textual dataset. *Neural Computing and Applications*, 34(1):721–744.

Shweta Yadav, Jainish Chauhan, Joy Prakash Sain, Krishnaprasad Thirunarayan, Amit Sheth, and Jeremiah Schumm. 2020. Identifying depressive symptoms from tweets: figurative language enabled multitask learning framework. *arXiv preprint arXiv:2011.06149*.

Andrew Yates, Arman Cohan, and Nazli Goharian. 2017. Depression and self-harm risk assessment in online forums. In *Proceedings of the 2017 Conference on Empirical Methods in Natural Language Processing*, pages 2968–2978, Copenhagen, Denmark. Association for Computational Linguistics.

Amir Hossein Yazdavar, Hussein S Al-Olimat, Monireh Ebrahimi, Goonmeet Bajaj, Tanvi Banerjee, Krishnaprasad Thirunarayan, Jyotishman Pathak, and Amit Sheth. 2017. Semi-supervised approach to monitoring clinical depressive symptoms in social media. In *Proceedings of the 2017 IEEE/ACM International Conference on Advances in Social Networks Analysis and Mining 2017*, pages 1191–1198.

Li Zhou, Amy W Baughman, Victor J Lei, Kenneth H Lai, Amol S Navathe, Frank Chang, Margarita Sordo, Maxim Topaz, Feiran Zhong, Madhavan Murrali, et al. 2015. Identifying patients with depression using free-text clinical documents. In *MEDINFO 2015: eHealth-enabled Health*, pages 629–633. IOS Press.

Ayah Zirikly, Philip Resnik, Ozlem Uzuner, and Kristy Hollingshead. 2019. Clpsych 2019 shared task: Predicting the degree of suicide risk in reddit posts. In *Proceedings of the sixth workshop on computational linguistics and clinical psychology*, pages 24–33.

Identifying stable speech-language markers of autism in children: Preliminary evidence from a longitudinal telephony-based study

Sunghye Cho[1], Riccardo Fusaroli[2], Maggie Rose Pelella[3], Kimberly Tena[3], Azia Knox[3], Aili Hauptmann[3], Maxine Covello[3], Alison Russell[3], Judith Miller[3], Alison Hulick[3], Jennifer Uzokwe[3], Kevin Walker[1], James Fiumara[1], Juhi Pandey[3], Christopher Chatham[4], Christopher Cieri[1], Robert T. Schultz[3], Mark Liberman[1], Julia Parish-Morris[3]

[1]Linguistic Data Consortium, University of Pennsylvania, [2]Interacting Minds Center, School of Culture and Society, Aarhus University, [3]Center for Autism Research, Children's Hospital of Philadelphia, [4]F Hoffman-La Roche Ltd

Abstract

This study examined differences in linguistic features produced by autistic and neurotypical (NT) children during brief picture descriptions, and assessed feature stability over time. Weekly speech samples from well-characterized participants were collected using a telephony system designed to improve access for geographically isolated and historically marginalized communities. Results showed stable group differences in certain acoustic features, some of which may potentially serve as key outcome measures in future treatment studies. These results highlight the importance of eliciting semi-structured speech samples in a variety of contexts over time, and adds to a growing body of research showing that fine-grained naturalistic communication features hold promise for intervention research.

1 Introduction

Natural sampling is a rich approach to investigating speech and language in autistic children. Previous studies have shown that language behavior in autism differs from neurotypical (NT) patterns in a number of ways. For example, autistic children who are more severely impacted have been shown to produce less speech (Bone et al., 2014), slower speech (Parish-Morris et al., 2016; Bonneh et al., 2011), and speech with atypical voice quality compared to NT peers (Paul et al., 2005; Shriberg et al., 2001), including heightened jitter, increased jitter variability, but reduced harmonic-to-noise ratio (Bone et al., 2014). It has also been observed that autistic children's prosody differs from NT children, with qualitative observations ranging from "sing-songy" and exaggerated to monotonous, machine-like, or hollow (Bonneh et al., 2011;DePape et al., 2012; Lord et al., 1994; Wehrle et al., 2020; Fusaroli2017; Fusaroli2021). In the lexical domain, prior research has shown that autistic children use more nouns than NT peers when narrating a story from a picture, suggesting that the storytelling of children with autism is more object-focused (Boorse et al., 2019). Also, children with autism use fewer filler words during clinical assessments than NT children (Parish-Morris et al., 2017), and they talk less about social topics during get-to-know-you conversations compared to NT children (Song et al., 2021). It is also observed that children with autism have difficulties in using words in non-literal ways (Bara et al., 1999; Rutherford et al., 2012). Research in this domain continues to emerge, but samples remain small and results occasionally conflict, as in the description of prosody being either "sing-songy" or monotonous, or fail to replicate (See Fusaroli et al., 2017 for a meta-analysis of previous findings).

Prior studies of natural language in autism used a variety of data collection and analysis methods that could critically affect results and may have led to conflicting findings. For example, the presence of an unfamiliar adult during in-person or remote elicitations could adversely impact the behavior of autistic children, thus reducing the quality and informativeness of their language samples (Barokova and Tager-Flusberg, 2020). Also, children's linguistic behavior might differ depending on the specifics of the elicitation task in a given study, i.e., whether natural conversations or semi-structured speech tasks are used, and the characteristics of certain elicitation stimuli.

In order to develop scalable, cost-effective, reliable intervention progress monitoring systems of autistic symptoms using speech as a primary target, it is necessary to understand how contextual and testing factors affect children's behavior. Then, it will be possible to identify robust features that reliably index autism symptoms across heterogeneous testing conditions. Toward this goal, we developed a telephony protocol to examine how various factors affect speech performance in autistic children and adolescents. Telephony has particular potential to address service and monitoring gaps for

Proceedings of the Eighth Workshop on Computational Linguistics and Clinical Psychology, pages 40 - 46
July 15, 2022

autistic and NT children from historically marginalized and/or low-resource communities(Omer et al., 2022), and is a useful alternative to in-person data collection during the COVID-19 pandemic. The final battery of our protocol consisted of seven versions of seven tasks that a parent or legal guardian could independently facilitate. In this preliminary report from an on-going study, we assessed children's speech and language features during one of the seven tasks (picture descriptions) collected in the first and second phone sessions. Our goals were to (1) identify diagnostic group differences in automated speech and language features that are stable over time, and (2) examine potential effects of staff vs. parent administration in each diagnostic group.

2 Methods

2.1 Participants

Study inclusion and exclusion criteria are included in the Appendix. In this report, we analyzed data from 29 children who successfully completed two sessions. Participant groups were matched on age, full-scale IQ, and self-reported race. Groups were not matched on sex (p=0.015), which is expected due to the prevalence of ASD in boys (Baio et al., 2018), and we are currently addressing with targeted recruitment. One autistic participant identified as non-binary. Autism and NT groups differed in several clinical ratings (Table 1).

	Autism (n=13)	NT (n=16)	p-value
Age (years)	9.8 (2.5)	9.6 (2.6)	0.767
Sex (%)	10 boys (76.9%)	6 boys (37.5%)	0.015
Full scale IQ	115.1 (15.4)	119.1 (13.7)	0.469
Race	4 non-whites	5 non-whites	0.69
SCQ (total)	17 (6.6)	1.2 (1.1)	<0.001
SRS-2 (total)	70.5 (7)	42.1 (3.5)	<0.001
CCC-2 (speech)	9.2 (2.5)	11.8 (0.8)	<0.001
CCC-2 (non-speech)	5.5 (2.2)	11.8 (1.3)	<0.001

Table 1: Demographic and clinical characteristics of the participants. Groups were compared with t-tests, except the sex ratio, where a chi-square test was used. SCQ: Social communication questionnaire (Rutter et al., 2003), SRS: Social responsiveness scale (Constantino, 2011), CCC: Children's communication checklist (Bishop, 2006).

2.2 Data collection and annotation

We developed a telephony platform to support single and dual speaker modes. This platform consisted of a high-availability server, voice over internet protocol (VoIP) service by Vonage, telephony software framework (Asterisk 13.18.3), a relational database, and telephony applications.

The seven sessions included seven age-appropriate tasks, and the picture description task was included in all sessions. Children described different pictures in all seven sessions, and four sessions were administered by study staff and the other three sessions were proctored by children's caregivers. The data collection is on-going, and we only analyzed the first and second sessions in this study. Prior to the first official data collection call, study staff held an "informational call" with the participating parent to review standard elicitation methods to be utilized across sessions. During the first session with the child, study staff remained on the line and facilitated tasks with the parent and child. During the second session, children and parents independently completed all seven tasks on their own. The second session was collected approximately one week after the first session was completed. The study was reviewed by the institutional review board at the Children's Hospital of Philadelphia. Written informed consent was obtained from parents and children provided a verbal assent before study enrollment.

Recordings were transcribed by trained annotators using a web-based transcription tool with a built-in speech activity detector (SAD) function. Data were stored in secured HIPAA-compliant servers, and all annotators were trained to protect patients' identities and identifiable information. For dual speaker mode recordings, SAD ran on each channel separately. Annotators also corrected speech segment boundary errors.

2.3 Acoustic and text features

Words were automatically tagged for part-of-speech (POS) categories using spaCy (Honnibal and Johnson, 2015). POS categories, fillers, partial words, repetitions, and "hm" were counted separately and converted to counts per 100 words. Content words were rated for word frequency (Brysbaert and New, 2009), concreteness (Brysbaert et al., 2014), ambiguity (Hoffman et al., 2013), age of acquisition (AoA) (Brysbaert et al., 2018), and familiarity (Brysbaert et al., 2018). We also ran the Language Inquiry and Word Count program (Pennebaker et al., 2015) to calculate additional word-level measures found to be useful in clinical population.

For acoustic processing, stereo recordings were split into single channels for precise audio processing. We extracted low-level descriptors of pitch, jitter, shimmer, harmonic-to-noise ratio (HNR), and four spectral moments (1st order: centroid, 2nd order: standard deviation, 3rd order: skewness, 4th order: kurtosis) from participants' picture descriptions per 10 ms using openSMILE with the ComParE13 configuration file (Eyben et al., 2013). Pitch values in hertz were converted to semitones (st) using individuals' 10th percentiles to normalize physiological differences among participants ($St = \log_2(f0 / 10th\ percentile) \times 12$). Since this method used each speaker's baseline (i.e., the 10th percentile of individual's pitch range) to convert raw hertz values to semitones, it allowed us to compare the groups directly despite the significant difference in sex ratio and the wide age range. Several durational measures were computed from SAD timestamps.

2.4 Statistical considerations

Preliminary analyses revealed that our variable residuals met the assumptions of parametric tests, so we employed analysis of covariance (ANCOVA) models. Speech/language features were included as dependent variables, with group, session, and the interaction of group and session as independent variables. Sex was covaried in all models. Since this was a first exploratory analysis, with findings that would be considered reliable only once the data collection is over, we did not currently correct p-values for multiple comparisons.

3 Results

3.1 Acoustic measures

Median shimmer and jitter values were higher for autistic children than NT children (shimmer: $F(1,52)=4.17$, $p=0.046$; jitter: $F(1,52)=3.96$, $p=0.052$, Figure 1A-B). Mean, standard deviation (SD), and interquartile range (IQR) of jitter and shimmer did not differ by group. Autistic children also had higher mean (skewness: $F(1,52)=13.46$, $p<0.001$; kurtosis: $F(1,52)=12.98$, $p<0.001$), median (skewness: $F(1,52)=6.17$, $p=0.016$; kurtosis: $F(1,52)=4.7$, $p=0.035$, Figure 1C-D), SD (skewness: $F(1,52)=9.89$, $p=0.003$; kurtosis: $F(1,52)=13.86$, $p<0.001$), and IQR values (skewness: $F(1,52)=7$, $p=0.011$; kurtosis: $F(1,52)=8.26$, $p=0.006$) of spectral skewness and kurtosis than NT children. Groups did not differ in pitch and HNR, and Session had no significant effect on any acoustic variables.

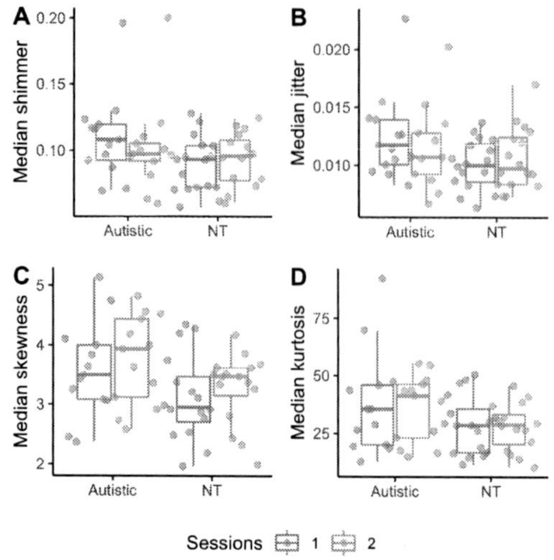

Figure 1: Acoustic features during picture description tasks. Shimmer refers to variability in signal amplitude, whereas jitter represents variability in signal frequency (A, B). Spectral skewness and kurtosis refers to the third and fourth order of spectral moments, which are known to characterize voice timber (C, D). Only median values are plotted for an illustration purpose.

3.2 Durational measures

Autistic children produced longer ($F(1,52)=7.79$, $p=0.007$) and more variable ($F(1,52)=8.49$, $p=0.005$) speech segment durations than NT children (Figure 2A-B). The difference in total speech duration between the first and second sessions was larger for autistic children than NT children ($F(1,52)=4.34$, $p=0.042$). Total pause duration was shorter in autistic participants than NT children ($F(1,52)=5.14$, $p=0.028$, Figure 2C-D), and children paused longer during the first session compared to the second ($F(1,52)=4.82$, $p=0.033$). Autistic children paused less frequently than NT children ($F(1,52)=6.33$, $p=0.015$).

3.3 Textual measures

Autistic participants produced fewer conjunctions ($F(1,52)=5.06$, $p=0.029$) and pronouns ($F(1,52)=$

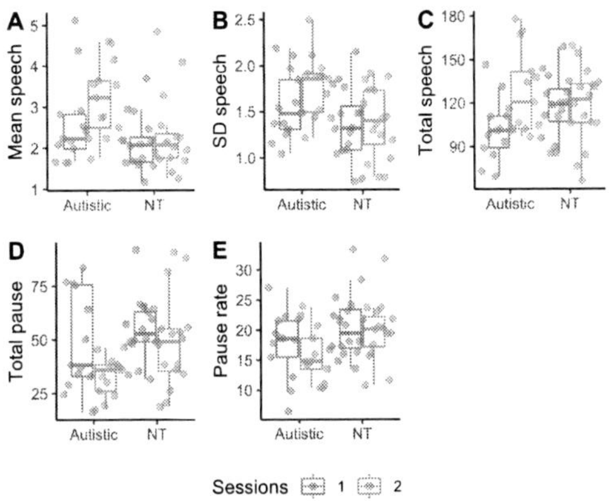

Figure 2: Durational measures during picture descriptions. The units of the y-axis are seconds, except the pause rate, where pause rate per minute was plotted.

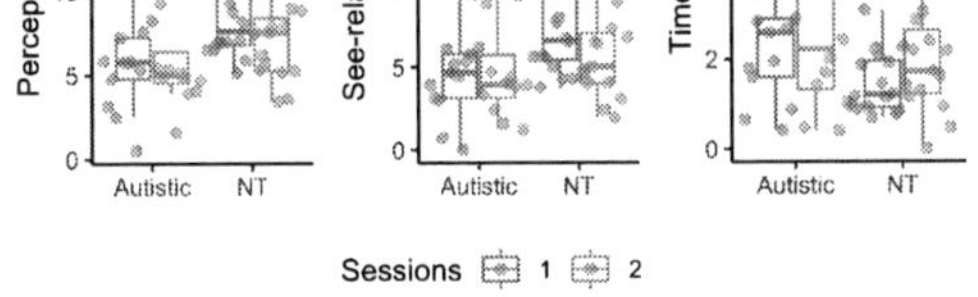

Figure 3: Lexical measures during picture description tasks. All POS counts are per 100 words, and the age of acquisition was averaged across all content words produced by each child. The counts of LIWC categories were also normalized.

4.75, p=0.034) than NT children, and their content words had a higher AoA than those of NT children (F(1,52)=6.35, p=0.015, Figure 3A-C). Also, autistic children produced fewer perception (F(1, 52)=9.17, p=0.004) and see-related words (F(1,52) =7.1, p=0.01) and more time-related words (F(1, 52)=4.79, p=0.033) than NT children (Figure 3).

Regardless of diagnostic status, children produced more adverbs (F(1,52)=9.08, p=0.003) and prepositions (F(1,52)=6.47, p=0.014) during the second session than the first (not shown in the figure). Children also produced content words that were more ambiguous (F(1,52)=10.82, p=0.002), later acquired (F(1,52)=54.9, p<0.001), and familiar (F(1,52)=14.85, p<0.001) during the second session than the first session. Finally, several LIWC categories, including anger (F(1,52) =4.69, p=0.035), difference (F(1,52)=5.55, p=0.023), feeling (F(1,52)=4.06, p=0.049), bio (F(1, 52)=4.99, p=0.03), and ingestion (F(1,52)=19, p<0.001), showed significant effects of Session.

4 Discussion

In this study, we elicited picture descriptions from autistic and NT children using a telephony platform, and tested for the presence of diagnostic group differences in a variety of acoustic and lexical features over two sessions. Results showed that autistic children produced greater local jitter, shimmer and the third and fourth orders of spectral moments, as well as shorter and less frequent pauses compared to NT children, across two sessions and with different stimuli. Autistic children produced more speech during the second session when parents administered the task without study staff, compared to the first session. In contrast, NT children's speech duration did not differ by session. Lexically, autistic children produced fewer conjunctions and pronouns than NT children, and used later-acquired content words compared to NT peers. Our results also showed that autistic children used fewer see- or perception-related words and more time-related words than NT children. However, many other lexical features differed by session without significant group differences, suggesting that the picture stimuli may have had more influence than diagnostic group on lexical production.

Given that the acoustic features described here remained stable from the first to the second telephony session, and also distinguished diagnostic groups, they might hold potential as reliable speech markers of autism. Higher jitter (variability in frequency) and shimmer (variability in amplitude) are perceived as harsh, hoarse, or breathy voice (Tsanas et al., 2011). The observation that autistic children's jitter and jitter variability were higher than NT peers is consistent with a previous study that also showed positive correlations between jitter and autism symptomology (ADOS scores; Bone et al., 2014). Yet, a recent meta-analysis study found that jitter was lower in autistic children than NT children in US and it did not differ in Denmark, so future study is needed to resolve this mixed finding (Fusaroli et al., 2022). Also, previous studies showed consistently lower HNR values for autistic children compared to NT peers, with mixed findings in shimmer (Fusaroli et al., 2022); this differs from our pattern of results, where we found no difference in HNR but higher shimmer in children with autism. Spectral moments in autism have rarely been studied, even though these measures are known to characterize individuals' voice timbre (Lerch, 2012). We plan to study these features further in a larger sample after completing the data collection, to explore whether they could serve as validated speech markers of autism.

Children on the autism spectrum spoke longer and paused less frequently during the second session than the first session, whereas TD children's duration measures did not differ by session. This finding is in line with prior research where fewer pauses were consistently observed in children with autism (Fusaroli et al., 2022). This finding has at least two potential explanations: First, autistic individuals experience social-communicative challenges which might have hindered their willingness to speak freely in the presence of unfamiliar study staff. In this case, they may have spoken longer in the second session because their parent administered the task. Thus, it is important to consider the presence or absence of study staff when interpreting studies of speech and language in autism. Alternatively, children's greater speaking duration in the second session could simply be due to task familiarity; by week 2, children knew what to expect and had already completed the picture description once.

Finally, our study also found that autistic children produced fewer conjunctions, pronouns, see- and perception-related words with high AoA than NT children. We also observed that many word-level features differed by session in both the autistic and NT groups, suggesting that picture selection has an outsized effect on lexical features. In this study, we selected seven different pictures to prevent boredom and practice effects across multiple sessions. However, since different pictures include unique objects that children are likely to list in their descriptions, this will result in significant session-based differences in word-level features. Picture selection and objects in pictures need to be carefully designed in future research, potentially with less weight placed on specific content words as stable outcome measures. As data collection continues in the current study, we will investigate whether group differences in more abstract lexical features (e.g., pronoun use) might remain stable over all seven sessions.

5 Conclusion

Telephony carries great potential as a low-cost and scalable platform for monitoring intervention responses from afar, as well as measuring longitudinal developmental changes in individual children. Acoustic features extracted from data collected using a telephony system, which delivered consistent, high-quality recordings, could be important tools for identifying speech markers of autism.

References

Jon Baio, Lisa Wiggins, Deborah L. Christensen, Matthew J. Maenner, Julie Daniels, Zachary Warren, Margaret Kurzius-Spencer, Walter Zahorodny, Cordelia Robinson Rosenberg, Tiffany White, Maureen S. Durkin, Pamela Imm, Loizos Nikolaou, Marshalyn Yeargin-Allsopp, Li Ching Lee, Rebecca Harrington, Maya Lopez, Robert T. Fitzgerald, Amy Hewitt, Sydney Pettygrove, John N. Constantino, Alison Vehorn, Josephine Shenouda, Jennifer Hall-Lande, Kim Van Naarden Braun, and Nicole F. Dowling. 2018. Prevalence of autism spectrum disorder among children aged 8 Years - Autism and developmental disabilities monitoring network, 11 Sites, United States, 2014. *MMWR Surveillance Summaries*, 67(6):1–23.

Bruno G. Bara, Francesca M. Bosco, and Monica Bucciarelli. 1999. Developmental pragmatics in normal and abnormal children. *Brain and Language*, 68(3):507–528.

Mihaela Barokova and Helen Tager-Flusberg. 2020. Commentary: Measuring Language Change Through Natural Language Samples. *Journal of Autism and Developmental Disorders*, 50(7):2287–2306.

Dorothy Bishop. 2006. *Children's Communication Checklist-2 U.S. Edition*. Psychological Corporation, San Antonio, TX.

Daniel Bone, Chi Chun Lee, Matthew P. Black, Marian E. Williams, Sungbok Lee, Pat Levitt, and Shrikanth Narayanan. 2014. The psychologist as an interlocutor in Autism Spectrum Disorder assessment: insights from a study of spontaneous prosody. *Journal of Speech, Language, and Hearing Research*, 57:1162–1177.

Yoram S. Bonneh, Yoram Levanon, Omrit Dean-Pardo, Lan Lossos, and Yael Adini. 2011. Abnormal speech spectrum and increased pitch variability in young autistic children. *Frontiers in Human Neuroscience*, 4(JANUARY):1–7.

Jaclin Boorse, Meredith Cola, Samantha Plate, Lisa Yankowitz, Juhi Pandey, Robert T. Schultz, and Julia Parish-Morris. 2019. Linguistic markers of autism in girls: Evidence of a "blended phenotype" during storytelling. *Molecular Autism*, 10(1):1–12.

Marc Brysbaert, Paweł Mandera, and Emmanuel Keuleers. 2018. Word prevalence norms for 62 ,000 English lemmas. *Behavior Research Methods*, (July 2018):467–479.

Marc Brysbaert and Boris New. 2009. Moving beyond Kučera and Francis: A critical evaluation of current word frequency norms and the introduction of a new and improved word frequency measure for American English. *Behavior Research Methods*, 41(4):977–990.

Marc Brysbaert, Amy Beth Warriner, and Victor Kuperman. 2014. Concreteness ratings for 40 thousand generally known English word lemmas. *Behavior Research Methods*, 46(3):904–911.

John N. Constantino. 2011. *Social Responsiveness Scale, Second Edition*. Western Psychological Services, Los Angeles, CA.

Anne Marie R. DePape, Aoju Chen, Geoffrey B.C. Hall, and Laurel J. Trainor. 2012. Use of prosody and information structure in high functioning adults with Autism in relation to language ability. *Frontiers in Psychology*, 3(MAR):1–13.

Florian Eyben, Felix Weninger, Florian Gross, and Björn Schuller. 2013. Recent developments in openSMILE, the munich open-source multimedia feature extractor. In *Proceedings of the 21st ACM international conference on Multimedia - MM '13*, pages 835–838, New York, New York, USA. ACM Press.

Riccardo Fusaroli, Ruth Grossman, Niels Bilenberg, Cathriona Cantio, Jens Richardt Møllegaard Jepsen, and Ethan Weed. 2022. Toward a cumulative science of vocal markers of autism: A cross-linguistic meta-analysis-based investigation of acoustic markers in American and Danish autistic children. *Autism Research*, 15(4):653–664.

Riccardo Fusaroli, Anna Lambrechts, Dan Bang, Dermot M. Bowler, and Sebastian B. Gaigg. 2017. "Is voice a marker for Autism spectrum disorder? A systematic review and meta-analysis". *Autism Research*, 10(3):384–407.

Paul Hoffman, Matthew A. Lambon Ralph, and Timothy T. Rogers. 2013. Semantic diversity: A measure of semantic ambiguity based on variability in the contextual usage of words. *Behavior Research Methods*, 45(3):718–730.

Matthew Honnibal and Mark Johnson. 2015. An improved non-monotonic transition system for dependency parsing. In *EMNLP 2015: Conference on Empirical Methods in Natural Language Processing*, pages 1373–1378.

Alexander Lerch. 2012. *An Introduction to Audio Content Analysis*. John Wiley Sons, Inc., Hoboken, NJ, USA.

Catherine Lord, Michael Rutter, and Ann Le Couteur. 1994. Autism Diagnostic Interview-Revised: A revised version of a diagnostic interview for caregivers of individuals with possible pervasive developmental disorders. *Journal of Autism and Developmental Disorders*, 24(5):659–685.

Khalid Omer, Umaira Ansari, Amar Aziz, Khalid Hassan, Lami Aminati Bgeidam, Muhd Chadi Baba, Yagana Gidado, Neil Andersson, and Anne Cockcroft. 2022. Participatory health research under COVID-19 restrictions in Bauchi State, Nigeria: Feasibility of cellular teleconferencing for virtual discussions with community groups in a low-resource setting. *Digital Health*, 8.

Julia Parish-Morris, Mark Liberman, Neville Ryant, Christopher Cieri, Leila Bateman, Emily Ferguson, and Robert T Schultz. 2016. Exploring Autism Spectrum Disorders Using HLT. *Computational Linguistics and Clinical Psychology,*, pages 74–84.

Julia Parish-Morris, Mark Y. Liberman, Christopher Cieri, John D. Herrington, Benjamin E. Yerys, Leila Bateman, Joseph Donaher, Emily Ferguson, Juhi Pandey, and Robert T. Schultz. 2017. Linguistic camouflage in girls with autism spectrum disorder. *Molecular Autism*, 8(1):1–12.

Rhea Paul, Lawrence D. Shriberg, Jane McSweeny, Domenic Cicchetti, Ami Klin, and Fred Volkmar. 2005. Brief report: Relations between prosodic performance and communication and socialization ratings in high functioning speakers with autism spectrum disorders. *Journal of Autism and Developmental Disorders*, 35(6):861–869.

James W. Pennebaker, Roger J. Booth, Ryan L. Boyd, and Martha E. Francis. 2015. Linguistic inquiry and word count: Liwc2015.

Helena J. V. Rutherford, Justin D. Wareham, Ioanna Vrouva, Linda C. Mayes, Peter Fonagy, and Marc N. Potenza. 2012. Sex differences moderate the relationship between adolescent language and mentalization. *Personality Disorders: Theory, Research, and Treatment*, 3(4):393–405.

Michael Rutter, Anthony Bailey, and Catherine Lord. 2003. *SCQ: The Social Communication Questionnaire*. Western Psychological Services, Los Angeles, CA.

L. D. Shriberg, R. Paul, J. L. McSweeny, A. Klin, D. J. Cohen, and F. R. Volkmar. 2001. Speech and prosody characteristics of adolescents and adults with high-functioning autism and Asperger syndrome. *Journal of Speech, Language, and Hearing Research*, 44(5):1097–1115.

Amber Song, Meredith Cola, Samantha Plate, Victoria Petrulla, Lisa Yankowitz, Juhi Pandey, Robert T. Schultz, and Julia Parish-Morris. 2021. Natural language markers of social phenotype in girls with autism. *Journal of Child Psychology and Psychiatry and Allied Disciplines*, 62(8):949–960.

Athanasios Tsanas, Max A. Little, Patrick E. McSharry, and Lorraine O. Ramig. 2011. Nonlinear speech analysis algorithms mapped to a standard metric achieve clinically useful quantification of average Parkinson's disease symptom severity. *Journal of The Royal Society Interface*, 8(59):842–855.

Simon Wehrle, Francesco Cangemi, Harriet Hanekamp, Kai Vogeley, and Martine Grice. 2020. Assessing the intonation style of speakers with autism spectrum disorder. *Proceedings of the International Conference on Speech Prosody*, 2020-May(May):809–813.

A Inclusion and Exclusion Criteria

Inclusion criteria for participants were the following:

- Subjects age 6 – 17.99
- English is participant's first language
- Verbally fluent – language on grade level/consistent with chronological age
- Strongly suspected/confirmed diagnosis of autism or typical development
- Full-scale and verbal IQ > 75
- For autistic children, current SCQ score >= 11
- For the NT group, current SCQ scores < 11

Exclusion criteria for participants were the following:

- Known genetic condition that impacts neurodevelopment or vocal production/language
- History of persistent language deficits that are currently affecting child's language abilities such that it impacts their ability to have a conversation
- Extreme prematurity (<32 weeks)
- History of severe neurological injury likely to affect expressive language and communication behavior
- If NT, no first-degree family members with autism
- Plan to begin or change medication during study duration
- Plan to begin or change an intervention during study duration
- Diagnosis of hearing impairment or cochlear implant

Psychotherapy is Not One Thing: Simultaneous Modeling of Different Therapeutic Approaches

Maitrey Mehta[1], Derek D. Caperton[2], Katherine Axford[3], Lauren Weitzman[4], David Atkins[5], Vivek Srikumar[1], Zac E. Imel[3]
[1]School of Computing, University of Utah
[2]Calgary Counseling Centre
[3]Department of Educational Psychology, University of Utah
[4]University Counseling Center, University of Utah
[5]University of Washington

Abstract

There are many different forms of psychotherapy. Itemized inventories of psychotherapeutic interventions provide a mechanism for evaluating the quality of care received by clients and for conducting research on how psychotherapy helps. However, evaluations such as these are slow, expensive, and are rarely used outside of well-funded research studies. Natural language processing research has progressed to allow automating such tasks. Yet, NLP work in this area has been restricted to evaluating a single approach to treatment, when prior research indicates therapists used a wide variety of interventions with their clients, often in the same session. In this paper, we frame this scenario as a multi-label classification task, and develop a group of models aimed at predicting a wide variety of therapist talk-turn level orientations. Our models achieve F1 macro scores of 0.5, with the class F1 ranging from 0.36 to 0.67. We present analyses which offer insights into the capability of such models to capture psychotherapy approaches, and which may complement human judgment.

1 Introduction

A typical psychotherapy session involves a client–therapist dialog with the aim of diagnosing and assuaging a client's mental health condition. Psychotherapists, generally, rely on certain approaches (e.g., Cognitive Behavioral or Interpersonal Therapy) and interventions differ across these approaches.[1]. For example, a therapist might focus on a client's interpersonal relationships, their emotions, or help develop behavioral activities designed to reduce symptoms (or all of the above). A key goal of psychotherapy research is to categorize such approaches and study them to determine the effectiveness of each approach in any given scenario. We refer to this process of categorizing and detecting approaches based on an overarching theory as 'evaluation'.

In this paper, we study an application of Natural Language Processing (NLP) to mental health, and focus on therapists' approach to psychotherapy (Imel et al., 2015). Past NLP research has developed tools for evaluating specific types of interventions like Motivational Interviewing (Cao et al., 2019) or Cognitive Behavioral therapy (Flemotomos et al., 2021). However, psychotherapists differ from each other in the approaches they take. Furthermore, they can also vary in the interventions they use within and between sessions. The lines of work mentioned before assume that a session is comprised of exactly one approach, and consequently do not attempt to automatically evaluate different psychotherapy approaches that may coexist in the same session.

McCarthy and Barber (2009) proposed one multiple-approach evaluation methodology—the *Multitheoretical List of Therapeutic Interventions (MULTI)*, which is a list of 60 interventions (or, items) against which a psychotherapy session as a whole is evaluated post-session. The *MULTI* items are grouped into eight approaches. Note the *MULTI* is a session-level measure and thereby limited in specificity because it does not record therapist language that informs a given item's presence. Caperton (2021) extend the scheme to the evaluation of therapist monologues, talk-turn by talk-turn, in addition to the session-level evaluation. Such a scheme provides additional detail over time in a session.

Evaluating sessions with the *MULTI* requires a certain amount of time to be set aside post-session. Evaluating talk-turns manually for every session would be even more onerous and inefficient. This calls for a better automatic/semi-automatic method(s) to evaluate talk-turns. These methods serve two advantages: i) reducing the amount of

[1]We use the words 'approach' and 'orientation' interchangeably. Later in this paper, we use 'subscales' to align with practical usage.

Proceedings of the Eighth Workshop on Computational Linguistics and Clinical Psychology, pages 47 - 58
July 15, 2022

effort required in manual classification for research and quality assurance, and ii) creating applications to analyze approaches deemed helpful on out-of-session platforms (e.g., social media).

To that end, we present a neural machine learning model which aims to automate talk-turn level approach annotation. The task is set up in the following fashion: Given a therapist input talk-turn, does the input (or part of the input) correspond to one or more approaches. A talk-turn might only represent one approach, or might have different parts that correspond to different approaches. It is also possible that a therapist talk-turn does not fall within a specific therapeutic approach (e.g., minimal encouragers, small talk, etc.). Examples are shown in Table 1. This problem posits itself perfectly as a multi-label classification task.

The state-of-the-art in natural language processing (NLP) has seen significant improvements with the advent of transformer-based models (Vaswani et al., 2017; Devlin et al., 2019). In this paper, we show the performance of one such pre-trained transformer based language model on three paradigms, and experiment with changing context windows. Our models achieve around 0.5 F1 macro scores with the class F1 ranging from 0.36 to 0.67. Our analyses reveal that while our models mispredict on certain talk-turns during a session, they capture the dominant approaches when viewed from a session-level perspective. Furthermore, we show that certain decisions rely on inter-session context, and even common-sense knowledge which sets up a challenge for current models.

2 Talk-turn Level MULTI-30 Coding

MULTI-60 and MULTI-30. The *Multitheoretical List of Therapeutic Interventions* (*MULTI*) was originally developed as a list of 60 interventions (McCarthy and Barber, 2009). The 60-items belonged to eight different coarse-grained *subscales*, each representing a therapeutic approach. Each item was rated on a 5-point Likert scale for how prevalent the intervention was over the course of a psychotherapy session. The *MULTI*-60 was later re-evaluated through an item reduction procedure to create the more parsimonious *MULTI*-30 (Solomonov et al., 2019), comprised of the same eight subscales. In this work, we use focus on the eight coarse-grained approaches.

Each subscale was defined by a psychotherapeutic theoretical orientation. We describe each subscale briefly here.

1. *Psychodynamic* (**PD**) items focus on addressing nonconscious content from the client's psyche to alleviate distress.

2. *Process-experiential* (**PE**) items emphasize what is happening in the moment during a therapy session with the understanding that what happens in-session mirrors processes in the client's life outside of session.

3. *Interpersonal* (**IP**) items focus on relationship issues with other people in the client's life.

4. *Person-centered* (**PC**) interventions focus on elucidating client experiences and opinions to gain clarity on distress.

5. *Behavioral* (**BT**) items encourage adaptive behavioral activation strategies, assuming that productive actions will produce changes in mental wellbeing.

6. *Cognitive* (**CT**) items address possible distortions or unhelpful patterns in client thinking.

7. *Dialectical-behavioral* (**DBT**) interventions emphasize the client's non-judgment of present experience and the balance between accepting themselves as they are while believing they can be better.

8. *Common factors* (**CF**) items are purportedly transtheoretical and include interventions where the therapist demonstrates encouraging, sympathetic, and attentive listening behaviors.

Data Source. Psychotherapy audio data was collected from a university counseling center at large public school in the western United States. There were 243 unique sessions transcribed, some of which were annotated more than once, totaling to 473 sessions. These sessions were annotated using a talk-turn level version of the MULTI-30 (Caperton, 2021).

Coding Procedure and Reliability. Seven graduate students in mental health fields annotated session content for their varying use of theoretical interventions. Each coder received approximately 18 hours of training during in-person meetings and practiced coding sessions for an additional 36 hours before annotating session data used in this study. To minimize coder drift over time,

Case	Example Talk-turns	Approach(es)
Non-Approach	Okay let's set you up with an appointment.	No Code
	So, All of us trainees get to have a break as well.	No Code
Single Approach	You're scared.	Process-Experiential
	I definitely notice a lot of progress that you've made.	Common Factors
Multiple Approaches	Unfortunately, it's very normal. But I want you to continue practicing that exercise.	Common Factors Behavioral
	When you say that he's better off without you, what do you mean by that? It seems like he still has you.	Person-Centered, Cognitive

Table 1: Examples of talk-turns which have a single, multiple or no approach categories assigned. In the Multiple Approaches examples, colored text snippets correspond to their respective approach categories with the same color.

coders met together with their team leader every two weeks to discuss difficult talk-turns, items, and areas of disagreement.

Coders were tasked with identifying the presence or absence of theory-derived content in therapists' language at every therapist talk-turn (i.e., a string of words or statements uninterrupted by client speech). A given talk-turn could be identified with one, multiple, or no interventions.

Of the 243 unique sessions, 102 were annotated by multiple coders, resulting in 270 codings for interrater analysis. The statement-level interrater reliability of the eight theoretical orientations (subscales) was calculated using Cohen's kappa. Kappa was calculated for every possible coder pair who rated the same session and weighted according to the number of comparisons. Subscale kappa scores ranged from .37 ('fair' reliability; Landis and Koch (1977)) to .63 ('substantial').

The dataset was split by client randomly into train/dev/test sets containing 70%, 15% and 15% of the clients respectively. The splits contain 338, 66, and 76 sessions respectively containing 74k, 14k, and 17k talk-turns in total. Dataset statistics for the training split are presented in Table 2.

Class Name	Counts
No Code	58584
Psychodynamic	1024
Process-Experiential	3865
Interpersonal	1446
Person-centered	4810
Common Factors	5931
Behavioral	1531
Cognitive	1940
Dialectical-Behavioral	1765

Table 2: Training Data Statistics

3 Models

While we want to model the eight subscales (plus the 'No Code' class) conventionally used in literature, we deviate from these eight classes for the implementation. The Behavioral, Cognitive, and Dialectical-Behavioral subscales contain overlapping items (e.g., items 1 and 10 are shared by all three subscales). We break these subscales into four categories such that each of these categories contains mutually exclusive items. Note that the other subscales(Psychodynamic, Interpersonal, etc.) remain the same. Hence, in total, we obtain ten modified model classes (including the 'No Code' class). We refer the reader to Tables 7 and 8 in Appendix A for further details on the breakdown. This method can aid downstream analysis by allowing credit/blame assessment on a smaller set of items.

In our setup, each therapist talk-turn u_i has a corresponding binary label vector y_i. The binary label vector is ten dimensional, one decision each for the nine model classes (i.e., modified subscales) and one additional class indicating the absence of any code (NC). For all our experiments, we consider the RoBERTa-base (Liu et al., 2019b) model as the language model of choice. This model takes in a talk-turn u_i as input to produce contextual representations for its words. We take the pooler output of these contextual representations which gives us a vector representation h_i for the talk-turn.

$$h_i = Pooler(RoBERTa(u_i)) \quad (1)$$

We consider three modeling paradigms for our experiments.

Stand-Alone (SA) Model. This model is the vanilla multi-label classifier. Talk-turn representations are passed through a linear layer with the number of output nodes equal to the model classes. The result is passed through a sigmoid layer resulting in a vector of presence probabilities for each label $\hat{y}_i$. That is

$$\hat{y}_i = \sigma(w^T h_i + b) \quad (2)$$

where w and b are the weights of the linear layer. For inference, a probability of 0.5 or above indicates label presence for a particular class.

Pipeline Model. A heavily imbalanced dataset can hinder model performance for the under-represented categories. As seen in Table 2, the number of examples with a "No Code" class highly skews the dataset, potentially leading to performance bias towards the class. To alleviate this problem, we define a pipeline model that uses a separate binary classifier to determine whether a talk-turn deserves an orientation category or not. If this binary classifier predicts that the talk-turn supports at least one orientation, then the talk-turn is given to a multi-label model to predict over the nine model classes. The multi-label model will be similar to the one mentioned in the Stand-Alone Model, except nine classes are considered since predicting a "No Code" would be redundant. The multi-label model has the flexibility, nonetheless, to predict an absence of orientation by predicting that none of the codes are present (i.e., a zero vector). Note that two separate RoBERTa models are used for the binary and the multi-label classifiers.

Multi-Task Model. The Pipeline Model trains two separate RoBERTa models — one for the binary classifer and one for the multi-label model. A major drawback of this system is that training two RoBERTa models is computationally expensive and memory-intensive. An alternative method is to share the RoBERTa layer between the two tasks and have two separate linear layers for the respective binary and multi-label classification. This strategy of multi-task or joint learning has shown to be of promise in literature (Liu et al., 2019a; Stickland and Murray, 2019) and allows for better shared representation. The losses for both the tasks are combined as a weighted sum for learning. We consider two variants of the model based on the number of output classes for the multi-label classifier. The **MultiTask**$_{10}$ variant considers all the classes including "No Code" while **MultiTask**$_9$ excludes the "No Code" class. The inference is identical to the Pipeline Model. The Multi-Task and Stand-Alone model paradigm can be thought of as fairly similar architectures. However, the Multi-Task model assigns a higher loss weight to the binary classifier, uses a different optimization metric and utilizes a pipelined inference approach as opposed to the one-shot prediction by the Stand-Alone model.

So far, we explained that we break the conventional eight subscales into nine which have mutually exclusive items. While this approach allows us for better analysis, it is essential to present performance on the original theoretical subscales. To that end, we aggregate binary vector model predictions to the conventional eight *MULTI* subscales during evaluation. The output of the model, a ten-dimensional vector, will be mapped to a nine-dimensional vector(eight subscales plus 'No Code'). We use these nine-dimensional vectors to perform model evaluation. Table 8 is a guide for mapping model classes to the *MULTI* subscales.

4 Results

4.1 Experimental Setup

All the models use the RoBERTa-base implementation in HuggingFace's Transformers library (Wolf et al., 2020) for obtaining contextual representations. We utilize the pooler output as defined by the library which uses the embedding of the classification token passed through a pre-trained linear layer followed by a tanh activation. We use weighted losses to account for class-imbalance in all cases. The loss weight for a label *i* is determined by $1 - \frac{n_i}{n}$, where n_i are the number of talk-turns where label i is coded, and n is the total number of talk-turns in the training data. This choice ensures that rarer classes are given greater importance during learning.

Hyperparameters. All the models use a learning rate of 10^{-5} and the RoBERTa layer is fine-tuned is each case. We use the early stopping mechanism set at 5 epochs to avoid overfitting. The macro-averaged F1 score on a held-out validation is used to choose the best multi-label classification model. We use macro-averaged F2 score, instead, for the binary classification models since it favors recall on the positive label. This metric is ideal since the multi-label classifier would have the opportunity to correct false positives leaking from the binary classifier. Hyperparameters are tuned based on experimental results on a smaller dataset. All results are averages across three random seeds.

4.2 Results and Discussion

How do our models perform on the dataset? The comparative performance of our models are shown in Table 3. We report the model performance in terms of exact accuracy, micro and macro-averaged F1 scores across the label set, including the No Code (NC) label, and excluding it. We see that all the modeling paradigms perform almost similarly and to our surprise, the Pipeline or the MultiTask models do not produce substantial gains. Furthermore, we investigate the performance of the models on individual approach categories to understand the results further. These are reported in Table 4. We observe that model performances for categories do not deviate substantially between paradigms. By comparing to the number of training examples per label in Table 2, we observe that the performance closely correlates to the amount of data seen by the model.

Does added context help? For the results in Table 3, we consider just the therapist talk-turn and not the context surrounding it, i.e., the client and therapist talk-turns before or after it. We investigate whether adding additional context helps. We consider the following two approaches in addition to the previously shown approach:

1. Client talk-turn immediately preceding the therapist talk-turn in question can help determine the subscale. Take, for example, the Person-Centered subscale items. In these interventions, therapists often paraphrase statements which clients had just made. Hence, we concatenate the previous client **(PrevC)** talk-turn to the therapist talk-turn.
2. We observe from the training data that subscales tend to occur in chunks with the therapist opting for a certain orientation for a period of the session. We experiment with added therapist talk-turn context (**TC**) preceding and following the talk-turn in question.

We choose the MultiTask$_9$ model for this comparison which achieves the best performance. The results are in Table 5. We see that there is a small increase observed when therapist contexts are added. However, these gains are not substantial ($< 2\%$). Client context does not help the performance.
We also show some example predictions of a session snapshot in Table 6.

5 Analysis

In this section, we present analyses on the development set. We choose the best performing MultiTask$_9$ model for our analysis.

Do our models capture the global prevalence of approaches? The *MULTI*, to begin with, was intended to capture approaches at the session level. We investigate whether our models replicate the trends at a session-level. The comparative analysis for a randomly chosen session is shown in Figure 1. We see that despite making mistakes locally, the model captures approaches over therapist talk-turns. In this case, we see that the therapist scarcely uses a Psychodynamic or Interpersonal intervention and the model prediction shows similar behavior. On the other hand, the other subscale interventions are used almost uniformly over the length of the session. The model again captures this pattern.

Which categories are confused with each other? Figure 2a presents which categories tend to co-occur with each other. We observe a category Process-Experiential (PE) co-occurs with Person-Centered (PC) almost every third instance. Similarly, Psychodynamic (PD) approach almost always co-occurs with Process-Experiential (PE). Note that this is not commutative, i.e., PE co-occurs with PD about every fourth instance. Figure 2b shows the same, however, between gold labels and model prediction. Here we ask the question: for a certain category that exists in the gold data, what are the categories predicted by the model? Figures 2a and 2b should be identical if our model is ideal. Studying these figures in conjunction, gives us an idea of where the model confuses predictions the most. For example, a lot of Cognitive (CT) instances get misclassified as Person-Centered (PC), a trend which is not reflected in Figure 2a. We also observe that Psychodynamic (PD) items get significantly mispredicted as Process-Experiential (PE). A large number of approach-labeled instances get classified as 'No Code'. We expected this observa-

Test Labels	Metrics (in %)	SA	Pipeline	MultiTask_9	MultiTask_{10}
All	Exact Accuracy	76.84	74.63	**78.14**	75.86
	F1_{Macro}	48.24	48.52	**49.35**	47.79
	F1_{Micro}	**79.06**	75.43	78.63	78.17
Non-NC	F1_{Macro}	42.79	43.32	**44.06**	42.32
	F1_{Micro}	47.03	46.90	**47.64**	46.26

Table 3: Experimental results for all the classes (top half) and the eight subscales excluding 'No Code' (bottom-half)

Class	Class Abbrv.	SA	Pipeline	MultiTask_9	MultiTask_{10}
No Code	**NC**	**91.88**	90.07	91.65	91.55
Psychodynamic	**PD**	32.11	32.64	30.65	**32.97**
Process-Experiential	**PE**	67.20	65.30	**67.53**	67.32
Interpersonal	**IP**	33.25	35.21	**38.16**	34.34
Person-centered	**PC**	43.95	**44.86**	43.13	43.77
Common Factors	**CF**	**48.99**	48.87	48.06	47.30
Behavioral	**BT**	41.25	43.00	**43.90**	38.96
Cognitive	**CT**	33.95	33.12	**36.41**	34.26
Dialectical-Behavioral	**DBT**	41.62	43.60	**44.65**	39.67

Table 4: Class-wise F1 Results (in %)

Labels	Metrics	Va	PrevC	TC
All	Acc	78.14	78.34	**78.69**
	F1_{Macro}	49.35	49.12	**50.17**
	F1_{Micro}	78.63	78.67	**79.00**
Non-NC	F1_{Macro}	44.06	43.80	**44.96**
	F1_{Micro}	47.64	47.25	**48.00**

Table 5: Comparison of model performance(in %) with added contexts as compared to the MultiTask_9 model with just the therapist talk-turn (**Va**). This table shows results for all labels (top half) and the eight subscales excluding 'No Code' (bottom half)

tion given the skew in the training data.

6 Qualitative Analysis

F1 scores and Cohen's kappa scores cannot be compared directly. We analyze some model error examples to assess examples in a fair manner. We selected 22 examples at random with the constraint of selecting different combinations of labels. Out of the 22 examples chosen, five were ones which had an 'NC' gold label and a non-'NC' model prediction, while five had the opposite. The remaining twelve examples were mis-predictions between approach classes. Of the twelve, four were cases in which the talk-turn had a single gold approach and a single model prediction which did not match, while four each were cases in which there were multiple gold approaches but a single model predicted approach, and vice-versa. We made sure that the cases were diverse. We present five of these examples. We consider the best MutliTask_9 model which is trained on just the therapist talk-turn (**Va**) for this analysis.

Example 1
"*Interaction with your ex, like that's better for you*"
Human Annotation: NC
Model Prediction: IP
Here the human assessed that the talk-turn was not structured or specific enough to earn a code, despite the presence of interpersonal content. However, the model identified interpersonal language which may or may not be linked to client distress. In this case, the human seems to have been more conservative than the model in applying a code.

Example 2
"*And did you journal? Or keep a log?*"
Human Annotation: BT, CT, DBT
Model Prediction: NC
Here, journaling and log-keeping likely refers to reviewing homework, so the annotator marked an Item 10. This item, subsequently, maps onto three

Speaker	Talk-turn	Gold	SA	Pipeline	MultiTask$_9$
Client	Okay, sounds good, thank you.	-	-	-	-
Therapist	Yeah, so I just want to check in again, see how you're feeling in the room.	PD,PEI	PEI	PEI	PEI,PC
Client	Um I still feel fine, um, yeah I feel pretty good I guess.	-	-	-	-
Therapist	Okay, and that's also okay if you don't feel good, if you feel anxious. I still feel a little anxious as we're getting to know each other.	PEI, PC, CF	PEI, CF	PEI, CF	PEI
Client	Yeah.	-	-	-	-
Therapist	I just want to acknowledge that we have about twenty minutes left in our session. I'm curious is there anything you want to bring up, anywhere you want to start exploring?	CF	CF	CF	CF

Table 6: Example model predictions

subscales (BT, CT, and DBT). The model, in contrast, would not have known the homework context from this statement alone, resembling a case of atheoretical information gathering, hence an NC.

Example 3
"*Yeah and it sound sounds to me like you've already been incredibly patient with him, waiting for him to do those things, and recently he's just been letting you down over and over.*"
Human Annotation: IP
Model Prediction: IP, PC, CF
Both human and model identify clear evidence of client distress linked to an inter-personal relationship. However, the model detects justifiable PC and CF codes, explained by the emotion-added paraphrase and support for the client.

Example 4
"*I would guess that, I mean, that that's a really hard place for her to figure out.*"
Human Annotation: PE, PC
Model Prediction: CF
There is no clear argument for PE with only the context from this talk-turn. The human coder likely saw that the therapist made a paraphrase to justify the PC code. The model's CF coding is likely linked to the phrase 'really hard', which often arises from therapists providing empathic support for their client.

Example 5
"*So how was that experience, this last week of paying attention to your thoughts?*"
Human Annotation: PC, BT, CT, DBT
Model Prediction: PC
The therapist clearly asks about the client's experience, justifying a PC label. The phrase "last week of paying attention to your thoughts", however, sounds like a homework check-in (Item 10). Similar to example 2, Item 10 triggers three subscales and the human annotation of BT, CT, and DBT subscales seems appropriate and highlights a case which the model does not capture. This is an interesting case of annotation based on common-sense knowledge with which NLP models still struggle.

We should emphasize again that the humans do not annotate eight subscales directly; rather, they annotate based on the 30-item inventory. For instance, in example 2, the human annotator does not annotate the BT, CT, and DBT categories individually. They, instead, might have just annotated a single item (item 10) which maps to the three subscales. Hence, it should not be misconstrued that the human has over-labeled in that scenario.

In general, after analyzing the 22 examples, we find that in many such erroneous cases, prior intra-session (short or long range) and even inter-session contextual information might be relevant to determine the correct context. We leave this as a possible direction for future research.

7 Related Work

Artificial Intelligence and its sub-domains are being increasingly discussed as possible sources of

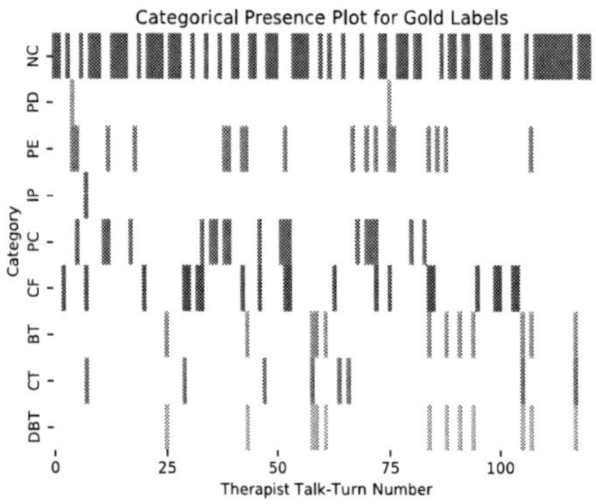

(a)

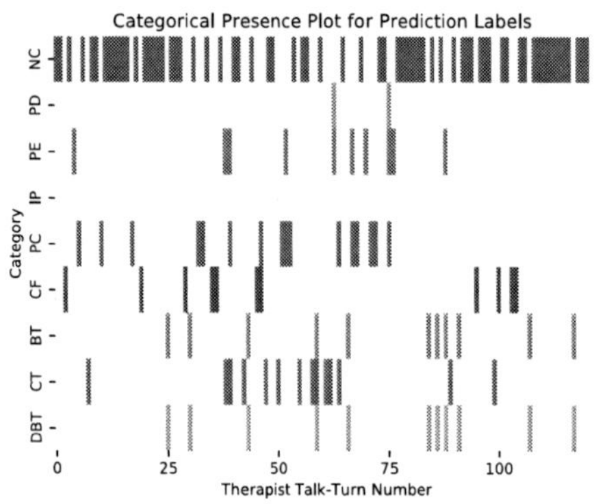

(b)

Figure 1: Predictions over a therapy session. Session proceeds left to right with a colored bar indicating the presence of an approach (or lack thereof) for the respective category. Plot (a) shows the approaches in gold annotations, (b) shows the same for model predictions.

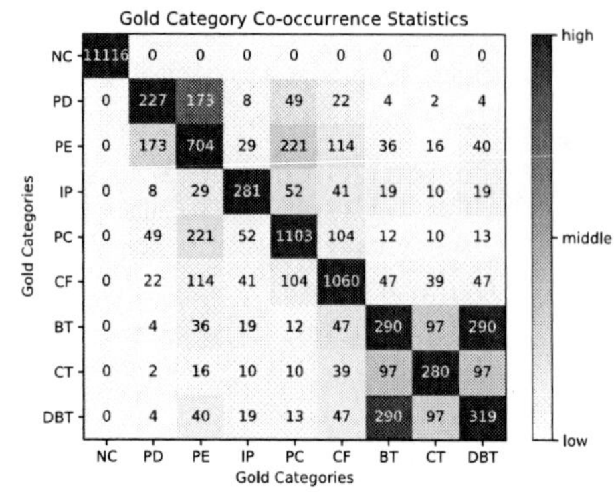

(a)

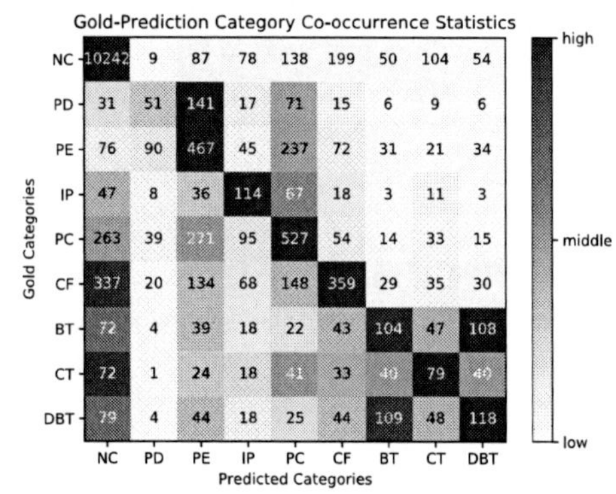

(b)

Figure 2: Co-occurrence Statistics. Figure (a) describes co-occurrence between approaches in therapist talk-turns in the human-annotated gold data. E.g., out of 227 talk-turns where PD is annotated, 173 talk-turns also had PE annotation. Figure (b) describes co-occurrence between approaches in the gold data and the model predictions. E.g., out of 227 talk-turns where PD is annotated as mentioned in (a), only 51 talk-turns had a PD model prediction. The color gradients are normalized on rows.

improvements in mental health conversations (Lee et al., 2021; Aafjes-van Doorn et al., 2021). Moreover, transcribed therapy data from counselling centres, and public mental health forums have encouraged interest in the NLP community (Goharian et al., 2021; Le Glaz et al., 2021). NLP tools have since been used to help automate Motivational Interviewing (Tanana et al., 2016; Pérez-Rosas et al., 2017), suicide ideation detection (Huang et al., 2014; Sawhney et al., 2018), etc. to name a few. More recently, pre-trained language models have been increasing finding use in various facets like qualitative session content analysis (Grandeit et al., 2020), detecting (Wu et al., 2021) and determining the direction of empathy (Hosseini and Caragea, 2021b,a). Li et al. (2022) use transformer-based pre-trained language models to evaluate interventions from a client perspective. Client talk-turn responses to therapist interventions are evaluated based on 3-class response type and a 5-class experience type adapted from TCCS (Ribeiro et al., 2013). To the best of our knowledge, this will be the first work to automate the *MULTI* subscale assignment of therapist talk-turns.

8 Conclusion

The expanding awareness and need for mental health improvement demands the ubiquity of such resources. Therapeutic evaluation becomes increasingly important as more people leverage mental health resources. We consider one such evaluation strategy — a talk-turn level adaptation of the *MULTI* — which evaluates therapist orientations. A major downside of such strategies remains their time-intensive nature. In this paper, we propose using pre-trained language models, which have proven to be high performance systems, to automate this evaluation. We experiment across three modeling paradigms using a pre-trained language model — RoBERTa. In addition, we show substantial analyses to understand the results. Our experiments are encouraging, however, we stress that substantial gaps in performance remain. We see this work as a significant stepping stone towards improving therapeutic feedback using NLP tools.

9 Ethics Statement

We note that the gold data used for this project was collected at a university counseling center at a university in the western United States. This induces a demographic bias in the data. It is highly possible that this data is neither representative of the various dialects of the English language spoken around the globe, nor of mental health concerns in the broader population. Our models are built using pre-trained language models, which, by design, are opaque. Consequently, our results are not interpretable.

The data was anonymized to protect information disclosures. Text snippets have been paraphrased by a Psychology graduate to mask stylistic cues.

10 Acknowledgements

We would like to thank the members of the Utah NLP group, and Utah Laboratory for Psychotherapy Science for their invaluable suggestions through the course of this work. We would also like to thank the anonymous reviewers for their insightful feedback. The authors acknowledge the support of NIH/NIAAA award R01 AA018673 and NSF award #1822877 (Cyberlearning).

References

Katie Aafjes-van Doorn, Céline Kamsteeg, Jordan Bate, and Marc Aafjes. 2021. A Scoping Review of Machine Learning in Psychotherapy Research. *Psychotherapy Research*, 31(1):92–116.

Jie Cao, Michael Tanana, Zac Imel, Eric Poitras, David Atkins, and Vivek Srikumar. 2019. Observing Dialogue in Therapy: Categorizing and Forecasting Behavioral Codes. In *Proceedings of the 57th Annual Meeting of the Association for Computational Linguistics*, pages 5599–5611, Florence, Italy. Association for Computational Linguistics.

Derek D Caperton. 2021. Development of a Multitheoretical, Statement-level Measure of Psychotherapeutic Interventions. . *Dissertation*.

Jacob Devlin, Ming-Wei Chang, Kenton Lee, and Kristina Toutanova. 2019. BERT: Pre-training of Deep Bidirectional Transformers for Language Understanding. In *Proceedings of the 2019 Conference of the North American Chapter of the Association for Computational Linguistics: Human Language Technologies, Volume 1 (Long and Short Papers)*, pages 4171–4186, Minneapolis, Minnesota. Association for Computational Linguistics.

Nikolaos Flemotomos, Victor R Martinez, Zhuohao Chen, Torrey A Creed, David C Atkins, and Shrikanth Narayanan. 2021. Automated Quality Assessment of Cognitive Behavioral Therapy Sessions Through Highly Contextualized Language Representations. *PloS one*, 16(10):e0258639.

Nazli Goharian, Philip Resnik, Andrew Yates, Molly Ireland, Kate Niederhoffer, and Rebecca Resnik, editors. 2021. *Proceedings of the Seventh Workshop on Computational Linguistics and Clinical Psychology: Improving Access*. Association for Computational Linguistics, Online.

Philipp Grandeit, Carolyn Haberkern, Maximiliane Lang, Jens Albrecht, and Robert Lehmann. 2020. Using BERT for Qualitative Content Analysis in Psychosocial Online Counseling. In *Proceedings of the Fourth Workshop on Natural Language Processing and Computational Social Science*, pages 11–23.

Mahshid Hosseini and Cornelia Caragea. 2021a. Distilling Knowledge for Empathy Detection. In *Findings of the Association for Computational Linguistics: EMNLP 2021*, pages 3713–3724, Punta Cana, Dominican Republic. Association for Computational Linguistics.

Mahshid Hosseini and Cornelia Caragea. 2021b. It Takes Two to Empathize: One to Seek and One to Provide. *Proceedings of the AAAI Conference on Artificial Intelligence*, 35(14):13018–13026.

Xiaolei Huang, Lei Zhang, David Chiu, Tianli Liu, Xin Li, and Tingshao Zhu. 2014. Detecting Suicidal Ideation in Chinese Microblogs with Psychological Lexicons. In *2014 IEEE 11th Intl Conf on Ubiquitous Intelligence and Computing and 2014 IEEE 11th Intl Conf on Autonomic and Trusted Computing and 2014 IEEE 14th Intl Conf on Scalable Computing and Communications and Its Associated Workshops*, pages 844–849. IEEE.

Zac E Imel, Mark Steyvers, and David C Atkins. 2015. Computational Psychotherapy Research: Scaling up the Evaluation of Patient–Provider Interactions. *Psychotherapy*, 52(1):19.

J Richard Landis and Gary G Koch. 1977. The Measurement of Observer Agreement for Categorical Data. *Biometrics*, pages 159–174.

Aziliz Le Glaz, Yannis Haralambous, Deok-Hee Kim-Dufor, Philippe Lenca, Romain Billot, Taylor C Ryan, Jonathan Marsh, Jordan Devylder, Michel Walter, Sofian Berrouiguet, et al. 2021. Machine Learning and Natural Language Processing in Mental Health: Systematic Review. *Journal of Medical Internet Research*, 23(5):e15708.

Ellen E Lee, John Torous, Munmun De Choudhury, Colin A Depp, Sarah A Graham, Ho-Cheol Kim, Martin P Paulus, John H Krystal, and Dilip V Jeste. 2021. Artificial Intelligence for Mental Health Care: Clinical Applications, Barriers, Facilitators, and Artificial Wisdom. *Biological Psychiatry: Cognitive Neuroscience and Neuroimaging*, 6(9):856–864.

Anqi Li, Jingsong Ma, Lizhi Ma, Pengfei Fang, Hongliang He, and Zhenzhong Lan. 2022. Towards Automated Real-time Evaluation in Text-based Counseling. *arXiv preprint arXiv:2203.03442*.

Xiaodong Liu, Pengcheng He, Weizhu Chen, and Jianfeng Gao. 2019a. Multi-Task Deep Neural Networks for Natural Language Understanding. In *Proceedings of the 57th Annual Meeting of the Association for Computational Linguistics*, pages 4487–4496, Florence, Italy. Association for Computational Linguistics.

Yinhan Liu, Myle Ott, Naman Goyal, Jingfei Du, Mandar Joshi, Danqi Chen, Omer Levy, Mike Lewis, Luke Zettlemoyer, and Veselin Stoyanov. 2019b. Roberta: A Robustly Optimized BERT Pretraining Approach. *arXiv preprint arXiv:1907.11692*.

Kevin S McCarthy and Jacques P Barber. 2009. The Multitheoretical List of Therapeutic Interventions (MULTI): Initial Report. *Psychotherapy research*, 19(1):96–113.

Verónica Pérez-Rosas, Rada Mihalcea, Kenneth Resnicow, Satinder Singh, Lawrence An, Kathy J Goggin, and Delwyn Catley. 2017. Predicting Counselor Behaviors in Motivational Interviewing Encounters. In *Proceedings of the 15th Conference of the European Chapter of the Association for Computational Linguistics: Volume 1, Long Papers*, pages 1128–1137.

Eugénia Ribeiro, Antonio P Ribeiro, Miguel M Gonçalves, Adam O Horvath, and William B Stiles. 2013. How Collaboration in Therapy Becomes Therapeutic: The Therapeutic Collaboration Coding System. *Psychology and Psychotherapy: Theory, Research and Practice*, 86(3):294–314.

Ramit Sawhney, Prachi Manchanda, Puneet Mathur, Rajiv Shah, and Raj Singh. 2018. Exploring and Learning Suicidal Ideation Connotations on Social Media with Deep Learning. In *Proceedings of the 9th Workshop on Computational Approaches to Subjectivity, Sentiment and Social Media Analysis*, pages 167–175, Brussels, Belgium. Association for Computational Linguistics.

Nili Solomonov, Kevin S McCarthy, Bernard S Gorman, and Jacques P Barber. 2019. The Multitheoretical List of Therapeutic Interventions–30 Items (MULTI-30). *Psychotherapy Research*, 29(5):565–580.

Asa Cooper Stickland and Iain Murray. 2019. BERT and PALs: Projected Attention Layers for Efficient Adaptation in Multi-Task Learning. In *International Conference on Machine Learning*, pages 5986–5995. PMLR.

Michael Tanana, Kevin A Hallgren, Zac E Imel, David C Atkins, and Vivek Srikumar. 2016. A Comparison of Natural Language Processing Methods for Automated Coding of Motivational Interviewing. *Journal of substance abuse treatment*, 65:43–50.

Ashish Vaswani, Noam Shazeer, Niki Parmar, Jakob Uszkoreit, Llion Jones, Aidan N Gomez, Łukasz Kaiser, and Illia Polosukhin. 2017. Attention Is All You Need. In *Advances in neural information processing systems*, pages 5998–6008.

Thomas Wolf, Lysandre Debut, Victor Sanh, Julien Chaumond, Clement Delangue, Anthony Moi, Pierric Cistac, Tim Rault, Remi Louf, Morgan Funtowicz, Joe Davison, Sam Shleifer, Patrick von Platen, Clara Ma, Yacine Jernite, Julien Plu, Canwen Xu, Teven Le Scao, Sylvain Gugger, Mariama Drame, Quentin Lhoest, and Alexander Rush. 2020. Transformers: State-of-the-Art Natural Language Processing. In *Proceedings of the 2020 Conference on Empirical Methods in Natural Language Processing: System Demonstrations*, pages 38–45, Online. Association for Computational Linguistics.

Zixiu Wu, Rim Helaoui, Diego Reforgiato Recupero, and Daniele Riboni. 2021. Towards Low-Resource Real-Time Assessment of Empathy in Counselling. In *Proceedings of the Seventh Workshop on Computational Linguistics and Clinical Psychology: Improving Access*, pages 204–216, Online. Association for Computational Linguistics.

A Subscales, Items and Model Classes

MULTI-30 Items	Model Classes
ϕ	No Code
2,6,12,14,15	Psychodynamic
5,7,18,23	Process-Experiential
25,26,27,30	Interpersonal
4,21,22	Person-Centered
3,11,16,17	Common Factors
8,9,19	Behavioral_{only}
13,20,24	Cognitive_{only}
28,29	$\text{Dialectical-Behavioral}_{only}$
1,10	$\text{Cognitive-Behavioral}_{shared}$

Table 7: Mapping between model classes and the *MULTI*-30 item codes (Solomonov et al., 2019). We use these classes for model training to facilitate flexibility in at a finer level. The author-defined model classes which are not part of the conventional *MULTI* subscale are highlighted.

MULTI Subscales	*MULTI*-30 Items	Constituent Model Classes
No Code	ϕ	No Code
Psychodynamic	2,6,12,14,15	Psychodynamic
Process-Experiential	5,7,18,23	Process-Experiential
Interpersonal	25,26,27,30	Interpersonal
Person-Centered	4,21,22	Person-Centered
Common Factors	3,11,16,17	Common Factors
Behavioral	8,9,19,1,10	Behavioral$_{only}$,Cognitive-Behavioral$_{shared}$
Cognitive	13,20,24,1,10	Cognitive$_{only}$,Cognitive-Behavioral$_{shared}$
Dialectical-Behavioral	28,29,1,10,8,9,19	Dialectical-Beh.$_{only}$,Cognitive-Beh.$_{shared}$,Beh.$_{only}$

Table 8: The conventional subscales and their constituent *MULTI*-30 items are shown here. Note that the Behavioral, Cognitive and Dialectical-Behavioral subscales (highlighted) have overlapping items. The constituent model classes from Table 7 are shown. Note that all our evaluations are presented on the conventional *MULTI* sub-scales by aggregating performance on their constituent model classes.

Then and Now: Quantifying the Longitudinal Validity of Self-Disclosed Depression Diagnoses

Keith Harrigian and **Mark Dredze**
Johns Hopkins University
kharrigian@jhu.edu, mdredze@cs.jhu.edu

Abstract

Self-disclosed mental health diagnoses, which serve as ground truth annotations of mental health status in the absence of clinical measures, underpin the conclusions behind most computational studies of mental health language from the last decade. However, psychiatric conditions are dynamic; a prior depression diagnosis may no longer be indicative of an individual's mental health, either due to treatment or other mitigating factors. We ask: to what extent are self-disclosures of mental health diagnoses actually relevant over time? We analyze recent activity from individuals who disclosed a depression diagnosis on social media over five years ago and, in turn, acquire a new understanding of how presentations of mental health status on social media manifest longitudinally. We also provide expanded evidence for the presence of personality-related biases in datasets curated using self-disclosed diagnoses. Our findings motivate three practical recommendations for improving mental health datasets curated using self-disclosed diagnoses:

1. Annotate diagnosis dates and psychiatric comorbidities
2. Sample control groups using propensity score matching
3. Identify and remove spurious correlations introduced by selection bias

1 Introduction

The ability to provide equitable access to psychiatric healthcare has become more difficult than ever, inhibited by an entanglement of lingering public policy effects (Miranda et al., 2020), heightened levels of physician burnout (Johnson et al., 2018), and infrastructural challenges arising from global crisis (Davis et al., 2021). Meanwhile, social media platforms have become the predominant means of communication for much of the population, providing the opportunity to share personal experiences and seek support from others (Mueller et al., 2021). Noting these parallel timelines, computational scientists have devoted substantial effort to engineering statistical models capable of translating social media data into reliable insights regarding mental health. Core objectives of this work include optimizing psychiatric treatment, identifying early stages of mental illness, and measuring the effect of public policy on a population's well-being (Losada et al., 2017; Fine et al., 2020).

The most significant advances in computational mental health research have not come from improved modeling architectures (Benton et al., 2017b), but from methods for curating large-scale datasets which contain robust and clinically-relevant ground truth annotations of mental health status (Coppersmith et al., 2014). Use of regular expressions to identify genuine self-disclosures of a psychiatric diagnosis remains one of the most widely adopted annotation mechanisms by the research community (Chancellor and De Choudhury, 2020; Harrigian et al., 2021), offering a relatively reliable proxy in place of clinical measures which are not only costly to collect, but also often unable to be shared beyond a single institution due to patient privacy policies (Macavaney et al., 2021). Datasets leveraging self-disclosed diagnoses as annotations of mental health status have yielded a variety of insights that align with clinical knowledge and psychological theory (Mowery et al., 2017; Lee et al., 2021). However, a growing body of work has raised questions about whether such datasets provide sufficient information to train statistical models that generalize to new populations (Harrigian et al., 2020; Aguirre et al., 2021).

Despite the prevalence of datasets dependent on self-disclosure, no analyses have considered how associating a single self-disclosed diagnosis label with data from a variable-length period of time may inhibit the learning of robust statistical relationships. If a user tweets a depression diagnosis in 2015, is their data from 2018 still representative

Proceedings of the Eighth Workshop on Computational Linguistics and Clinical Psychology, pages 59 - 75
July 15, 2022

of the condition? Presentation of several mental health conditions change dynamically and (sometimes) precipitously over time (Collishaw et al., 2004). Yet, it remains common in the computational research community to treat mental health conditions as a static attribute with equal relevance at multiple time points (MacAvaney et al., 2018). In reality, it is likely that only a small fraction of an individual's social media activity is appropriate for training optimal classifiers. Moreover, that a mental health status label may be appropriate for only a subset of time suggests that evaluations of longitudinal model generalization as they are traditionally structured in the community may be insufficient (Sadeque et al., 2018).

We ask: to what extent do mental health diagnosis self-disclosures remain valid over time? We focus specifically on extended durations (i.e., multiple years), a setting which has particular relevance to those who wish to estimate generalization strength of their statistical classifiers for use in longitudinal monitoring applications, as well as those interested in updating existing models with new data to mitigate the effects covariate shift (Agarwal and Nenkova, 2021). In reviewing recent online activity from individuals in the 2015 CLPsych Shared Task dataset who disclosed a depression diagnosis on Twitter over five years ago (Coppersmith et al., 2015), we not only acquire a new understanding of how presentations of mental health status on social media present over time, but also find new evidence to support prior claims regarding the presence of personality-related confounds in datasets curated using self-disclosures (Preoţiuc-Pietro et al., 2015; Vukojevic and Šnajder, 2021). Our analysis provides critical guidance to practitioners as they curate mental health social media datasets, while also elucidating factors which inhibit robustness in a dataset that remains one of the most widely adopted by the research community.

2 Background

The majority of mental health research based on social media leverages the same experimental design—assume individuals have a fixed mental health status and attempt to infer this latent attribute using historical online activity traces (e.g., posts, follower network dynamics) (Guntuku et al., 2017; Chancellor and De Choudhury, 2020). This training setting is convenient given the inherent complexities of acquiring temporally-granular psychiatric measures at scale (Canzian and Musolesi, 2015). However, the setting implicitly relies on assumptions that are not supported by clinical knowledge regarding psychiatric dynamics (Johnson and Nowak, 2002; Schoevers et al., 2005). Some work has been done to incorporate time-based priors into mental health models, which allow practitioners to train statistical classifiers using a static label while also explicitly accounting for longitudinal variation in label relevance (Wongkoblap et al., 2019; Uban et al., 2021). Others have eschewed the use of a static label altogether and instead curated datasets that contain multiple points of ground truth mental health status, albeit still with some element of historical data aggregation (Chancellor et al., 2016).

Dataset	Dates	# Users	# Posts
Original	2012 – 2015	D: 477 C: 872	D: 1,121,388 C: 1,907,508
Updated	2012 – 2021	D: 444 C: 172	D: 1,372,868 C: 546,826

Table 1: Summary statistics for the original and updated versions of the 2015 CLPsych Shared Task dataset, further stratified by [C]ontrol and [D]epression groups.

Temporally-aware classifiers have achieved better performance benchmarks than their static counterparts in some cases (Rao et al., 2020), though these evaluations remain limited by the dearth of data with mental health status annotations at multiple time points. Meanwhile, datasets which do support dynamic evaluation are curated almost exclusively using protected clinical measures (Reece et al., 2017), cost-intensive interviews (Nobles et al., 2018), or non-trivial shifts in non-language-based online behavior (De Choudhury et al., 2016).

Computational studies that have focused on self-disclosed diagnoses have not comprehensively reviewed how individual activity evolves over long periods of time (Saha et al., 2021). Our study thus fulfills an important void in the research space by providing a new understanding of long term mental health dynamics in social media, and more particularly, within convenience samples curated using self-disclosed diagnoses.

3 Data

We support our study using a newly updated version of the 2015 CLPsych Shared Task dataset (Coppersmith et al., 2015). The original Twitter dataset was constructed in a two-stage process, with regular-expressions first being used to identify candidate self-disclosures of a depression diagnosis and experts manually verifying the authenticity of the match thereafter. Individuals in the control group were sampled randomly from the 1% public Twitter stream such that the joint distribution of inferred age and gender attributes (Sap et al., 2014) was in alignment with the depression group. Up to 3,000 tweets were acquired for each individual in the resulting sample using Twitter's public API. The dataset has not only become one of the most widely adopted social media datasets for mental health (Harrigian et al., 2021), but also inspired the annotation procedures for numerous successors across various platforms and languages (Cohan et al., 2018; Shen et al., 2018).

In line with guidance from Benton et al. (2017a), individual identifiers in the official version of the CLPsych dataset have been anonymized, with linkages between anonymized and de-anonymized identifiers erased in entirety. However, the original de-anonymized identifiers remain available under explicit permission from Coppersmith et al. (2015), who provided this information to reverse engineer the original anonymization mapping. To do so, we first query up to 3,200 of the most recent tweets from each de-anonymized user identifier using Twitter's public API and further isolate all relevant tweets found in our institution's cache of Twitter's 1% data stream. We identify candidate pairs of anonymized and de-anonymized accounts based on overlap of raw timestamps within the original dataset's collection window. Normalized text (i.e., punctuation removal, case standardization) from candidate pairs is compared using exact matching to verify final linkages.

Statistics for the original dataset and its updated counterpart are provided in Table 1. We find that a majority of accounts which were unable to be linked had significantly smaller activity traces in the original dataset. These accounts are likely to either have been deleted in entirety or to have tweeted with a small enough frequency such that the 1% stream does not contain any samples. The discrepancy in match rates between individuals in the depression and control groups is unfortunately not fully-understood, though discussions with the dataset's authors suggest this may just be an artifact of the original archival process.

Preprocessing. Twitter's language tags and automatic language identification (Lui and Baldwin, 2012) are used to isolate English text. Retweets are excluded to most acutely highlight personal experiences with depression over time. Unless specified otherwise, keyword-based tweet filtering is applied to preemptively mitigate sampling-induced biases which can artificially inflate estimates of predictive performance. Some of these biases have been recognized and addressed by the research community (e.g., filtering tweets which include diagnosis disclosures and/or mental health related keywords/hashtags) (De Choudhury and De, 2014), while others have been traditionally overlooked.

A preliminary qualitative analysis of influential n-grams and their source tweets reveals a previously unrecognized surplus of "fan accounts" (e.g., supporters of Harry Styles and Demi Lovato) and tweets containing account statistics (e.g., new followers) within the depression cohort. Meanwhile, daily horoscope tweets were identified with an anomalous frequency within the control group. The latter two sources of noise do not have a clear clinical explanation, while the former (i.e., fan accounts) arises in the context of discussion regarding the mental health of young celebrities. Although some of these motifs represent genuine behavioral correlates of depression, their importance in prediction tends to be inflated due to context of the original collection time period.

4 Inference Under Latent Dynamics

Enabling reliable use of statistical models to evaluate change in mental health status remains a core objective for computational researchers (Choi et al., 2020; Fine et al., 2020). Our success in this task domain critically depends on access to ground truth at multiple time points, not only for evaluating generalization error (DeMasi et al., 2017; Tsakalidis et al., 2018), but also for mitigating the effects of covariate shift (Sugiyama and Kawanabe, 2012). As discussed above, it is often trivial to update activity traces for individuals with a prior mental health diagnosis disclosure. Nonetheless, clinical knowledge suggests original disclosure-based labels may not be relevant over the course of time, either due to a condition's episodic presentations (Angst et al., 2009) or the effects of psychiatric treatment (Saha et al., 2021). We ask whether the CLPsych Shared Task dataset supports this theory.

Methods. A natural framework for answering this inquiry emerges from computational research regarding label noise (Frénay and Verleysen, 2013).

Under such a perspective, we can view changes in mental health status as a stochastic process which blindly alters the correctness of class labels over time. The implications of this mechanism allow us to reason about predictive performance of a statistical classifier within and outside of the time period in which it is trained. Differences in within-time-period performance for two different time periods may be caused by two factors—different levels of label noise and/or different signal-to-noise ratios. Meanwhile, degradation in performance when transferring a classifier from one time period to another may be caused by three possible factors—label noise in the source time period, label noise in the target time period, or distributional shift between the time periods. Although isolated differences in predictive performance in a longitudinal setting do not implicate a single causal factor, multiple comparisons taken together may allow us to reason about underlying changes in the data.

This logic guides our search for evidence in support of the hypothesis that mental health annotations cannot be treated as fixed attributes. We consider a standard longitudinal domain transfer setup (Huang and Paul, 2019), chunking the CLPsych dataset into three discrete three-year periods[1] (2012–2015, 2015–2018, 2018–2021) and evaluating within- and between-time-period predictive performance for all available pairs. We use Monte Carlo Cross Validation (Xu and Liang, 2001) to obtain estimates of predictive generalization, chosen over alternative protocols that would be unreliable given the limited sample size of the updated CLPsych dataset (Varoquaux, 2018).

Each iteration of the cross validation procedure (1,000 total) begins by randomly splitting individuals into a 60/40 train/test split, with control and depression groups demographically aligned[2] using propensity scores (Imbens and Rubin, 2015). To control for differences in data availability between time periods, we not only constrain the sampling process such that splits have an *equal class balance*, but also that individual-level representations are constructed using an *equal document history size* (250 randomly-sampled posts from each time period). A single binary logistic regression classifier provided with document-term TF-IDF representations (Baeza-Yates et al., 1999) is fit for each

	Test		
Train	2012-2015	2015-2018	2018-2021
2012-2015	$.71_{(.70,.72)}$	$.66_{(.65,.66)}$	$.69_{(.68,.70)}$
2015-2018	$.66_{(.65,.67)}$	$.66_{(.65,.66)}$	$.68_{(.67,.69)}$
2018-2021	$.65_{(.65,.66)}$	$.67_{(.66,.68)}$	$.68_{(.67,.69)}$

Table 2: Average test-set area under the curve (AUC) and 95% confidence intervals across 1,000 Monte Carlo Cross Validation iterations. Within-time-period performance is significantly higher around the original disclosure window than in subsequent time periods.

time period using data from individuals in the training set. Each classifier is applied to all three time periods, evaluating performance using individuals in the sampled test set.

Results. We report the average test set area under the curve (AUC) and 95% confidence intervals for each discrete time period pairing in Table 2. Focusing first on *within-time-period* performance (top left to bottom right diagonal), we find that within-time-period performance is significantly higher in the dataset's original time period (2012-2015) than within subsequent time periods. This holds true even when running experiments only with individuals that have sufficiently-sized post histories in the new time periods, demonstrating that the outcome is not an artifact of survivor bias. At a high level, the differences in within-time-period performance suggest that either label noise has increased or that the signal-to-noise ratio has decreased over time.

Unfortunately, examination of *between-time-period* generalization does not conclusively resolve which of these two factors are responsible for the variation. Focusing first on models trained using data from older time periods (top right triangle), we do not observe any significant difference in predictive performance compared to the benchmarks established by models trained and deployed during the same time period. This serves as a contrast to models deployed on older data (bottom left triangle), where we note that classifiers trained on both of the new time periods incur a loss when being applied to the original CLPsych dataset time period. Interestingly, the absolute differences in performance are minimal. We note that the coefficients of the logistic regression classifiers from each independent time period exhibit significantly positive Pearson correlations, ranging from 0.47 to 0.52, and in turn promote stable performance.

Discussion. Although these experiments have

[1]Time periods were chosen to maximize the number of discrete windows while ensuring enough posts were available to construct informative individual-level representations.

[2]Aligned on gender and age dimensions.

not conclusively answered our primary research question regarding longitudinal label validity, they have provided evidence that not all time periods of data are equally informative for training a robust depression classifier. Critically, these results suggest that practitioners cannot assume it better to train a depression classifier using new data, which may be more relevant to their deployment scenario, if it means potentially compromising the temporal relevance of the original ground truth annotations.

What remains to be understood is *why* the predictive task appears to become more difficult in the updated time periods at a statistically significant level, but not one that would necessarily raise immediate concerns to a practitioner. Had underlying dynamics significantly changed since the original data collection period, we would have expected to see a more dramatic loss in predictive performance. Has the mental health status for these individuals genuinely remained static, or is there a spurious confound in the data inflating our performance estimates?

5 Interpreting Model Performance

We attempt to better understand the variation in predictive performance estimated above by comparing language within the updated dataset to the original CLPsych sample. In particular, we adopt a mixed methods approach that allows us to estimate changes in the proportion of depression labels which remain relevant in the updated dataset, and to qualitatively summarize drivers of model decision-making across time periods. We support our analysis by manually coding content-related motifs within a large sample of document histories in the updated dataset, focusing primarily on criteria for diagnosing depression as defined within the DSM-5 (APA, 2013). We draw inspiration from the growing literature on "train-set debugging" (Koh and Liang, 2017; Han et al., 2020), which leverages instance attribution and other diagnostics to succinctly interpret the relationship between training data, learned model parameters, and downstream predictions.

Methods. An annotator is presented with up to 30 anonymized tweets made by a single individual during one of the time periods and asked to indicate whether the individual exhibits evidence of depression. The annotator must mark one of four options — Uncertain, No Evidence, Some Evidence (Moderate Confidence), Strong Evidence (High Confidence). Explicit disclosures of a depression diagnosis and references to living with depression are automatically assigned to the Strong Evidence category. Otherwise, the annotator is instructed to indicate their confidence based on the nine DSM-5 criteria for diagnosing depression (APA, 2013) and their prior knowledge regarding the presentation of mental health conditions within social media. If at least some evidence of a depression diagnosis is indicated, the annotator is asked to identify whether the depression appears to be in remission (e.g., discussion of overcoming depression). They are also asked to indicate which DSM-5 criteria and/or prior knowledge was used to inform their decision, along with any other notable thematic content.

Our goal of this analysis is *not* to make diagnostic claims regarding the mental health status of individuals in our dataset, but rather to broadly understand what the statistical classifiers are learning. Accordingly, tweets presented to the annotator are those which had the largest positive effect on the classifier's estimated probability of depression, as measured by their influence on user-level predictions within a given time period τ. Formally, we define the influence of a tweet $I(x)$ amongst a set of tweets $x \in X_\tau$ as follows:

$$I(x) = \sum_{k=1}^{K} P_{k,\tau}(y = 1|X_\tau) - P_{k,\tau}(y = 1|X_\tau^{\neg x})$$

where $P_{k,\tau}(\cdot)$ is the probability of depression estimated by a classifier trained on the k-th random sample of data from time period τ, out of K total samples. As was the case in the classification experiments above, each training sample contains 60% of the available data, with the learned classifiers only being applied to the remaining 40% of individuals at each iteration. We refrain from filtering mental health related tweets and those containing explicit diagnosis disclosures, as the goal in this experiment is not to quantify predictive ability, but rather to identify evidence of depression over time. Note that we control for distributional shift over time by estimating influence using a model trained during the time period in which a tweet was posted.

Data. A total of 300 individuals (574 total instances) were selected randomly for annotation. One author, a doctoral student in computer science with multiple years experience working with the CLPsych dataset, was responsible for all coding. They consulted one additional co-author, an expert in computational modeling of social media and mental health, to develop a common mental model for identifying DSM-5 criteria and other common linguistic motifs in the text. During a pilot round of coding, 16 thematic patterns were identified within the annotated instances to complement the original DSM-5 criteria. Exemplary tweets (paraphrased non-trivially to preserve anonymity (Ayers et al., 2018)) for each of the DSM-5 criteria and alternative thematic categories are provided in Appendix A.3. A breakdown of annotation results is presented in the Table 3. We provide a distribution of the top 20 most common evidence categories amongst individuals who displayed at least some evidence of a depression diagnosis in Appendix A.3.

Reliability. Two non-authors with a background in computational psychology independently annotated a subsample of the coded instances to assess the primary coder's reliability. Agreement regarding whether an individual exhibits evidence of depression was fair to moderate; we observe Krippendorff's α measures of 0.438 and 0.499 for the four-class (Uncertain, No Evidence, Some Evidence, Strong Evidence) and three-class (Uncertain, No Evidence, Some or Strong Evidence) scenarios, respectively (Krippendorff, 2011). Agreement regarding remission status varied significantly between pairs of annotators and was generally weaker than agreement regarding evidence of depression ($\alpha = 0.356$). We include an analysis of the disagreements in Appendix A.2 to better contextualize observations from the primary coder's annotations. Succinctly, we identify two reasons for the variation: 1) each annotator's propensity to select the "Uncertain" category, and 2) each annotator's sensitivity to displays of emotion as an indicator of depression.

5.1 What proportion of labels in the updated sample remain relevant?

In line with underlying clinical knowledge regarding the dynamic nature of depression, we observe a significant decrease in linguistic evidence of depression over the course of time. Roughly 76% of individuals in the original depression group displayed at least some clear evidence of a depression diagnosis during the first time period (2012-2015), in comparison to 45% and 39% of individuals in the 2015-2018 and 2018-2021 time periods, respectively. Across all time periods, only a small number

	Dates	Total	Some Evi.	Strong Evi.	Not Active
Con.	2012-2015	83	15	3	1
	2015-2018	50	10	2	0
	2018-2021	40	5	0	0
Dep.	2012-2015	215	164	136	10
	2015-2018	107	49	28	2
	2018-2021	79	31	16	1

Table 3: Breakdown of coding labels as a function of time period and labels from the original CLPsych dataset. Clinically aligned evidence of a depression diagnosis becomes less prevalent over time.

of affirmative instances of depression appear to be in remission. That said, the non-zero level of inactive depression annotations in the original time period highlights an important consideration for practitioners who would like to leverage disclosure-based mechanisms to annotate mental health data moving forward.

The presence of evidence for a depression diagnosis in a subset of the original control group is quite striking. Other studies have raised questions regarding the possible risk of introducing such label noise when curating a control group using a random sampling protocol (Wolohan et al., 2018), though none have provided tangible evidence of this contamination to the best of our knowledge. We see that approximately 4% of individuals in the control group display strong evidence of a depression diagnosis within the original time period. Although relatively small, it is an important reminder of the pitfalls of random control group sampling for health-related social media modeling tasks.

Discussion. The decrease in evidence of a depression diagnosis over time lends support to the introduction of label noise in the updated dataset. Furthermore, it would explain the decrease in predictive performance observed in our previous classification experiments. However, the proportional drop in evidence of a depression diagnosis over time appears too large given the relatively minor reduction in classification accuracy.

We identify two possible explanations for this inconsistency. First, we recognize the possibility that our annotation procedure is insufficient to provide an annotator with appropriate information and comprehensive criteria for indicating evidence of a depression diagnosis. Only a small subset of an individual's entire post history is displayed to the annotator, a subset chosen using an inherently error-prone statistical ranking method. It is possible that stronger indicators of a depression diagnosis lie outside the 30-tweet sample size window for some individuals. Moreover, the annotator was instructed to rely predominantly on DSM-5 criteria to inform their decision, though several prior computational studies have shown language informative of depression may stray from explicit diagnostic criteria and be difficult for humans to recognize altogether (e.g., increased personal pronoun usage (Holtzman et al., 2017)).

More concerning is the possible presence of non-trivial confounds introduced by the original dataset's sampling/annotation procedure which may artificially inflate predictive performance estimates. Similar types of bias have been identified in prior work when attempting to transfer statistical mental health models trained using proxy-based annotations to new populations of individuals (e.g., demographics, patient populations) (Ernala et al., 2019; Aguirre et al., 2021). Although sampling-based artifacts may be causally-related to the original diagnosis disclosure (e.g., a coping mechanism that becomes a hobby, heightened levels of neuroticism), they may be serve as a red herring in place of primary indicators of depression.

5.2 Do presentations of depression provide evidence of sampling-related confounds?

Personality-related attributes are prominent features in all periods of the updated dataset. For example, indications of a depressed and/or irritable mood were the most common form of evidence in support of an individual having a depression diagnosis. In many cases, anger and irritation were displayed in the form of interpersonal confrontation (passively and actively) with other Twitter profiles. Negative emotions such as loneliness, fear, and existential dread were also displayed readily amongst those showing signs of a depression diagnosis. This result aligns with knowledge regarding the relationship between personality and depression, with elevated levels of neuroticism (negative affectivity and vulnerability to stress) being common in those living with depression (Bagby et al., 2008; Lahey, 2009; Bondy et al., 2021). Although etiologically relevant, this heightened level of emotional affect emerges as one possible artifact which may confound displays of depression and serve as a nuisance variable in linguistic models of the condition (Tackman et al., 2019).

We also found it common for individuals to mention comorbid psychiatric conditions—such as obsessive compulsive disorder, bipolar disorder, and general anxiety. Many of these conditions share similar underlying symptoms and causes with depressive disorders (Franklin and Zimmerman, 2001; Goodwin, 2015), but tend to assume a different temporal profile (Schoevers et al., 2005). The significant overlap often makes it difficult for trained physicians to properly diagnose individuals (Bowden, 2001) and for language-based algorithms to achieve appropriate discriminative sensitivity (Ive et al., 2018). We recognize the possibility that these comorbid conditions are active during the updated time periods for some individuals and may assume a proxy role in place of depression.

Although not captured by any single evidence category in isolation, there emerged a distinct propensity for "oversharing" amongst individuals from the original dataset's depression group. More specifically, we identified ample discussion of topics that are typically considered socially inappropriate in public discourse spaces (e.g., sexual activity, familial conflict, use of controlled substances). On one hand, this is an interesting finding given that individuals living depression often demonstrate lower levels of emotional self-disclosure (Wei et al., 2005; Kahn and Garrison, 2009). On the other hand, we note that prior work in clinical psychology has recognized a similar propensity for depressed and anxious individuals to engage in oversharing within social media (Radovic et al., 2017; Law et al., 2020).

The theory behind that latter is that social media offers an opportunity to discuss the oft stigmatized challenges of mental health (Betton et al., 2015) and increase feelings of connectedness in a less personal environment (Luo and Hancock, 2020). With this in mind, perhaps it is not surprising that those who have openly disclosed their experience with depression also feel comfortable discussing the aforementioned "taboo" topics. Nonetheless, this personal comfort remains relatively unique amongst the larger social media population. The unfortunate effect of this nuance is that it transforms the primary depression inference task into, essentially, a topic-classification task.

Discussion. Our analysis affirms what other recent studies on proxy-based mental health annotations have claimed — individuals who disclose a mental health condition systematically differ from the larger population of individuals living with that condition (Ernala et al., 2019; Saha et al., 2021). As a research community, we must be careful to disambiguate 1) training a language classifier to identify individuals who live with a mental health condition, and 2) training a language classifier to identify individuals who live with a mental health condition *and* disclose their diagnosis. Inappropriately equating the two creates an opportunity to erroneously estimate population-level dynamics (Amir et al., 2019) and ignore underrepresented voices from communities who tend to possess conservative ideologies regarding mental health (Loveys et al., 2018; Aguirre et al., 2021).

6 Discussion

Demand for computational methods to quantify mental health dynamics within social media data is at an all time high (Galea et al., 2020). However, the potential impact of these methods remains bounded by the robustness of datasets used for their development. Spanning nearly a decade of online activity, our study uniquely identifies evidence of these limitations as they currently manifest in non-clinically derived mental health social media datasets. This evidence leads us to offer three recommendations for enhancing data curation and model evaluation.

Annotate Diagnosis Date & Comorbidities. We identified several instances within our dataset where a diagnosis disclosure was made in reference to a condition that had since entered remission. In other cases, depression diagnoses were either supplanted by or augmented with alternative psychiatric diagnoses. Indicators regarding the time a diagnosis was made, many of which can be identified using inexpensive algorithms (MacAvaney et al., 2018), can provide important signal regarding the temporal relevance of a psychiatric diagnosis. Meanwhile, inclusion of comorbidities may provide researchers an opportunity to model psychiatric heterogeneity (Arseniev-Koehler et al., 2018) and interpret longitudinal generalization.

Sample Control Groups using Propensity Matching. Control group selection is influential in both training and evaluation of statistical models of mental health (Pirina and Çöltekin, 2018). Prior work has leveraged a myriad of criteria to match individuals who have disclosed a psychiatric diagnosis with suitable counterparts—demographics (Coppersmith et al., 2014), online behavior (Cohan et al., 2018), and language (De Choudhury et al., 2016). Though use of inconsistent matching criteria is less than ideal, the absence of any protocol is potentially more problematic (Shen et al., 2018; Wolohan et al., 2018). We recommend practitioners leverage propensity-based matching (Imbens and Rubin, 2015) to reduce the effect of self-disclosure biases (e.g., personality, interests, demographics). In addition to the aforementioned dimensions, researchers may augment their criteria using classifiers to infer relevant latent attributes (Preoţiuc-Pietro et al., 2015) or neural models to derive user-level embeddings (Amir et al., 2017).

Identify and Filter Sampling Biases. Our analysis benefited from context that emerged when attempting to train classifiers that generalize over long time periods. However, access to supplementary data is not necessary to understand whether artifacts may exist in a dataset. Algorithmic approaches, such as those from Le Bras et al. (2020), may be used to identify instances containing spurious correlations. These approaches should be used to augment insights derived from manual annotation and review. We found our technique for ranking the influence of individual posts on user-level predictions began yielding insights after only a few dozen examples, though alternative ranking methodologies are available (Uban et al., 2021). Outcomes should be used to inform preprocessing decisions, construct fair evaluations (Poliak et al., 2018), and inform the description of a dataset within documentation/datasheets (Gebru et al., 2021).

6.1 Limitations and Qualifiers

Though our analysis identified data attributes that may inhibit statistical generalization, we also found evidence in support of the validity of self-disclosed diagnoses for annotating mental health status. The majority of individuals within the CLPsych dataset's original time window showed clear evidence of depression that aligns with clinical criteria. Many of these indicators remained stable over the course of time. Moreoever, the 2015 CLPsych Shared Task dataset is just one of many resources in this research community, all of which are likely to exhibit varying degrees of noise depending on their respective sampling protocols. Conclusive statements regarding the validity of self-disclosed diagnoses require evidence from multiple social media platforms, cultural groups, and time periods.

7 Ethical Considerations

Ethical challenges emerging from use of public social media data to analyze an individual's mental health have been examined extensively by members of both computational and clinical/public health communities (Conway and O'Connor, 2016; Chancellor et al., 2019). Privacy-related concerns are the most poignant for our study, which relies both on de-anonymizing records from a vulnerable population and manually reviewing/analyzing individual posts.

Indeed, many individuals who publicly discuss their mental health or disclose a psychiatric condition within social media admit that they worry about harmful repercussions of sharing such sensitive information with the public (Ford et al., 2019; Naslund and Aschbrenner, 2019). Primary fears include risking occupational stability, damaging interpersonal relationships, and being subjected to hostile communications. Whether potential positive outcomes (e.g., development of systems for recommending mental health care, fiduciary aid to address population-level crises) offset these threats remains largely dependent on an individual's personal life experience. For example, psychiatric patients have expressed stronger approval toward analysis of their social media than members of the general public (Mikal et al., 2017). The same holds true amongst younger individuals (Naslund and Aschbrenner, 2019).

Recognizing these viewpoints, we are careful to mitigate privacy-related risks to the greatest extent possible given our primary research aim. For example, account identifiers distributed within the 2015 CLPsych Shared Task dataset are de-anonymized only temporarily to link updated records with existing post histories. We also redact account handles and URLs from the text analyzed during our manual coding procedure (§5). In line with protocols enumerated by Benton et al. (2017a), all data is stored on a remote server and secured using OS-level group permissions. We perform our analysis under the external guidance of clinical psychologists and psychiatrists. Our study is also reviewed by our Institutional Review Board (IRB), obtaining exempt status under 45 CFR §46.104.

Critically, our intention is not to develop a public-facing system for algorithmic analysis of mental health. Rather, our goal is to evaluate the validity of an existing and widely-adopted data curation practice (Chancellor and De Choudhury, 2020; Harrigian et al., 2021). Failure to comprehensively understand biases that arise under this methodology can have severe detrimental effects in downstream systems. In the case of estimating population-level health trends, for instance, we have already seen machine learning classifiers produce outcomes that are inconsistent across computational studies (Wolohan, 2020; Biester et al., 2021; Harrigian and Dredze, 2022) and in conflict with traditional measurement techniques (Amir et al., 2019). Continuing to pursue this line of research without questioning the validity of its underlying data has the potential to irreparably damage the public's trust in this domain, and worse, enable ill-informed decision making in highly-sensitive circumstances.

Acknowledgements

We thank Ayah Zirikly and Carlos Aguirre for contributing annotations to use for evaluating inter-rater reliability. We also thank the anonymous reviewers for providing additional clinical grounding of our study and highlighting opportunities to improve our technical approach.

References

Oshin Agarwal and Ani Nenkova. 2021. Temporal effects on pre-trained models for language processing tasks. *arXiv preprint arXiv:2111.12790*.

Carlos Aguirre, Keith Harrigian, and Mark Dredze. 2021. Gender and racial fairness in depression research using social media. In *16th Conference of the European Chapter of the Association for Computational Linguistics: Main Volume*.

Silvio Amir, Glen Coppersmith, Paula Carvalho, Mário J Silva, and Bryon C Wallace. 2017. Quantifying mental health from social media with neural user embeddings. In *Machine Learning for Healthcare Conference*. PMLR.

Silvio Amir, Mark Dredze, and John W Ayers. 2019. Mental health surveillance over social media with digital cohorts. In *Sixth Workshop on Computational Linguistics and Clinical Psychology*.

Jules Angst, Alex Gamma, Wulf Rössler, Vladeta Ajdacic, and Daniel N Klein. 2009. Long-term depression versus episodic major depression: results from the prospective zurich study of a community sample. *Journal of affective disorders*, 115(1-2).

APA. 2013. Diagnostic and statistical manual of mental disorders: Dsm-5. *Arlington, VA*.

Alina Arseniev-Koehler, Sharon Mozgai, and Stefan Scherer. 2018. What type of happiness are you looking for?-a closer look at detecting mental health from language. In *Fifth Workshop on Computational Linguistics and Clinical Psychology: From Keyboard to Clinic.*

John W Ayers, Theodore L Caputi, Camille Nebeker, and Mark Dredze. 2018. Don't quote me: reverse identification of research participants in social media studies. *NPJ digital medicine.*

Ricardo Baeza-Yates, Berthier Ribeiro-Neto, et al. 1999. *Modern information retrieval.* ACM press New York.

R Michael Bagby, Lena C Quilty, and Andrew C Ryder. 2008. Personality and depression. *The Canadian Journal of Psychiatry.*

Adrian Benton, Glen Coppersmith, and Mark Dredze. 2017a. Ethical research protocols for social media health research. In *First ACL Workshop on Ethics in Natural Language Processing.*

Adrian Benton, Margaret Mitchell, and Dirk Hovy. 2017b. Multitask learning for mental health conditions with limited social media data. In *15th Conference of the European Chapter of the Association for Computational Linguistics: Volume 1, Long Papers.*

Victoria Betton, Rohan Borschmann, Mary Docherty, Stephen Coleman, Mark Brown, and Claire Henderson. 2015. The role of social media in reducing stigma and discrimination. *The British Journal of Psychiatry.*

Laura Biester, Katie Matton, Janarthanan Rajendran, Emily Mower Provost, and Rada Mihalcea. 2021. Understanding the impact of covid-19 on online mental health forums. *ACM Transactions on Management Information Systems (TMIS).*

Erin Bondy, David AA Baranger, Jared Balbona, Kendall Sputo, Sarah E Paul, Thomas F Oltmanns, and Ryan Bogdan. 2021. Neuroticism and reward-related ventral striatum activity: Probing vulnerability to stress-related depression. *Journal of Abnormal Psychology*, 130(3):223.

Charles L Bowden. 2001. Strategies to reduce misdiagnosis of bipolar depression. *Psychiatric Services.*

Luca Canzian and Mirco Musolesi. 2015. Trajectories of depression: unobtrusive monitoring of depressive states by means of smartphone mobility traces analysis. In *2015 ACM international joint conference on pervasive and ubiquitous computing.*

Stevie Chancellor, Michael L Birnbaum, Eric D Caine, Vincent MB Silenzio, and Munmun De Choudhury. 2019. A taxonomy of ethical tensions in inferring mental health states from social media. In *conference on fairness, accountability, and transparency.*

Stevie Chancellor and Munmun De Choudhury. 2020. Methods in predictive techniques for mental health status on social media: a critical review. *NPJ digital medicine*, 3(1).

Stevie Chancellor, Zhiyuan Lin, Erica L Goodman, Stephanie Zerwas, and Munmun De Choudhury. 2016. Quantifying and predicting mental illness severity in online pro-eating disorder communities. In *19th ACM Conference on Computer-Supported Cooperative Work & Social Computing.*

Daejin Choi, Steven A Sumner, Kristin M Holland, John Draper, Sean Murphy, Daniel A Bowen, Marissa Zwald, Jing Wang, Royal Law, Jordan Taylor, et al. 2020. Development of a machine learning model using multiple, heterogeneous data sources to estimate weekly us suicide fatalities. *JAMA network open.*

Arman Cohan, Bart Desmet, Andrew Yates, Luca Soldaini, Sean MacAvaney, and Nazli Goharian. 2018. Smhd: a large-scale resource for exploring online language usage for multiple mental health conditions. In *27th International Conference on Computational Linguistics.* ACL.

Jacob Cohen. 1960. A coefficient of agreement for nominal scales. *Educational and psychological measurement*, 20(1):37–46.

Stephan Collishaw, Barbara Maughan, Robert Goodman, and Andrew Pickles. 2004. Time trends in adolescent mental health. *Journal of Child Psychology and psychiatry*, 45(8).

Mike Conway and Daniel O'Connor. 2016. Social media, big data, and mental health: current advances and ethical implications. *Current opinion in psychology.*

Glen Coppersmith, Mark Dredze, and Craig Harman. 2014. Quantifying mental health signals in twitter. In *workshop on computational linguistics and clinical psychology: From linguistic signal to clinical reality.*

Glen Coppersmith, Mark Dredze, Craig Harman, Kristy Hollingshead, and Margaret Mitchell. 2015. Clpsych 2015 shared task: Depression and ptsd on twitter. In *2nd Workshop on Computational Linguistics and Clinical Psychology: From Linguistic Signal to Clinical Reality.*

Cassandra R Davis, Jevay Grooms, Alberto Ortega, Joaquin Alfredo-Angel Rubalcaba, and Edward Vargas. 2021. Distance learning and parental mental health during covid-19. *Educational Researcher.*

Munmun De Choudhury and Sushovan De. 2014. Mental health discourse on reddit: Self-disclosure, social support, and anonymity. In *Eighth international AAAI conference on weblogs and social media.*

Munmun De Choudhury, Emre Kiciman, Mark Dredze, Glen Coppersmith, and Mrinal Kumar. 2016. Discovering shifts to suicidal ideation from mental health content in social media. In *2016 CHI conference on human factors in computing systems.*

Orianna DeMasi, Konrad Kording, and Benjamin Recht. 2017. Meaningless comparisons lead to false optimism in medical machine learning. *PloS one.*

Sindhu Kiranmai Ernala, Michael L Birnbaum, Kristin A Candan, Asra F Rizvi, William A Sterling, John M Kane, and Munmun De Choudhury. 2019. Methodological gaps in predicting mental health states from social media: triangulating diagnostic signals. In *2019 chi conference on human factors in computing systems*.

Alex Fine, Patrick Crutchley, Jenny Blase, Joshua Carroll, and Glen Coppersmith. 2020. Assessing population-level symptoms of anxiety, depression, and suicide risk in real time using nlp applied to social media data. In *Fourth Workshop on Natural Language Processing and Computational Social Science*.

Elizabeth Ford, Keegan Curlewis, Akkapon Wongkoblap, and Vasa Curcin. 2019. Public opinions on using social media content to identify users with depression and target mental health care advertising: mixed methods survey. *JMIR mental health*.

C Laurel Franklin and Mark Zimmerman. 2001. Posttraumatic stress disorder and major depressive disorder: Investigating the role of overlapping symptoms in diagnostic comorbidity. *The Journal of nervous and mental disease*.

Benoît Frénay and Michel Verleysen. 2013. Classification in the presence of label noise: a survey. *IEEE transactions on neural networks and learning systems*.

Sandro Galea, Raina M Merchant, and Nicole Lurie. 2020. The mental health consequences of covid-19 and physical distancing: the need for prevention and early intervention. *JAMA internal medicine*.

Timnit Gebru, Jamie Morgenstern, Briana Vecchione, Jennifer Wortman Vaughan, Hanna Wallach, Hal Daumé Iii, and Kate Crawford. 2021. Datasheets for datasets. *Communications of the ACM*.

Guy M Goodwin. 2015. The overlap between anxiety, depression, and obsessive-compulsive disorder. *Dialogues in clinical neuroscience*.

Sharath Chandra Guntuku, David B Yaden, Margaret L Kern, Lyle H Ungar, and Johannes C Eichstaedt. 2017. Detecting depression and mental illness on social media: an integrative review. *Current Opinion in Behavioral Sciences*.

Xiaochuang Han, Byron C Wallace, and Yulia Tsvetkov. 2020. Explaining black box predictions and unveiling data artifacts through influence functions. *arXiv preprint arXiv:2005.06676*.

Keith Harrigian, Carlos Aguirre, and Mark Dredze. 2020. Do models of mental health based on social media data generalize? In *2020 conference on empirical methods in natural language processing: findings*.

Keith Harrigian, Carlos Aguirre, and Mark Dredze. 2021. On the state of social media data for mental health research. In *Seventh Workshop on Computational Linguistics and Clinical Psychology: Improving Access*.

Keith Harrigian and Mark Dredze. 2022. The problem of semantic shift in longitudinal monitoring of social media. *Proceeedings of the 14th ACM Web Science Conference*.

Nicholas S Holtzman et al. 2017. A meta-analysis of correlations between depression and first person singular pronoun use. *Journal of Research in Personality*.

Xiaolei Huang and Michael Paul. 2019. Neural temporality adaptation for document classification: Diachronic word embeddings and domain adaptation models. In *57th Annual Meeting of the Association for Computational Linguistics*.

Guido W Imbens and Donald B Rubin. 2015. *Causal inference in statistics, social, and biomedical sciences*. Cambridge University Press.

Julia Ive, George Gkotsis, Rina Dutta, Robert Stewart, and Sumithra Velupillai. 2018. Hierarchical neural model with attention mechanisms for the classification of social media text related to mental health. In *Fifth Workshop on Computational Linguistics and Clinical Psychology: From Keyboard to Clinic*. Association for Computational Linguistics.

Judith Johnson, Louise H Hall, Kathryn Berzins, John Baker, Kathryn Melling, and Carl Thompson. 2018. Mental healthcare staff well-being and burnout: A narrative review of trends, causes, implications, and recommendations for future interventions. *International journal of mental health nursing*.

Sheri L Johnson and Andrzej Nowak. 2002. Dynamical patterns in bipolar depression. *Personality and Social Psychology Review*.

Jeffrey H Kahn and Angela M Garrison. 2009. Emotional self-disclosure and emotional avoidance: Relations with symptoms of depression and anxiety. *Journal of counseling psychology*, 56(4):573.

Pang Wei Koh and Percy Liang. 2017. Understanding black-box predictions via influence functions. In *International conference on machine learning*, pages 1885–1894. PMLR.

Klaus Krippendorff. 2004. Reliability in content analysis: Some common misconceptions and recommendations. *Human communication research*, 30(3):411–433.

Klaus Krippendorff. 2011. Computing krippendorff's alpha-reliability.

Benjamin B Lahey. 2009. Public health significance of neuroticism. *American Psychologist*, 64(4):241.

Danielle M Law, Jennifer D Shapka, and Rebecca J Collie. 2020. Who might flourish and who might languish? adolescent social and mental health profiles and their online experiences and behaviors. *Human Behavior and Emerging Technologies*, 2(1):82–92.

Ronan Le Bras, Swabha Swayamdipta, Chandra Bhagavatula, Rowan Zellers, Matthew Peters, Ashish Sabharwal, and Yejin Choi. 2020. Adversarial filters of dataset biases. In *International Conference on Machine Learning*. PMLR.

Andrew Lee, Jonathan K Kummerfeld, Larry An, and Rada Mihalcea. 2021. Micromodels for efficient, explainable, and reusable systems: A case study on mental health. In *Findings of the Association for Computational Linguistics: EMNLP 2021*.

David E Losada, Fabio Crestani, and Javier Parapar. 2017. erisk 2017: Clef lab on early risk prediction on the internet: experimental foundations. In *International Conference of the Cross-Language Evaluation Forum for European Languages*. Springer.

Kate Loveys, Jonathan Torrez, Alex Fine, Glen Moriarty, and Glen Coppersmith. 2018. Cross-cultural differences in language markers of depression online. In *Proceedings of the fifth workshop on computational linguistics and clinical psychology: from keyboard to clinic*, pages 78–87.

Marco Lui and Timothy Baldwin. 2012. langid. py: An off-the-shelf language identification tool. In *ACL 2012 system demonstrations*.

Mufan Luo and Jeffrey T Hancock. 2020. Self-disclosure and social media: motivations, mechanisms and psychological well-being. *Current Opinion in Psychology*, 31:110–115.

Sean MacAvaney, Bart Desmet, Arman Cohan, Luca Soldaini, Andrew Yates, Ayah Zirikly, and Nazli Goharian. 2018. Rsdd-time: Temporal annotation of self-reported mental health diagnoses. In *Fifth Workshop on Computational Linguistics and Clinical Psychology: From Keyboard to Clinic*.

Sean Macavaney, Anjali Mittu, Glen Coppersmith, Jeff Leintz, and Philip Resnik. 2021. Community-level research on suicidality prediction in a secure environment: Overview of the clpsych 2021 shared task. In *Seventh Workshop on Computational Linguistics and Clinical Psychology: Improving Access*.

Jude Mikal, Samantha Hurst, and Mike Conway. 2017. Investigating patient attitudes towards the use of social media data to augment depression diagnosis and treatment: a qualitative study. In *fourth workshop on computational linguistics and clinical psychology—from linguistic signal to clinical reality*.

Jeanne Miranda, Lonnie R Snowden, and Rupinder K Legha. 2020. Policy effects on mental health status and mental health care disparities. In *The palgrave handbook of American mental health policy*. Springer.

Danielle Mowery, Hilary Smith, Tyler Cheney, Greg Stoddard, Glen Coppersmith, Craig Bryan, and Mike Conway. 2017. Understanding depressive symptoms and psychosocial stressors on twitter: a corpus-based study. *Journal of medical Internet research*, 19(2).

Aaron Mueller, Zach Wood-Doughty, Silvio Amir, Mark Dredze, and Alicia Lynn Nobles. 2021. Demographic representation and collective storytelling in the me too twitter hashtag activism movement. *ACM on Human-Computer Interaction*.

John A Naslund and Kelly A Aschbrenner. 2019. Risks to privacy with use of social media: understanding the views of social media users with serious mental illness. *Psychiatric services*.

Alicia L Nobles, Jeffrey J Glenn, Kamran Kowsari, Bethany A Teachman, and Laura E Barnes. 2018. Identification of imminent suicide risk among young adults using text messages. In *2018 CHI Conference on Human Factors in Computing Systems*.

Inna Pirina and Çağrı Çöltekin. 2018. Identifying depression on reddit: The effect of training data. In *2018 EMNLP Workshop SMM4H: The 3rd Social Media Mining for Health Applications Workshop & Shared Task*.

Adam Poliak, Jason Naradoesky, Aparajita Haldar, Rachel Rudinger, and Benjamin Van Durme. 2018. Hypothesis only baselines in natural language inference. In *Seventh Joint Conference on Lexical and Computational Semantics*.

Daniel Preoţiuc-Pietro, Johannes Eichstaedt, Gregory Park, Maarten Sap, Laura Smith, Victoria Tobolsky, H Andrew Schwartz, and Lyle Ungar. 2015. The role of personality, age, and gender in tweeting about mental illness. In *2nd workshop on computational linguistics and clinical psychology: From linguistic signal to clinical reality*.

Ana Radovic, Theresa Gmelin, Bradley D Stein, and Elizabeth Miller. 2017. Depressed adolescents' positive and negative use of social media. *Journal of adolescence*, 55:5–15.

Guozheng Rao, Yue Zhang, Li Zhang, Qing Cong, and Zhiyong Feng. 2020. Mgl-cnn: A hierarchical posts representations model for identifying depressed individuals in online forums. *IEEE Access*.

Andrew G Reece, Andrew J Reagan, Katharina LM Lix, Peter Sheridan Dodds, Christopher M Danforth, and Ellen J Langer. 2017. Forecasting the onset and course of mental illness with twitter data. *Scientific reports*.

Farig Sadeque, Dongfang Xu, and Steven Bethard. 2018. Measuring the latency of depression detection in social media. In *Eleventh ACM International Conference on Web Search and Data Mining*.

Koustuv Saha, John Torous, Emre Kiciman, and Munmun De Choudhury. 2021. Understanding side effects of antidepressants: Large-scale longitudinal study on social media data. *JMIR mental health.*

Maarten Sap, Gregory Park, Johannes Eichstaedt, Margaret Kern, David Stillwell, Michal Kosinski, Lyle Ungar, and H Andrew Schwartz. 2014. Developing age and gender predictive lexica over social media. In *2014 Conference on Empirical Methods in Natural Language Processing (EMNLP).*

Robert A Schoevers, DJH Deeg, W Van Tilburg, and ATF Beekman. 2005. Depression and generalized anxiety disorder: co-occurrence and longitudinal patterns in elderly patients. *The American Journal of Geriatric Psychiatry.*

Tiancheng Shen, Jia Jia, Guangyao Shen, Fuli Feng, Xiangnan He, Huanbo Luan, Jie Tang, Thanassis Tiropanis, Tat Seng Chua, and Wendy Hall. 2018. Cross-domain depression detection via harvesting social media. In *2018 International Joint Conferences on Artificial Intelligence (IJCAI).* International Joint Conferences on Artificial Intelligence.

Masashi Sugiyama and Motoaki Kawanabe. 2012. *Machine learning in non-stationary environments: Introduction to covariate shift adaptation.* MIT press.

Allison M Tackman, David A Sbarra, Angela L Carey, M Brent Donnellan, Andrea B Horn, Nicholas S Holtzman, To'Meisha S Edwards, James W Pennebaker, and Matthias R Mehl. 2019. Depression, negative emotionality, and self-referential language: A multi-lab, multi-measure, and multi-language-task research synthesis. *Journal of personality and social psychology*, 116(5):817.

Adam Tsakalidis, Maria Liakata, Theo Damoulas, and Alexandra I Cristea. 2018. Can we assess mental health through social media and smart devices? addressing bias in methodology and evaluation. In *Joint European Conference on Machine Learning and Knowledge Discovery in Databases.* Springer.

Ana Sabina Uban, Berta Chulvi, and Paolo Rosso. 2021. Understanding patterns of anorexia manifestations in social media data with deep learning. In *Seventh Workshop on Computational Linguistics and Clinical Psychology: Improving Access.*

Gaël Varoquaux. 2018. Cross-validation failure: small sample sizes lead to large error bars. *Neuroimage*, 180.

Matej Gjurkovic Mladen Karan Iva Vukojevic and Mihaela Bošnjak Jan Šnajder. 2021. Pandora talks: Personality and demographics on reddit. *SocialNLP 2021.*

Meifen Wei, Daniel W Russell, and Robyn A Zakalik. 2005. Adult attachment, social self-efficacy, self-disclosure, loneliness, and subsequent depression for freshman college students: A longitudinal study. *Journal of counseling psychology*, 52(4):602.

JT Wolohan. 2020. Estimating the effect of covid-19 on mental health: Linguistic indicators of depression during a global pandemic. In *1st Workshop on NLP for COVID-19 at ACL 2020.*

JT Wolohan, Misato Hiraga, Atreyee Mukherjee, Zeeshan Ali Sayyed, and Matthew Millard. 2018. Detecting linguistic traces of depression in topic-restricted text: Attending to self-stigmatized depression with nlp. In *First International Workshop on Language Cognition and Computational Models.*

Akkapon Wongkoblap, Miguel A Vadillo, and Vasa Curcin. 2019. Predicting social network users with depression from simulated temporal data. In *IEEE EUROCON 2019-18th International Conference on Smart Technologies.* IEEE.

Qing-Song Xu and Yi-Zeng Liang. 2001. Monte carlo cross validation. *Chemometrics and Intelligent Laboratory Systems.*

A Interpreting Model Performance

A.1 Data

Three individuals (one author A_1, two non-authors B_1, B_2) independently generated the annotations used to facilitate the analysis presented in §5. Statistics presented in the analysis are computed using the author's annotations, while reliability measures are computed using additional annotations from the non-authors. All annotators have several years of experience modeling language within social media to assess mental health, but do not claim to be experts in clinical psychology. Additionally, all annotators have prior experience with the CLPsych 2015 Shared Task data (Coppersmith et al., 2015) — e.g., A_1 and B_1 have worked with the original CLPsych dataset extensively over the prior three years. We include the distribution of instances reviewed by each of our annotators in Table 4.

	Time Period			
	2012-2015	2015-2018	2018-2021	**Total**
A_1	298	157	119	574
B_1	103	62	40	205
B_2	26	15	12	53

Table 4: Distribution of instances coded by each annotator across the three time periods. Note that the set of instances annotated follows the relationship: $B_2 \subseteq B_1 \subseteq A_1$.

A.2 Inter-rater Reliability

As a first look into inter-rater reliability, we consider three dimensions of agreement — evidence of depression (four-class and three-class)[3] and remission status (four-class). We present pairwise annotator agreement matrices for each of these dimensions in Figure 1. We use Cohen's kappa κ to evaluate pairwise annotator agreement (Cohen, 1960) and Krippendorff's alpha α to evaluate multi-annotator agreement (Krippendorff, 2011).

We observe fair to moderate agreement for the evidence-of-depression task: $\alpha = 0.4376$ and $\alpha = 0.4988$ for the four-class and three-class versions, respectively. Meanwhile, agreement on remission status is poor, reflected by a Krippendorff's α of 0.3561. In isolation, these agreement measures would suggest the results of our analysis should be accepted tentatively at best (Krippendorff, 2004). However, we argue these statistics are perhaps a bit conservative and skewed by the small sample size of annotations generated by B_2. A review of the underlying distributions provides us an opportunity to understand axes of disagreement and, in turn, contextualize the results presented in §5.

As shown in Figure 1, annotator B_2 exhibits a higher propensity to use the "Uncertain" label in the evidence-of-depression tasks compared to annotators A_1 and B_1. At the same time, while annotator B_2 is more inclined to indicate they are uncertain about an example than annotator A_1, we note that annotator B_1 appears to have a higher baseline threshold of what constitutes evidence of depression than annotator A_1. The latter is demonstrated by the fact that nearly all examples marked in the affirmative by B_1 were also marked as such by A_1, but a large number of examples marked in the affirmative by A_1 were marked as not containing evidence of depression by B_1.

With respect to the remission status task (bottom subplot of Figure 1), we note that annotator B_1 is more likely to mark an example as uncertain and more likely to mark an example as being in-remission than annotators A_1 and B_2. Broadly, this distribution highlights the difficulty of distinguishing active cases of clinical depression from prior experiences and lingering effects. It also serves as support for our recommendation in §6 that researchers should attempt to include the time a diagnosis was received by an individual when curating new datasets.

We acquire additional context for our results by examining the distribution of annotations as a func-

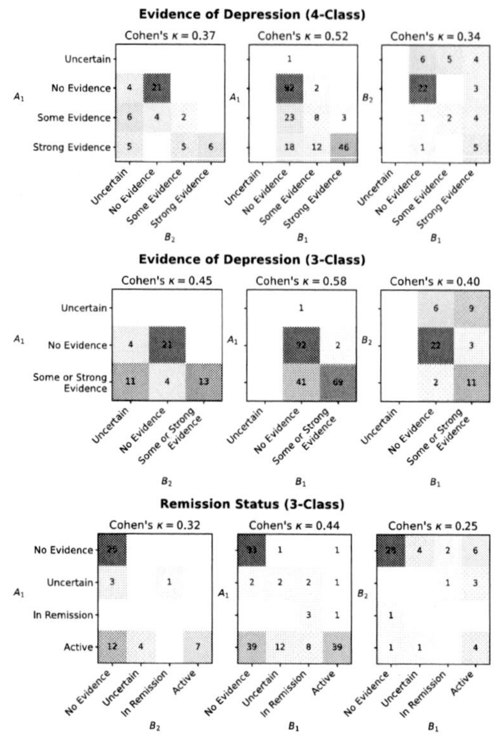

Figure 1: Pairwise agreement matrices for the annotation tasks. Underlying relationships reveal cognitive biases from annotator A_1 that may affect the outcomes presented in §5.

tion of the original CLPsych labels. Examining the results visualized in Figure 2, we first note that annotator A_1 classifies instances most accurately (under the assumption that ground truth is fixed over time). We believe this outcome to be a result of exposure bias; the annotation task was conducted *after* the completion of several modeling experiments, through which annotator A_1 was uniquely provided an opportunity to learn more about the presentation of depression by individuals in the 2015 CLPsych Shared Task dataset. We also note the distribution of "Uncertain" decisions from annotator B_2 concentrating within the original depression group. This seems to suggest annotator B_2 adopted a conservative coding approach when presented with instances that contained smaller degrees of evidence, whereas annotators A_1 and B_1 required a lower threshold of evidence to make a decision.

To conclude our reliability analysis, we examine agreement regarding the manner in which each annotator made their decision (i.e., evidence identi-

[3] Note that the three-class evidence-of-depression grouping simply merges the Some Evidence and Strong Evidence categories of the four-class version.

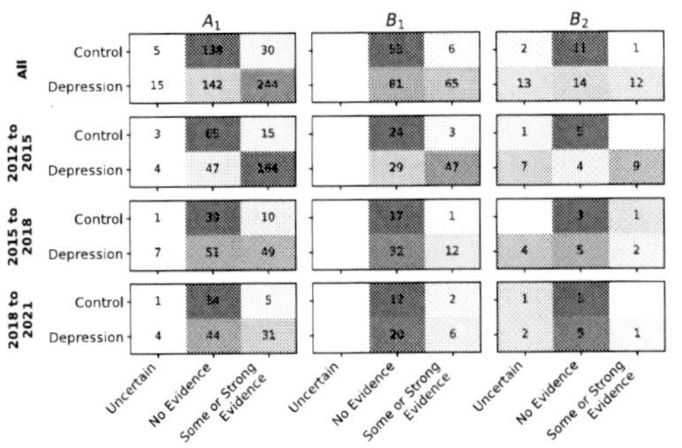

Figure 2: Distribution of annotations for the evidence of depression task (three-class) as a function of the original CLPsych labels. Affirmative evidence becomes less prevalent in the new time periods compared to the original time period for each annotator.

fication). We find that annotators A_1 and B_1 generally identify diagnosis disclosures within the same instances. Annotator B_2 often abstained from making a decision when presented with a disclosure due to uncertainty regarding the subject of the diagnosis. Annotator A_1 also indicated the presence of a depressed and/or irritable mood at a significantly higher rate than the other annotators, seemingly more sensitive to extreme negative emotions than the other annotators.

Discussion. Considering the difficulty of the annotation task, it is perhaps not surprising to have observed less than perfect annotator agreement. Machine learning classifiers often require hundreds of posts to make an accurate estimate of an individual's mental health status, while our annotators were only provided at maximum of 30 posts and encouraged to rely on varying levels of prior knowledge regarding the presentation of depression in social media. Critically, we emphasize that the goal of the analysis presented in §5 is *not* to curate ground truth labels of mental health status or act as clinical experts, but rather to understand biases that may exist in a depression dataset generated using self-disclosed diagnoses. The analysis of inter-rater reliability presented above provides an opportunity to further ground the results discussed in §5 and highlight areas that may benefit from future research.

A.3 Evidence Distribution

We include a breakdown of evidence annotations for individuals displaying some evidence of depression (§5) in Figure 3. Exemplary tweets for each of the evidence categories (paraphrased to maintain anonymity) are provided in Table 5. Both can be found on the following pages.

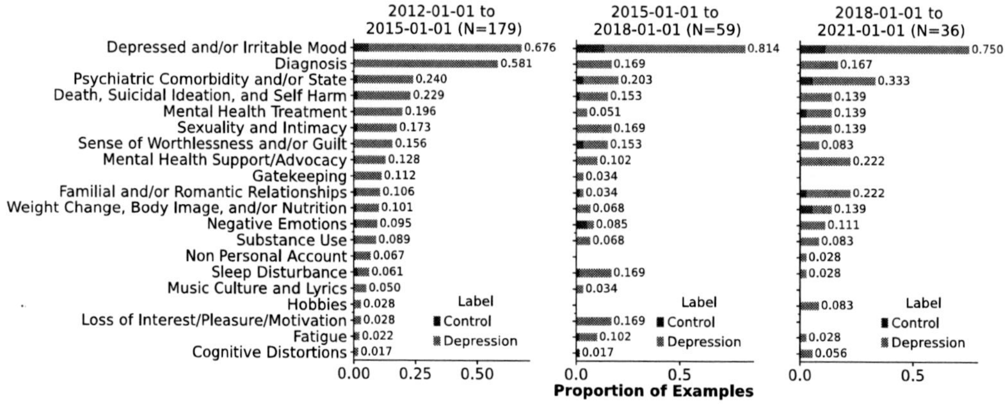

Figure 3: Distribution of evidence amongst individuals indicated as displaying at least some evidence of a depression diagnosis. Depressed and/or irritable mood is consistently the most common type of evidence within each of the three time periods.

Evidence	Exemplary Tweets
Diagnosis Disclosure	"Bipolar disorder and depression. My doctor finally agrees." "I have suffered from depression for several years now"
Depressed & Irritable Mood	"No one ever asks if I'm doing fine." "You don't understand what I'm dealing with. Get fucked."
Loss of Interest/Pleasure/Motivation	"...realizing you don't care about the things you used to enjoy" "cant get out of bed today"
Weight, Body Image, & Nutrition	"Not that anyone cares, but I'm almost at my goal weight." "I bought the dress I've always wanted, but still don't feel pretty."
Sleep Disturbance	"I CANT SLEEP. PAIN. JUST LIKE ALWAYS." "Shit! Surviving on only a couple of hours of sleep again :/"
Fatigue	"mentally drained from this pandemic" "This should be effortless but I can't work any harder"
Sense of Worthlessness & Guilt	"when you let someone do anything to you..." "It truly is always my fault. I probably suck."
Impaired Thought	"I'm failing my classes because I'm depressed." "at work. cant focus doe"
Death & Self Harm	"My scars are faded...unless you care to look close" "I wish you all never see a loved one fade away."
Cognitive Distortions	"Going to fail this exam. SCREWED." "I always think my bf is going to leave me"
Treatment	"Scared to tell a women that I'm in therapy" "Slowly weaning of the prozac."
Gatekeeping	"depression isn't just a bad day. fuck you all." "LET ME SHOW YOU WANT DEPRESSION IS"
Sexuality and Intimacy	"Who wants to come take some pics of me for only fans? ;)" "Every girl should watch porn with their bf"
Negative Emotions	"hi sunshine! Too bad no one to spend today with." "I feel like no one cares even though I know they do"
Coping Strategies	"Have you talked to anyone about it yet?" "Art is always the easiest way to distract me from my anxiety"
Psychiatric Comorbidity & State	"Really stressing today. Lots of built up anger" "I am anorexic and cut myself."
Non-psychiatric Comorbidity	"Could use a little bit of aid #DisabilityAid" "Lots of back pain ruining what should be a beautiful day."
Substance Use	"I really shouldn't be drunk this early." "Weed makes the dreams go away and thats a good thing."
Support & Advocacy	"If I can manage a smile, I believe you can too one day!" "RIP Chester. If you're going through pain, reach out to me."
Personality and Identity	"Girls say they love a man in uniform until they do their job" "Lol grandma still think I'm bringing a boy home"
Music Culture & Lyrics	"#FallingInReverse :D" "Scene doesn't mean emo idiots. I dont want to kill myself."
Familial/Romantic Relationships	"when bae dont answer the phone xx" "Mom: You'll never lose weight. Me: Is that why dad left?"
Political & Moral Beliefs	"look in the mirror if you're not upset a cop can murder" "Trump will kill us all"
Hobbies	"Missin the old days when eveyone played Pokemon yellow" "Boys that watch the Kardashians. Love."
Non-personal Accounts	"My life was about to fall apart until I found the Calm app..." "Breaking News: 5-alarm fire just outside Tulsa..."

Table 5: Exemplary tweets and phrases (modified to preserve anonymity) for each of the 25 evidence categories.

Tracking Mental Health Risks and Coping Strategies in Healthcare Workers' Online Conversations Across the COVID-19 Pandemic

Molly E. Ireland, Kiki Adams, and **Sean A. Farrell**
Receptiviti
{mireland, kiki, sfarrell}@receptiviti.com

Abstract

The mental health risks of the COVID-19 pandemic are magnified for medical professionals, such as doctors and nurses. To track conversational markers of psychological distress and coping strategies, we analyzed 67.25 million words written by self-identified healthcare workers (N = 5,409; 60.5% nurses, 39.5% physicians) on Reddit beginning in June 2019. Dictionary-based measures revealed increasing emotionality (including more positive and negative emotion and more swearing), social withdrawal (less affiliation and empathy, more "they" pronouns), and self-distancing (fewer "I" pronouns) over time. Several effects were strongest for conversations that were *least* health-focused and self-relevant, suggesting that long-term changes in social and emotional behavior are general and not limited to personal or work-related experiences. Understanding protective and risky coping strategies used by healthcare workers during the pandemic is fundamental for maintaining mental health among front-line workers during periods of chronic stress, such as the COVID-19 pandemic.

1 Introduction

The COVID-19 pandemic has magnified existing mental health disparities globally. Relative to people working in other fields, healthcare workers have experienced greater exposure to COVID-19 and, consequently, higher risk of death and illness as well as more time spent apart from loved ones during quarantine (Walton et al., 2020). An estimated 150,000-200,0000 healthcare workers have died globally from COVID-19 since the start of the pandemic, with higher rates of infection for nurses, women, and workers involved in COVID-19 screening, and higher mortality rates among doctors (Chutiyami et al., 2021; WHO, 2022). Deaths and illnesses among healthcare workers have led to severe understaffing in the hardest-hit areas, causing widespread overwork and burnout in the healthcare field (Andel et al., 2022). Healthcare workers experienced higher rates of depression and suicide than many other professions before the pandemic (Kalmoe et al., 2019), and suicidality, depression, and anxiety disorders have increased among healthcare workers in the last 2 years (Spoorthy et al., 2020; Young et al., 2021).

Beyond pandemic-related social isolation, personal health risks, and overwork, healthcare workers additionally cope with feeling responsible for the deaths and symptoms they witness firsthand in their patients (Zhang et al., 2020)—experiences exacerbated early in the pandemic by the fact that healthcare workers were often the only people permitted to be physically present in patients' final hours (Rabow et al., 2021). For many, the stress of the pandemic has been aggravated by widespread skepticism of vaccines and the medical system (Schneider et al., 2021). Others have reported survivors' guilt related to having early vaccine access, feelings of powerlessness with respect to limited COVID-19 patient treatment options, and the chronic stress of having insufficient personal protective equipment while working, particularly early in the pandemic (Rabow et al., 2021).

Dealing with chronic stress at the front line of an epidemic or pandemic requires extraordinary coping and emotion regulation skills—and, at the same time, likely compromises the mental health of even the most resilient nurses and doctors. In this project, we followed the linguistic trajectories of healthcare workers' risky and protective coping strategies over the course of the pandemic. The following sections first review past research on risk factors and resilience evident in language use following community traumas. We then describe a longitudinal study tracking social and emotional language used by several thousand self-labeled nurses and doctors on Reddit, a popular online social discussion platform, over a baseline period followed by roughly 2 years of the pandemic. Analyses focused on main

Proceedings of the Eighth Workshop on Computational Linguistics and Clinical Psychology, pages 76 - 88
July 15, 2022 ©2022 Association for Computational Linguistics

effects over time and moderator models exploring how language trajectories varied as a function of health-relevance, self-relevance, and role (nurse or physician). Finally, we explore the ethical, theoretical, and practical implications of the findings for clinical psychology and mental health technology.

1.1 Coping with Shared Trauma over Time

Tracking naturalistic language use on the internet is an effective method of measuring how people cope with trauma and experience emotions over time (Vine et al., 2020). Research has, for example, used both dictionary-based and open-vocabulary analyses of online language use (including social media, online forums, and search engine activity) to understand how individuals anticipate and then cope with traumatic events such as suicide attempts (De Choudhury et al., 2016; Ophir et al., 2020; Roy et al., 2020), relationship dissolution (Seraj et al., 2021), illnesses such as breast cancer (Verberne et al., 2019) and autoimmune disease (Jordan et al., 2019), and mental health conditions such as anxiety (Ireland and Iserman, 2018) and depression (Eichstaedt et al., 2018).

Several studies of community coping with shared traumas—such as the September 11th attacks and natural disasters—have found evidence of both distress and coping in naturalistic conversational language. Results show a common pattern of increasing affiliative and emotional language in the immediate 1-2 weeks after a traumatic event followed by a refractory period during which such communal coping indicators drop below baseline, theoretically reflecting social withdrawal (Cohn et al., 2004; Pennebaker and Harber, 1993; Stone and Pennebaker, 2002). For acute traumas, language typically returns to baseline after 4-6 weeks (Pennebaker and Chung, 2005).

Analyses of social media language use surrounding epidemics (e.g., Zika, Ebola) and sociopolitical movements (e.g., the Arab Spring) have focused primarily on the transmission of information about symptoms or events rather than psychological dimensions of messages (Hassan Zadeh et al., 2019; Howard et al., 2011). Previous analyses of psychological language use during epidemics or disease outbreaks have typically focused on tracking markers of distress over short spans of time. For example, Tausczik et al. (2012) tracked anxiety language in tweets about the H1N1 epidemic, revealing that fears about H1N1 were intense but short-lived, declining within weeks of the initial news about the disease.

At least one study has used dictionary-based measures to track coping across the first months of the COVID-19 pandemic. Based on a large Reddit sample of people posting in major U. S. city forums, Ashokkumar and Pennebaker (2021) found that anxious language spiked and positive emotional, angry, and analytic language dropped in March 2020. People also referred less to friends and more to family early in the pandemic. After roughly 6 weeks, these language categories plateaued but remained distinct from pre-pandemic levels in the previous year. It is unclear whether these patterns vary as a function of individuals' life stressors or will continue to shift over time.

1.2 Linguistic Markers of Distress

Overwork compromises mental health and has downstream consequences for the quality of individuals' close relationships and job performance. There are several potential indicators of burnout and work stress that may carry over from the workplace to online conversations. The clearest linguistic markers of distress and vulnerability to mental health conditions tend to be self-references (*I, me, my*) and negative emotional language, alone and particularly in combination (Baddeley et al., 2013; Coppersmith et al., 2015a; Tackman et al., 2019).

Work-related stress has disrupted healthcare workers' relationships throughout the COVID-19 pandemic. Long-term quarantining away from romantic partners and family members due to frequent exposure to the disease increases loneliness and relationship conflict (Murata et al., 2021). Relationship problems are closely linked with mental health; for example, breakups and relationship conflict are common triggers of suicide attempts (Bagge et al., 2013) and depressive episodes (Monroe et al., 1999). Thus, in tracking healthcare workers' conversational language use over the pandemic, it is critical to target linguistic markers of affiliation and social behavior.

1.3 Linguistic Markers of Coping

Just as self-directed negativity is a common indicator of psychological distress, the opposite pattern tends to reflect efforts to gain emotional distance from personal problems—a tactic that provides relief in the moment but may be risky long term. Research on self-talk and expressive writing has found that people tend to naturally self-distance, using less "I" and more "you," when recalling negative events or while discussing stressful events, with stronger effects for more distressing topics (Dolcos and Albarracin, 2014; Kross and Ayduk, 2017). The same strategy is effective experimentally as well, with people experiencing less distress when asked to write about negative life experiences or personal concerns using self-distancing (e.g., writing *you* instead of *I*). Psychological distance theoretically provides an emotional buffer, allowing people to consider the events that are causing them distress from the more objective perspective of an outside observer or friend. Thus, lower first-person singular pronoun usage may be a healthy emotion regulation strategy, especially when experiencing acute distress.

Despite the well-established body of work showing that self-distancing can help with emotional control and distress, decreased first-person singular pronoun is not an unambiguous sign of effective coping. In contexts where self-references indicate self-disclosure or self-reflection, using more "I"—or alternating between "I" and other personal pronouns—may be healthier. For example, people with ambiguous sexual self-concepts who used less first-person singular when discussing their sexuality were more likely to report drinking alcohol to cope with personal problems (Hancock et al., 2018). In expressive writing, where people repeatedly privately write about their deepest thoughts and feelings on a distressing topic, individuals tend to have better long-term mental and physical health after the writing intervention if their language indicates a perspective shift (moving from high to low self-references, or vice versa) across sessions (Pennebaker and Chung, 2007; Seih et al., 2008).

Separate research on compassion has found that discussing others' suffering in a less emotional, more socially distant way is associated with better mental health and greater likelihood of taking proactive steps to help the people who are suffering or need assistance (Buechel et al., 2018; Ministero et al., 2018). That is, people may be better able to provide assistance if they feel others' pain less acutely. These findings dovetail with research and practice regarding healthcare workers' bedside manner, where the goal is to show humanistic compassion for patients while maintaining enough distance to carry out complex and often risky and painful tasks (Weissmann et al., 2006).

Word category	Examples
Function Words	
First-person singular ("I")	*I, me, my*
Third-person plural ("they")	*they, them, their*
Negations	*no, not, never*
Affect	
Positive emotion	*lucky, love, happy*
Amusement	*haha, lol, funny*
Admiration	*cool, amazing, best*
Negative emotion	*hate, worry, sad*
Disgust	*creepy, vomit, ugh*
Social	
Affiliation	*call, party, together*
High empathy	*ally, rescue, we*
Low empathy	*yourself, asshat, waste*
Prosocial	*help, support, thanks*
Swear words	*dang, fuck, douche*

Table 1: Social and emotional language categories showing significant linear or curvilinear effects over time. Linguistic categories, affiliation, swear words, and prosocial are from LIWC-22 (Pennebaker et al., 2022). Affect categories are from SALLEE (Adams, 2022). High and low empathy are novel lexica adapted from Sedoc et al. (2020).

1.4 Hypotheses & Analytic Strategy

The current project took a quasi-exploratory approach, modeling the trajectories of a wide range of language variables that are theoretically relevant to risky and protective emotions and social behaviors (see Table 1). The main predictions were that healthcare workers would show signs of increasing distress (more negativity, less positivity), social detachment or isolation (more *I* and *they*, fewer social references, less empathetic language), and social problems (increased conflict and swearing, and decreased prosocial and polite language) over time. Linear, quadratic, and cubic associations were tested for all models. Finally, we tested three moderators for each model: professional role (nurse or doctor) and two aspects of linguistic context (first-person singular pronouns and references to health, e.g., *medicine, symptom, vaccine*).

2 Method

To obtain the initial sample, we first scraped a large sample of comments and submissions from medical-themed forums, or subreddits (r/medicine, 312,357 posts; r/nurses, 14,927 posts; r/emergencymedicine, 46,019 posts; r/AskDocs, 1,617,327 posts; r/StudentNurse, 191,525 posts), that appeared to be moderated by healthcare professionals and included "flair" indicating users' real-life qualifications or specializations. Initially, 2,182,155 messages posted between October 2018 and January 2021 were scraped using the Pushshift

Rank	Content	M (SD)	Function	M (SD)
High Empathy				
1	love	0.13 (0.98)	and	2.03 (1.78)
2	great	0.1 (0.68)	are	0.57 (1.14)
3	feel	0.1 (0.45)	your	0.4 (1.02)
4	thank	0.09 (0.91)	we	0.33 (0.84)
5	help	0.09 (0.67)	how	0.26 (0.88)
6	patients	0.08 (0.36)	her	0.15 (0.63)
7	hospital	0.07 (0.41)	our	0.13 (0.56)
8	home	0.07 (0.4)	us	0.1 (0.51)
9	patient	0.07 (0.37)	through	0.07 (0.34)
10	life	0.06 (0.49)	omg	0.01 (0.62)
Low Empathy				
1	time	0.22 (0.69)	the	3.29 (2.77)
2	think	0.18 (0.56)	in	1.31 (1.7)
3	new	0.14 (0.8)	that	0.97 (1.35)
4	going	0.12 (0.46)	but	0.62 (0.93)
5	same	0.11 (0.95)	be	0.54 (0.97)
6	better	0.09 (0.53)	not	0.53 (1.08)
7	thing	0.08 (0.41)	if	0.45 (0.78)
8	say	0.08 (0.39)	like	0.41 (1.06)
9	lol	0.07 (0.87)	one	0.29 (0.97)
10	long	0.07 (0.47)	up	0.27 (0.77)

Table 2: Frequency ranks and descriptive statistics for content and function words in the high and low empathy lexica. All numbers are % of total words per document (concatenated messages per user per month).

API (https://github.com/pushshift/api). Doctors and nurses were categorized via regular expression searches over the flair text of these messages, searching for commonly used phrases and acronyms used by medical professions (e.g., MD, M.D., MBBS for doctors; Nurse, PCCN, Nursing, NP, LPN, CAN, RN, R.N., BSN for nurses). A total of 2,585 doctors and 4,138 nurses were identified. Next, we downloaded all available comments and posts from the 6,723 self-labeled doctors or nurses on Reddit, totaling over 1.25 million texts, beginning in June 2019. The start date was selected in order to establish baseline norms for the sample, providing roughly 6 months of data from before the virus began spreading globally and 9 months before the WHO declared a pandemic.

Texts were concatenated by user and then by month, excluding months containing fewer than 100 words. We also excluded months for which fewer than 50% of the words were recognized by our dictionaries; given that over half of conversational language typically consists of function words ("stop words" such as pronouns and articles), texts containing half or more words that were not captured by our lexica are unlikely to be conversational English. Finally, we excluded months in which all punctuation made up 50% or more of the text (indicating, e.g., ASCII art). The final dataset included 5,409 unique users (n = 3,271 or 60.5% nurses; n = 2,138 or 39.5% medical doctors) and 67,247,147 words (M = 1,090, SD = 2,355, median = 434 words per user per month).

For mixed-effects regression modeling, we regressed language markers on time (linear, quadratic, and cubic effects), including random slopes for time, nested within authors, and specifying an autocorrelation structure of order 1 (corAR1) to account for the non-independence of adjacent (lag-1) months; all models used the nlme package (Pinheiro et al., 2021) and were plotted with splot (Iserman, 2022) in R version 4.2.0 (R Core Team, 2022). To simplify the time variable and make the regression coefficients more interpretable, we transformed months into quarters and then assigned each quarter a sequential number, starting with Q3 2019 as sequence 0 and ending with Q1 2022 as sequence 11. We then squared and cubed the sequence variable to create the polynomial predictors.

All references to statistical significance below use an adjusted p-value threshold rather than the traditional .05 in order to partly account for inflated false discovery rates, or Type I errors, due to multiple comparisons. Mixed-effects regression models tested effects for 30 language variables, each of which were explored in mixed-effects regression models including six tests (three linear and polynomial effects and three moderators). Thus, the corrected α level is .00028 using the Bonferroni method, a conservative but intuitive correction that is suitable for exploratory analyses in large samples (VanderWeele and Mathur, 2019).

Data collection methods and analytic strategies were approved by internal ethical review at Receptiviti, Inc. and meet federal guidelines for exempt research under the U. S. Department of Health and Human Services' (2017) revised Common Rule. Consistent with the Reddit API's User Agreement, all quantitative data are available online, and we will not profit from the use of these data.

Deidentified data, the high and low empathy lexica we developed, and R code for downloading individuals' messages can be accessed via the project's Open Science Framework (OSF) page.[1]

2.1 Language Measures

LIWC and SALLEE. Texts were analyzed using the latest version of the Linguistic Inquiry and

[1] https://osf.io/scmb7/?view_only=53e8bd3359b3460a907d19f5cb5a0ef6

Word category	*Linear β*	Quadratic β	*Cubic β*
Function Words			
I	-0.09	-0.18	**0.18**
They	-0.03	**0.42**	**-0.30**
Negations	0.01	0.25	**-0.20**
Affect			
Emotionality	-0.1	**0.55**	**-0.39**
Positive emo.	-0.12	**0.58**	**-0.42**
Amusement	-0.05	**0.41**	**-0.28**
Admiration	-0.14	**0.59**	**-0.43**
Negative emo.	0.01	0.13	-0.09
Disgust	-0.09	**0.41**	**-0.27**
Social			
Affiliation	**0.18**	**-0.63**	**0.42**
High empathy	**0.22**	**-0.66**	**0.42**
Low empathy	-0.06	**0.47**	**-0.34**
Prosocial	0.15	**-0.45**	**0.27**
Swear words	-0.07	**0.47**	**-0.32**
Question marks	0.01	**-0.45**	**0.34**

Table 3: Standardized β from the polynomial mixed-effects regression model controlling for role (doctor or nurse) and including linear, quadratic, and cubic sequence (time) effects, random slopes for time within authors, and random intercepts for authors. *I* = first-person singular pronouns, *they* = third-person plural pronouns, *emo.* = emotion. Bold = two-tailed $p \leq .0003$.

Word Count, LIWC-22 (Pennebaker et al., 2022; Boyd et al., 2022) and a sentiment analysis framework, SALLEE (Syntax-Aware LexicaL Emotion Engine; Adams 2022). LIWC is a widely used and well-validated dictionary-based text analysis tool that outputs the percentage of words in a given text that fall into one or more of several dozen grammatical (e.g., pronouns, articles), psychological (e.g., emotions, drives), and topical (e.g., work, health) categories. SALLEE is dictionary-based as well, providing fine-grained measures of specific emotions (e.g., curiosity, surprise, disgust) and summary affective states (e.g., emotionality, positive emotion) in addition to using syntax-based logic allowing words adjacent to emotion terms (e.g., swear words, negations, and intensifiers) to influence category weights (Adams, 2022).

Empathy lexica. The high and low-empathy lexica were both adapted from the data-driven empathy dictionary developed by Sedoc et al. (2020), which was initially trained on a gold-standard empathic reaction corpus (Buechel et al., 2018). For the revised dictionaries, we first took words in the highest and lowest-weighted quartiles of Sedoc et al.'s (2020) empathy dictionary. We then removed person and place names (e.g., *Abuja, Charles*; excepting names used synonymously with low or high empathy, such as *Bundy* and *Gandhi*, respectively), low-frequency misspellings (e.g., *entraprenerurship*), numerals, and other words that appeared

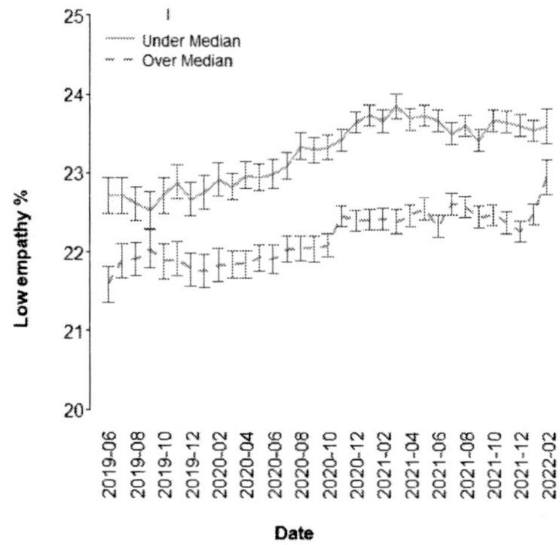

Figure 1: Low empathy language (% of total words) as a function of first-person singular pronoun ("I") usage. Error bars are standard errors.

to be highly contextual or time-specific. Removal judgments were made by the authors, with disagreements resolved through discussion. Wildcards were added sparingly to capture additional word variants where it was safe to do so (e.g., *ambulance**), and missing British English spellings (e.g., *analyse*) were added. Finally, we separated content and function words in order to explore whether effects were robust across both types of words. The final revised dictionaries included 4,059 words (2,105 low empathy, 1,954 high empathy).

Changes to the original empathy dictionary (Sedoc et al., 2020) were not intended not to improve measurement accuracy; rather, we aimed to increase interpretability and generalizability, with the long-term goal of making the lexica accessible to clinicians and mental health care providers. As in the original dictionary, high empathy words in the revised lexica focused primarily on suffering (e.g., *ravaged, hurt, lost*) using expressive (e.g., *emotions, feel*), prosocial language (e.g., *provide, reunite*), whereas low empathy language included unemotional or technical words (e.g., *acknowledge, result*) and disagreeable or insensitive language (e.g., *idgaf, trashy*). For examples used in the current sample, see Table 2. The two dictionaries were moderately negatively correlated, r = -.278.

3 Results

Regression results were consistent with our hypotheses with a few notable exceptions. Both effect sizes and AIC comparisons indicated that cubic models were the best fits for the data, and quadratic models were always a better fit than linear models, based on the ΔAIC > 2 criterion. The standard pattern was an approximately flat line at baseline followed by relatively sharp changes over the first year of the pandemic followed by another plateau or period of more gradual change in the same direction (see Table 3), similar to the overall patterns found by Ashokkumar and Pennebaker (2021).

For the social language categories, doctors and nurses both showed increasing rates of low-empathy words (a pattern that was strongest for less self-referential language; see Figure 1), swearing (Figure 2), and social detachment (more "they" pronouns, Figure 3) over the course of the pandemic. In parallel, healthcare workers showed decreasing rates of words reflecting or referring to social harmony and social engagement (high-empathy, prosocial, affiliation, and question marks) over time.[2]

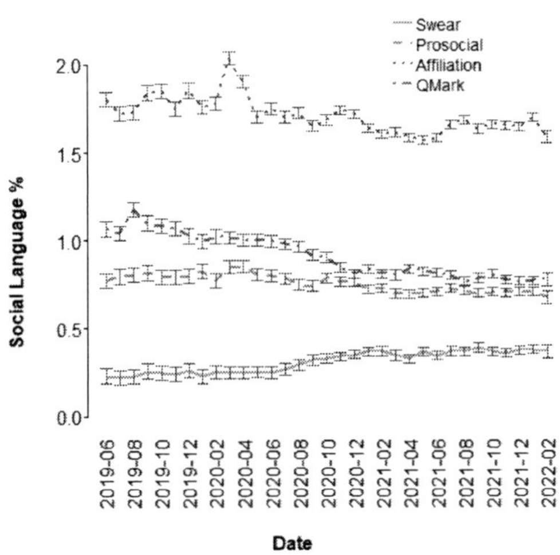

Figure 2: LIWC social categories (affiliation, prosocial, question marks, swear words) with significant polynomial effects over time.

Contrary to our predictions, first-person singular pronouns (e.g., *I, me, my*) decreased over the first year of the pandemic and then plateaued at a relatively low level (Figure 4). Nurses in particular used markedly less "I" (5.2% to 4.2%) from baseline to early 2022. Doctors' first-person singular usage was lower than nurses' at baseline (3.9%), perhaps reflecting physicians' relatively higher status (Kacewicz et al., 2014).

[2]For high empathy, effects were parallel and the conclusions of hypothesis tests were identical when function words were removed from the lexicon. For low empathy, results were not significant after removing function words, all $t < |2|$.

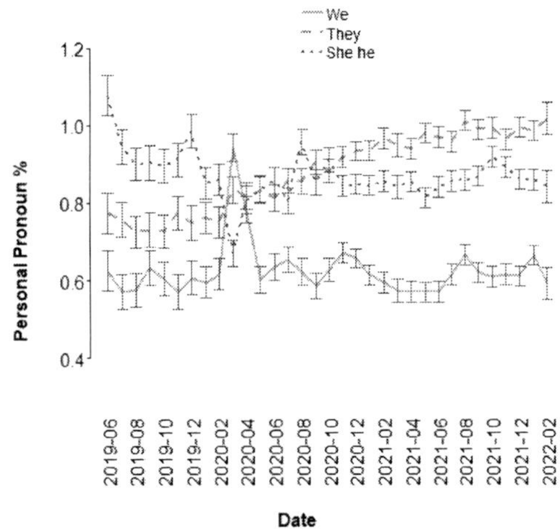

Figure 3: Other-focused pronouns over time. We = first-person plural, they = third-person plural, she/he = third-person singular.

The emotional language results were partly consistent with our predictions. As expected, emotionality and some negative emotions (namely disgust) increased over time. However, most negative emotion categories did not change significantly over time (e.g., sadness, fear). More surprisingly, overall positive emotional language increased over time, with amusement and admiration showing the strongest effects for specific emotions (Figure 5). Amusement is a low-frequency category ($M = 0.71$, $SD = 1.62$; 56.6% of months had 0% amusement) but showed robust quadratic and cubic effects.

Results for words referring to politeness and conflict from LIWC-22 were nonsignificant, despite showing the predicted trends (increasing conflict and decreasing politeness over time), both $ps > .10$, $ts < |3|$. Those categories' low base rates (M = 0.32% and 0.24%, respectively) may have limited our ability to detect subtle shifts over time.

3.1 Moderation by Health and Self-Relevance

For most variables, effects were not moderated by whether the conversations focused on health. There were a few exceptions: for swearing, positive emotions, and disgust, effects were strongest for conversations that were *not* about health. That is, changes in healthcare workers' language over time do not appear to be driven by online discussions of COVID-19 or challenges in their jobs as nurses and doctors; rather, linguistic changes were most evident in casual conversations about interests or hobbies, suggesting that the coping strategies that people have developed in response to the exigencies of the pandemic are carrying over into everyday conversations.

Moderation by self-referential language (*I, me* and *my* usage) was mixed. The overall pattern was for effects to be stronger for negative categories (negations, negative emotions, low empathy) when people were *not* talking about their own experiences; conversely, effects were strongest for positive or prosocial categories (affiliation, positive emotion, and high empathy) when people *were* talking about themselves. Such patterns are consistent with the self-protective tendency to distance oneself from negativity (Ayduk and Kross, 2010). People may feel more comfortable venting (e.g., expressing disgust) when not talking about themselves.

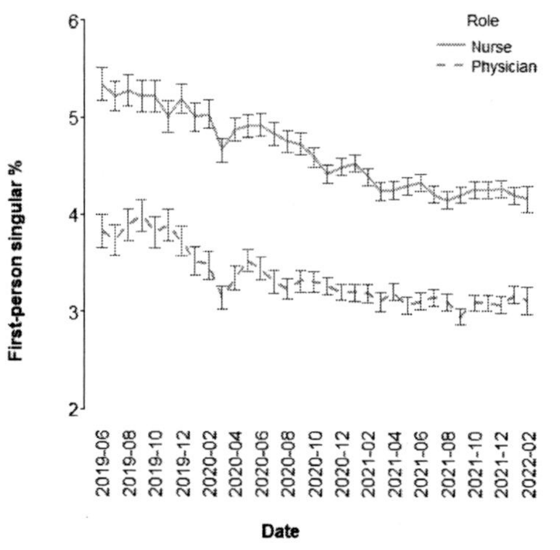

Figure 4: LIWC first-person singular pronoun usage (% of total words) as a function of role.

3.2 Additional Analyses

Early Pandemic Spikes. Many of the plots show deviations at the start of the pandemic followed by linear or flat patterns. First-person singular pronouns dipped sharply in March 2020 followed by a return to near baseline and then a gradual decrease over time. Affiliation language spiked in the first month of the pandemic, followed by a slow linear decline. Although there was no overall linear or curvilinear effect for first-person plural pronouns, there is a clear spike at the start of the pandemic where "we" increases and other pronouns drop before quickly returning to near baseline (Figure 3). Sadness and fear spiked in the same month, declined, and then increased gradually in the following months.

Word-Level Analyses. To better understand the results from the most data-driven (and thus least immediately intuitive) dictionaries, high and low empathy, we examined word-level frequencies. Table 2 shows that the most frequently used low-empathy content words are not rude or callous per se, but seem to reflect a degree of detachment (e.g., *lol, things, week, think*). Low empathy function words had some overlap with LIWC's composite analytic language category, including an article (*the*), impersonal pronouns (*that, there*), and prepositions (*up, in*)—all of which reflect more formal, categorical thinking—as well as negations (*no, not, never*).

New Case Rates. Monthly global new case rates (cases per million; Hannah Ritchie and Roser 2020) were largely uncorrelated with the language variables of interest in this study. In mixed-effects models regressing new case rates onto language variables, none met a $p < .001$ cut-off. The strongest effect was for first-person singular pronouns, quadratic effect $\beta = .013$, 95% CI [0.005,0.021], $SE = .004$, $p = .002$. However, controlling for new case rates as a covariate did not affect the conclusions for any models involving changes in first-person singular over time.

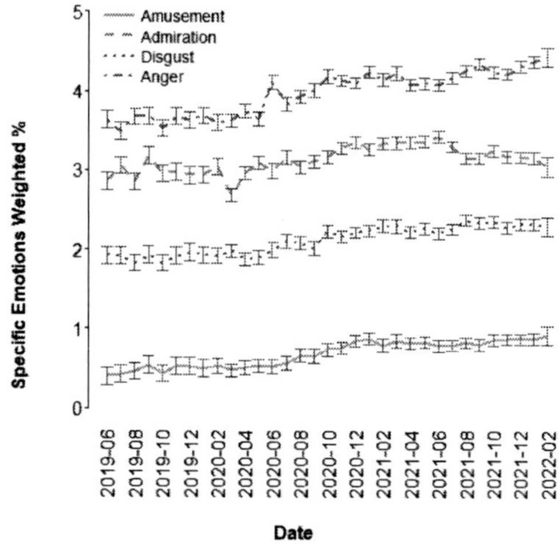

Figure 5: SALLEE emotions (amusement, admiration, disgust, and anger) that increased over time. All cubic effects except for anger are significant, $p < .001$; anger showed a nonsignificant but positive trend.

4 Discussion

The online conversational language of doctors and nurses over the course of the pandemic shows a coherent picture of people coping with chronic stress by self-distancing (fewer first-person singular pronouns) and adopting a more socially detached perspective (less empathic and affiliative language). At the same time, healthcare workers did not seem to be eschewing emotions; rather, emotional language increased over time, including more references to disgust and positive emotions in general.

The emotional effects should be qualified by the standard caveats of any language-based sentiment analysis: Affective words, when categorized correctly, indicate that a person is attending to and talking about an emotion—which sometimes but not always correlates with their emotional state at the time of speaking or writing (Sun et al., 2020; Eichstaedt et al., 2021). Thus, increases in positive emotional language may reflect emotion regulation attempts or coping strategies more than improvements in well-being or mood. What is most striking is not that positive emotional language increased near the end of our sample—which could be explained by decreasing case rates and a slow return of pre-pandemic freedom in much of the world—but that positive emotionality only dropped notably during the first month of the pandemic and did not decrease again during later spikes in global case or mortality rates (Figure 6). Indeed, post hoc analyses show that positive emotional language correlated weakly with global new case rates per million, $r = .015$. That pattern may support the supposition that positive language shifts reflected coping strategies (such as positive reframing) rather than overall well-being (Robbins et al., 2019).

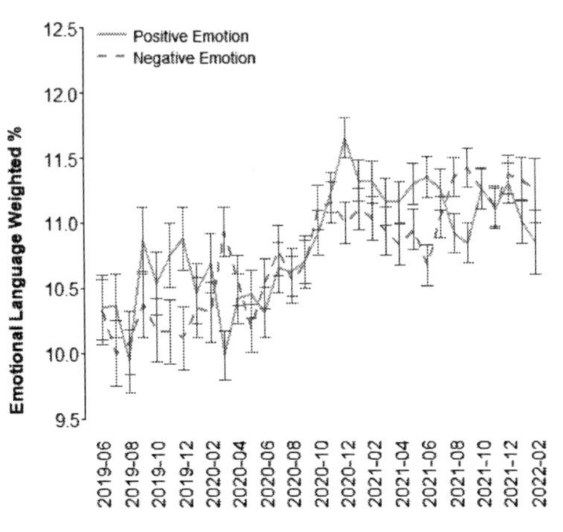

Figure 6: SALLEE composite positive and negative emotions. Only positive emotion showed significant linear, quadratic, and cubic effects; negative emotion is shown for context.

First-person singular pronouns (e.g., *I, me, my*) decreased across the pandemic, a pattern that was starker for nurses than for doctors. "I"-words typically indicate vulnerability to psychological distress (i.e., neuroticism or trait negative affectivity; Tackman et al. 2019) and mental health concerns related to affect dysregulation, including depression (Bucur et al., 2021; Holtzman et al., 2017), anxiety (Brockmeyer et al., 2015; Shen and Rudzicz, 2017), eating disorders (Coppersmith et al., 2015a), and suicidality (Coppersmith et al., 2015b; Stirman and Pennebaker, 2001). Although the results were inconsistent with general psychological distress, a pattern of decreasing first-person singular pronoun usage is consistent with using self-distancing as a self-regulation strategy during periods of chronic stress. People tend to naturally decrease "I" pronouns as a means of distancing themselves from distress and downregulating negative emotions (Ayduk and Kross, 2010; Dolcos and Albarracin, 2014). Coupled with less empathetic language over time, however, decreased "I" rates may represent an adaptation to chronic stress that helps preserve mental stability in the moment but leads to stress dysregulation and interpersonal problems in the future, after the period of severe stress has passed (Ellis et al., 2017).

4.1 Potential Applications

Occupational burnout has intensified throughout the pandemic, particularly for jobs that entail regular risk of exposure to the virus that causes COVID-19. The healthcare field has been among the most affected (Alrawashdeh et al., 2021), with women in particular experiencing more intense and debilitating burnout (Sriharan et al., 2021), as in other professions, partly as a result of gender inequality in the distribution of family responsibilities and household chores while working from home (Malisch et al., 2020). Being able to unobtrusively profile work-related stress or burnout in available texts (e.g., internal chats, emails) could help employers direct mental health resources to employees at risk of mental health crises before their symptoms become severe or their work is affected.

Before translating our findings to clinical or industrial/organizational practice, it will be necessary to disentangle which long-term or acute changes in language use are helpful or harmful. Some of healthcare workers' linguistic changes over time may be beneficial in the short-term but have long-term costs. For example, as already noted, self-distancing decreases distress in the moment (Kross and Ayduk, 2017) but may have long-term psychological costs (Hancock et al., 2018), parallel to the psychological and social toll of keeping major life secrets (Tausczik et al., 2016), refusing to discuss conflicts with romantic partners (Laursen and Hafen, 2010), or avoiding thoughts about traumatic experiences (Pennebaker, 1989, 2018). Indeed, people who use less authentic language (a composite measure that includes "I" pronouns) tend to be perceived as less likable and credible in social and entrepreneurial contexts, likely because first-person singular pronouns are a necessary part of self-disclosure and intimacy (Markowitz et al., 2022). Therefore, increasing self-distancing over time may lead to social and occupational fallout. Further research should confirm which linguistic markers of chronic stress may be harmful before implementing any language-based intervention.

4.2 Limitations

As with many archival samples of naturalistic conversations online, the current sample is limited by a lack of information about the users. It is not possible to verify each user's healthcare work experience, nor can we conclusively assess demographic characteristics or personality traits that may clarify or qualify our findings. Reddit users are diverse and global, but tend to skew American, young, and masculine (Gjurković et al., 2021). Although language-based models can estimate such individual differences (Eichstaedt et al., 2021), linguistic cues to mental health such as negative self-focus (Baddeley et al., 2013) are often confounded with gender, age, and culture. For example, younger people and women tend to use "I" more (Pennebaker and Stone, 2003; Tausczik and Pennebaker, 2010), and negative affect is less stigmatized in East Asian than in Western cultures (Park et al., 2020).

The results are also limited by the relatively short baseline period. Using a longer 1 or 2-year pre-pandemic sample would have more appropriately accounted for seasonality, i.e., cyclical patterns over time operating independently of but sometimes confounded with the variables of interest (Brendstrup et al., 2004).

Finally, our conclusions are limited by the relatively narrow focus on doctors and nurses. Coping strategies and emotional experiences over the course of the pandemic may differ for people in other workplaces (e.g., restaurants, public transit) who share doctors' and nurses' experiences with high-infectivity work environments and understaffing. However, we provisionally assume that doctors' and nurses' language patterns represent a microcosm of the global pandemic response, with people in all professions potentially showing the same linguistic changes over time to the degree that their lives have been disrupted by COVID-19.

4.3 Ethics and Privacy

Research on social media language is fraught with ethical ambiguity. All messages we analyzed are public, and Reddit norms encourage anonymity. Yet social media users often fail to realize the degree to which others may be able to triangulate personal information from messages they have posted online (Mneimneh et al., 2021). Furthermore, people who are comfortable disclosing private thoughts and feelings in a familiar online community may be less sanguine about researchers reading and republishing their messages. That is, despite the public nature of Reddit, users may have reasonably expected relative privacy (believing only fellow subreddit subscribers would see their messages) while writing.

To respect the individuals in this sample, texts and usernames will only be shared pending ethical review of the proposed research (see Bender et al. 2020). All deidentified, quantitative data are available at the OSF link referenced above.

4.4 Conclusion

Dictionary-based analyses of a large naturalistic, longitudinal sample of healthcare workers' online conversations revealed psychological strengths and vulnerabilities among people working in high-risk positions on the front lines of the COVID-19 pandemic. Understanding how people cope—adaptively and otherwise—with chronic stress can help to calibrate mental health treatment for not only doctors and nurses, but also other high-risk professions (Aulisio and May, 2020). In the workplace, such treatment improvements may decrease burnout, mitigate staffing shortages, and improve healthcare quality, thus lightening the global healthcare burden (Gandi et al., 2011). In terms of both theory and practice in clinical psychology, gaining a clearer picture of everyday coping strategies offers an opportunity to check and in some cases reject inaccurate assumptions about how chronic stress affects social and emotional behavior.

References

Kiki Adams. 2022. SALLEE documentation. Online Receptiviti Inc. documentation.

Social Security Administration, Department of Veterans Affairs, et al. 2017. Federal policy for the protection of human subjects. final rule. *Fed. Regist.*, 82:7149–7274.

Hamzeh Mohammad Alrawashdeh, Ala'a B Al-Tammemi, Mohammad Kh Alzawahreh, Ashraf Al-Tamimi, Mohamed Elkholy, Fawaz Al Sarireh, Mohammad Abusamak, Nafisa MK Elehamer, Ahmad Malkawi, Wedad Al-Dolat, et al. 2021. Occupational burnout and job satisfaction among physicians in times of covid-19 crisis: a convergent parallel mixed-method study. *BMC Public Health*, 21(1):1–18.

Stephanie A Andel, Archana M Tedone, Winny Shen, and Maryana L Arvan. 2022. Safety implications of different forms of understaffing among nurses during the covid-19 pandemic. *Journal of Advanced Nursing*, 78(1):121–130.

Ashwini Ashokkumar and James W Pennebaker. 2021. Social media conversations reveal large psychological shifts caused by covid-19's onset across us cities. *Science advances*, 7(39):eabg7843.

Mark P Aulisio and Thomas May. 2020. Why healthcare workers ought to be prioritized in asmr during the sars-cov-2 pandemic. *The American Journal of Bioethics*, 20(7):125–128.

Özlem Ayduk and Ethan Kross. 2010. Analyzing negative experiences without ruminating: The role of self-distancing in enabling adaptive self-reflection. *Social and Personality Psychology Compass*, 4(10):841–854.

Jenna L Baddeley, James W Pennebaker, and Christopher G Beevers. 2013. Everyday social behavior during a major depressive episode. *Social Psychological and Personality Science*, 4(4):445–452.

Courtney L Bagge, Catherine R Glenn, and Han-Joo Lee. 2013. Quantifying the impact of recent negative life events on suicide attempts. *Journal of Abnormal Psychology*, 122(2):359.

Emily M Bender, Dirk Hovy, and Alexandra Schofield. 2020. Integrating ethics into the nlp curriculum. In *Proceedings of the 58th Annual Meeting of the Association for Computational Linguistics: Tutorial Abstracts*, pages 6–9.

Ryan L Boyd, Ashwini Ashokkumar, Sarah Seraj, and James W Pennebaker. 2022. The development and psychometric properties of liwc-22.

Bjarne Brendstrup, Svend Hylleberg, Morten Ørregaard Nielsen, Lars Skipper, and Lars Stentoft. 2004. Seasonality in economic models. *Macroeconomic Dynamics*, 8(3):362–394.

Timo Brockmeyer, Johannes Zimmermann, Dominika Kulessa, Martin Hautzinger, Hinrich Bents, Hans-Christoph Friederich, Wolfgang Herzog, and Matthias Backenstrass. 2015. Me, myself, and i: self-referent word use as an indicator of self-focused attention in relation to depression and anxiety. *Frontiers in psychology*, 6:1564.

Ana-Maria Bucur, Ioana R Podină, and Liviu P Dinu. 2021. A psychologically informed part-of-speech analysis of depression in social media. In *Proceedings of the International Conference on Recent Advances in Natural Language Processing, RANLP*, pages 199–207.

Sven Buechel, Anneke Buffone, Barry Slaff, Lyle Ungar, and João Sedoc. 2018. Modeling empathy and distress in reaction to news stories. In *Proceedings of the 2018 Conference on Empirical Methods in Natural Language Processing*, pages 4758–4765.

Muhammad Chutiyami, Allen MY Cheong, Dauda Salihu, Umar Muhammad Bello, Dorothy Ndwiga, Reshin Maharaj, Kogi Naidoo, Mustapha Adam Kolo, Philomina Jacob, Navjot Chhina, et al. 2021. Covid-19 pandemic and overall mental health of healthcare professionals globally: A meta-review of systematic reviews. *Frontiers in Psychiatry*, 12.

Michael A Cohn, Matthias R Mehl, and James W Pennebaker. 2004. Linguistic markers of psychological change surrounding september 11, 2001. *Psychological Science*, 15(10):687–693.

Glen Coppersmith, Mark Dredze, Craig Harman, and Kristy Hollingshead. 2015a. From adhd to sad: Analyzing the language of mental health on twitter through self-reported diagnoses. In *Proceedings of the 2nd Workshop on Computational Linguistics and Clinical Psychology: From Linguistic Signal to Clinical Reality*, pages 1–10.

Glen Coppersmith, Ryan Leary, Eric Whyne, and Tony Wood. 2015b. Quantifying suicidal ideation via language usage on social media. In *Joint Statistics Meetings Proceedings, Statistical Computing Section, JSM*, volume 110.

Munmun De Choudhury, Emre Kiciman, Mark Dredze, Glen Coppersmith, and Mrinal Kumar. 2016. Discovering shifts to suicidal ideation from mental health content in social media. In *Proceedings of the 2016 CHI conference on human factors in computing systems*, pages 2098–2110.

Sanda Dolcos and Dolores Albarracin. 2014. The inner speech of behavioral regulation: Intentions and task performance strengthen when you talk to yourself as a you. *European Journal of Social Psychology*, 44(6):636–642.

Johannes C Eichstaedt, Margaret L Kern, David B Yaden, HA Schwartz, Salvatore Giorgi, Gregory Park, Courtney A Hagan, Victoria A Tobolsky, Laura K Smith, Anneke Buffone, et al. 2021. Closed-and open-vocabulary approaches to text analysis: A review, quantitative comparison, and recommendations. *Psychological Methods*, 26(4):398.

Johannes C Eichstaedt, Robert J Smith, Raina M Merchant, Lyle H Ungar, Patrick Crutchley, Daniel Preoţiuc-Pietro, David A Asch, and H Andrew Schwartz. 2018. Facebook language predicts depression in medical records. *Proceedings of the National Academy of Sciences*, 115(44):11203–11208.

Bruce J Ellis, Albertine J Oldehinkel, and Esther Nederhof. 2017. The adaptive calibration model of stress responsivity: An empirical test in the tracking adolescents' individual lives survey study. *Development and Psychopathology*, 29(3):1001–1021.

Joshua C Gandi, Paul S Wai, Haruna Karick, and Zubairu K Dagona. 2011. The role of stress and level of burnout in job performance among nurses. *Mental Health in Family Medicine*, 8(3):181.

Matej Gjurković, Mladen Karan, Iva Vukojević, Mihaela Bošnjak, and Jan Šnajder. 2021. Pandora talks: Personality and demographics on reddit. In *Proceedings of the Ninth International Workshop on Natural Language Processing for Social Media*, pages 138–152.

David W Hancock, Amelia E Talley, Jennifer Bohanek, Micah D Iserman, and Molly E Ireland. 2018. Sexual orientation self-concept ambiguity and alcohol use disorder symptomology: The roles of motivated psychological distancing and drinking to cope. *Journal of Studies on Alcohol and Drugs*, 79(1):96–101.

Lucas Rodés-Guirao Cameron Appel Charlie Giattino Esteban Ortiz-Ospina Joe Hasell Bobbie Macdonald Diana Beltekian Hannah Ritchie, Edouard Mathieu and Max Roser. 2020. Coronavirus pandemic (covid-19). *Our World in Data*. Https://ourworldindata.org/coronavirus.

Amir Hassan Zadeh, Hamed M Zolbanin, Ramesh Sharda, and Dursun Delen. 2019. Social media for nowcasting flu activity: Spatio-temporal big data analysis. *Information Systems Frontiers*, 21(4):743–760.

Nicholas S Holtzman et al. 2017. A meta-analysis of correlations between depression and first person singular pronoun use. *Journal of Research in Personality*, 68:63–68.

Philip N Howard, Aiden Duffy, Deen Freelon, Muzammil M Hussain, Will Mari, and Marwa Maziad. 2011. Opening closed regimes: what was the role of social media during the arab spring? *Project on Information Technology and Political Islam*.

Molly E Ireland and Micah D Iserman. 2018. Within and between-person differences in language used across anxiety support and neutral reddit communities. In *Proceedings of the Fifth Workshop on Computational Linguistics and Clinical Psychology: From Keyboard to Clinic*, pages 182–193.

Micah Iserman. 2022. *splot: Split Plot*. R package version 0.5.3.

Kayla N Jordan, James W Pennebaker, Keith J Petrie, and Nicola Dalbeth. 2019. Googling gout: exploring perceptions about gout through a linguistic analysis of online search activities. *Arthritis Care & Research*, 71(3):419–426.

Ewa Kacewicz, James W Pennebaker, Matthew Davis, Moongee Jeon, and Arthur C Graesser. 2014. Pronoun use reflects standings in social hierarchies. *Journal of Language and Social Psychology*, 33(2):125–143.

Molly C Kalmoe, Matthew B Chapman, Jessica A Gold, and Andrea M Giedinghagen. 2019. Physician suicide: a call to action. *Missouri medicine*, 116(3):211.

Ethan Kross and Ozlem Ayduk. 2017. Self-distancing: Theory, research, and current directions. In *Advances in Experimental Social Psychology*, volume 55, pages 81–136. Elsevier.

Brett Laursen and Christopher A Hafen. 2010. Future directions in the study of close relationships: Conflict is bad (except when it's not). *Social Development*, 19(4):858–872.

Jessica L Malisch, Breanna N Harris, Shanen M Sherrer, Kristy A Lewis, Stephanie L Shepherd, Pumtiwitt C McCarthy, Jessica L Spott, Elizabeth P Karam, Naima Moustaid-Moussa, Jessica McCrory Calarco, et al. 2020. Opinion: In the wake of covid-19, academia needs new solutions to ensure gender equity. *Proceedings of the National Academy of Sciences*, 117(27):15378–15381.

David M Markowitz, Maryam Kouchaki, Francesca Gino, Jeffrey T Hancock, and Ryan L Boyd. 2022. Authentic first impressions relate to interpersonal, social, and entrepreneurial success. *Social Psychological and Personality Science*, page 19485506221086138.

Lauren M Ministero, Michael J Poulin, Anneke EK Buffone, and Shane DeLury. 2018. Empathic concern and the desire to help as separable components of compassionate responding. *Personality and Social Psychology Bulletin*, 44(4):475–491.

Zeina Mneimneh, Josh Pasek, Lisa Singh, Rachel Best, Leticia Bode, Elizabeth Bruch, Ceren Budak, Pamela Davis-Kean, Katharine Donato, Nicole Ellison, andrew gelman, Erica Groshen, Libby Hemphill, William Hobbs, J. Jensen, George Karypis, Jonathan Ladd, Amy O'Hara, Trivellore Raghunathan, and Stefan Wojcik. 2021. Data acquisition, sampling, and data preparation considerations for quantitative social science research using social media data.

Scott M Monroe, Paul Rohde, John R Seeley, and Peter M Lewinsohn. 1999. Life events and depression in adolescence: relationship loss as a prospective risk factor for first onset of major depressive disorder. *Journal of Abnormal Psychology*, 108(4):606.

Stephen Murata, Taylor Rezeppa, Brian Thoma, Laura Marengo, Katie Krancevich, Elizabeth Chiyka, Benjamin Hayes, Eli Goodfriend, Meredith Deal, Yongqi Zhong, et al. 2021. The psychiatric sequelae of the covid-19 pandemic in adolescents, adults, and health care workers. *Depression and Anxiety*, 38(2):233–246.

Yaakov Ophir, Refael Tikochinski, Christa SC Asterhan, Itay Sisso, and Roi Reichart. 2020. Deep neural networks detect suicide risk from textual facebook posts. *Scientific Reports*, 10(1):1–10.

Jiyoung Park, Shinobu Kitayama, Yuri Miyamoto, and Christopher L Coe. 2020. Feeling bad is not always unhealthy: Culture moderates the link between negative affect and diurnal cortisol profiles. *Emotion*, 20(5):721.

James W Pennebaker. 1989. Confession, inhibition, and disease. In *Advances in experimental social psychology*, volume 22, pages 211–244. Elsevier.

James W Pennebaker. 2018. Expressive writing in psychological science. *Perspectives on Psychological Science*, 13(2):226–229.

James W Pennebaker, Ryan L Boyd, Roger J Booth, Ashwini Ashokkumar, and Martha E Francis. 2022. Linguistic inquiry and word count: Liwc-22.

James W Pennebaker and Cindy K Chung. 2005. Tracking the social dynamics of responses to terrorism: Language, behavior, and the internet. *NATO Security Through Science Series E: Human and Societal Dynamics*, 3:159.

James W Pennebaker and Cindy K Chung. 2007. Expressive writing, emotional upheavals, and health. In HS Friedman and RC Silver, editors, *Foundations of Health Psychology*. Oxford University Press, Oxford, England, UK.

James W Pennebaker and Kent D Harber. 1993. A social stage model of collective coping: The loma prieta earthquake and the persian gulf war. *Journal of Social Issues*, 49(4):125–145.

James W Pennebaker and Lori D Stone. 2003. Words of wisdom: language use over the life span. *Journal of Personality and Social Psychology*, 85(2):291.

Jose Pinheiro, Douglas Bates, Saikat DebRoy, Deepayan Sarkar, and R Core Team. 2021. *nlme: Linear and Nonlinear Mixed Effects Models*. R package version 3.1-153.

R Core Team. 2022. *R: A Language and Environment for Statistical Computing*. R Foundation for Statistical Computing, Vienna, Austria.

Michael W Rabow, Chao-Hui S Huang, Gloria E White-Hammond, and Rodney O Tucker. 2021. Witnesses and victims both: Healthcare workers and grief in the time of covid-19. *Journal of Pain and Symptom Management*, 62(3):647–656.

Megan L Robbins, Robert C Wright, Ana María López, and Karen Weihs. 2019. Interpersonal positive reframing in the daily lives of couples coping with breast cancer. *Journal of psychosocial oncology*, 37(2):160–177.

Arunima Roy, Katerina Nikolitch, Rachel McGinn, Safiya Jinah, William Klement, and Zachary A Kaminsky. 2020. A machine learning approach predicts future risk to suicidal ideation from social media data. *NPJ Digital Medicine*, 3(1):1–12.

Kristin E Schneider, Lauren Dayton, Saba Rouhani, and Carl A Latkin. 2021. Implications of attitudes and beliefs about covid-19 vaccines for vaccination campaigns in the united states: A latent class analysis. *Preventive medicine reports*, 24:101584.

João Sedoc, Sven Buechel, Yehonathan Nachmany, Anneke Buffone, and Lyle Ungar. 2020. Learning word ratings for empathy and distress from document-level user responses. In *Proceedings of the 12th Language Resources and Evaluation Conference*, pages 1664–1673.

Yi-Tai Seih, YC Lin, CL Huang, CW Peng, and SP Huang. 2008. The benefits of psychological displacement in diary writing when using different pronouns. *British Journal of Health Psychology*, 13(1):39–41.

Sarah Seraj, Kate G Blackburn, and James W Pennebaker. 2021. Language left behind on social media exposes the emotional and cognitive costs of a romantic breakup. *Proceedings of the National Academy of Sciences*, 118(7).

Judy Hanwen Shen and Frank Rudzicz. 2017. Detecting anxiety through reddit. In *Proceedings of the Fourth Workshop on Computational Linguistics and Clinical Psychology—From Linguistic Signal to Clinical Reality*, pages 58–65.

Mamidipalli Sai Spoorthy, Sree Karthik Pratapa, and Supriya Mahant. 2020. Mental health problems faced by healthcare workers due to the covid-19 pandemic–a review. *Asian Journal of Psychiatry*, 51:102119.

Abi Sriharan, Savithiri Ratnapalan, Andrea C Tricco, and Doina Lupea. 2021. Women in healthcare experiencing occupational stress and burnout during covid-19: a rapid review. *BMJ open*, 11(4):e048861.

Shannon Wiltsey Stirman and James W Pennebaker. 2001. Word use in the poetry of suicidal and nonsuicidal poets. *Psychosomatic medicine*, 63(4):517–522.

Lori D Stone and James W Pennebaker. 2002. Trauma in real time: Talking and avoiding online conversations about the death of princess diana. *Basic and Applied Social Psychology*, 24(3):173–183.

Jessie Sun, H Andrew Schwartz, Youngseo Son, Margaret L Kern, and Simine Vazire. 2020. The language of well-being: Tracking fluctuations in emotion experience through everyday speech. *Journal of Personality and Social Psychology*, 118(2):364.

Allison M Tackman, David A Sbarra, Angela L Carey, M Brent Donnellan, Andrea B Horn, Nicholas S Holtzman, To'Meisha S Edwards, James W Pennebaker, and Matthias R Mehl. 2019. Depression, negative emotionality, and self-referential language: A multi-lab, multi-measure, and multi-language-task research synthesis. *Journal of Personality and Social Psychology*, 116(5):817.

Yla Tausczik, Cindy Chung, and James Pennebaker. 2016. Tracking secret-keeping in emails. In *Proceedings of the International AAAI Conference on Web and Social Media*, volume 10, pages 388–397.

Yla Tausczik, Kate Faasse, James W Pennebaker, and Keith J Petrie. 2012. Public anxiety and information seeking following the h1n1 outbreak: blogs, newspaper articles, and wikipedia visits. *Health Communication*, 27(2):179–185.

Yla R Tausczik and James W Pennebaker. 2010. The psychological meaning of words: Liwc and computerized text analysis methods. *Journal of Language and Social Psychology*, 29(1):24–54.

Tyler J VanderWeele and Maya B Mathur. 2019. Some desirable properties of the bonferroni correction: is the bonferroni correction really so bad? *American journal of epidemiology*, 188(3):617–618.

Suzan Verberne, Anika Batenburg, Remco Sanders, Mies van Eenbergen, Enny Das, Mattijs S Lambooij, et al. 2019. Analyzing empowerment processes among cancer patients in an online community: A text mining approach. *JMIR Cancer*, 5(1):e9887.

Vera Vine, Ryan L Boyd, and James W Pennebaker. 2020. Natural emotion vocabularies as windows on distress and well-being. *Nature Communications*, 11(1):1–9.

Matthew Walton, Esther Murray, and Michael D Christian. 2020. Mental health care for medical staff and affiliated healthcare workers during the covid-19 pandemic. *European Heart Journal: Acute Cardiovascular Care*, 9(3):241–247.

Peter F Weissmann, William T Branch, Catherine F Gracey, Paul Haidet, and Richard M Frankel. 2006. Role modeling humanistic behavior: learning bedside manner from the experts. *Academic Medicine*, 81(7):661–667.

WHO. 2022. Who covid-19 dashboard.

Kevin P Young, Diana L Kolcz, David M O'Sullivan, Jennifer Ferrand, Jeremy Fried, and Kenneth Robinson. 2021. Health care workers' mental health and quality of life during covid-19: results from a mid-pandemic, national survey. *Psychiatric Services*, 72(2):122–128.

Yan Zhang, Lili Wei, Huanting Li, Yueshuai Pan, Jingyuan Wang, Qianqian Li, Qian Wu, and Holly Wei. 2020. The psychological change process of frontline nurses caring for patients with covid-19 during its outbreak. *Issues in Mental Health Nursing*, 41(6):525–530.

Are You Really Okay? A Transfer Learning-based Approach for Identification of Underlying Mental Illnesses

Ankit Aich and **Natalie Parde**
Natural Language Processing Laboratory
Department of Computer Science
University of Illinois at Chicago
{aaich2, parde}@uic.edu

Abstract

Evidence has demonstrated the presence of similarities in language use across people with various mental health conditions. In this work we investigate these relationships both as described in literature and as a data analysis problem. We also introduce a novel transfer learning based approach that learns from linguistic feature spaces of previous conditions and predicts unknown ones. Our model achieves strong performance, with F_1 scores of 0.75, 0.80, and 0.76 at detecting depression, stress, and suicidal ideation in a first-of-its-kind transfer task and offering promising evidence that language models can harness learned patterns from known mental health conditions to aid in their prediction of others that may lie latent.

1 Introduction

Mental health conditions are a pervasive but historically often overlooked societal and individual concern (Bertolote, 2008). In recent decades their study has gained increasing priority, and within the past decade this study has extended to techniques for automated analysis and detection of mental health conditions, including through patterns detected in written and spoken language (Resnik et al., 2014). Most work on automated assessment of mental health seeks to identify and possibly alleviate specific mental health conditions. Researchers have focused on a myriad of target illnesses and diagnoses such as depression (Schwartz et al., 2014a), schizophrenia (Gutiérrez et al., 2017), or even suicideal ideation[1] (Homan et al., 2014). However, to date they have not yet examined the overlap or interplay between these target illnesses. This overlap may present a valuable source of information, particularly in the resource-poor settings common in mental health and healthcare applications more generally.

In this paper we ask three important research questions centered on the interplay between the linguistic footprints of known and latent mental health conditions (MHCs),[2] and present answers to them with evidence.

- **RQ1:** *How do features relate across multiple MHCs?*
- **RQ2:** *Can we represent different MHCs under the same feature spaces and find relations?*
- **RQ3:** *Can we identify underlying MHCs using the language of known ones?*

Our first question relates to the linguistic markers of MHCs. We comprehensively examine existing psycholinguistic and mental health research to search for common underlying threads (§3). To answer our second question, we investigate the relation between the identified features using well defined and trusted NLP baselines (§4). Finally, we answer our last question by experimentally determining the success with which we can use similar and dissimilar linguistic feature spaces to predict the presence of latent MHCs (§5). To do so, we leverage transfer learning to achieve a strong benchmark accuracy of 85%.

2 Background

According to the National Institute of Mental Health, 43.6 million adults (nearly 18.1% of the

[1]Presence of *Suicidal Ideation* (SI) is not an illness, but a diagnosis which encompasses thoughts ranging from contemplation to preoccupations with death via suicide (Harmer et al., 2022).

[2]We define MHCs as any condition ranging along the spectrum from issues causing mental health concerns such as stress, to actual defined illnesses such as depression, or diagnoses such as SI.

Proceedings of the Eighth Workshop on Computational Linguistics and Clinical Psychology, pages 89 - 104
July 15, 2022

U.S. population) experience mental health conditions in a given year.[3] Oftentimes, symptoms may be recognizable when interacting with close relations (Insel, 2008) or even on social media (Berry et al., 2017). Berry et al. (2017) investigate the popularity of social media as an outlet for mental health discussion at length, finding reasons including anonymity, sense of empowerment, sense of community, and perceptions of the internet as a safe space. A growing number of approaches have sought to leverage social media data to aid in the automated identification of specific MHCs, with work to date including automated detection of depression (Yasaswini et al., 2021; Schwartz et al., 2014a; Tasnim and Stroulia, 2019; Rosenquist et al., 2010), post-traumatic stress disorder (Li et al., 2010), anxiety (Shen and Rudzicz, 2017), and stress (Naik et al., 2018). However, these approaches have lagged behind the state of the art in more fundamental NLP tasks. In particular, work harnessing high-powered transfer learning models has remained either scarce or singularly focused on one illness (Pegah et al., 2019; Howard et al., 2019).

We aim to fill this translational gap by synthesizing fundamental progress with the applied problem of detecting the presence of underlying MHCs. We follow Blodgett et al. (2020)'s lead and model our approach not only on existing NLP models, but on findings from pyscholinguistic and other domain-specific literature as well, including those correlating retention (Shen et al., 2009), cognitive attention and complexity (Vuilleumier, 2006; Tausczik and Pennebaker, 2010), reasoning (Jung et al., 2014), and problem-solving skills (Isen et al., 1987) with specific mental health conditions. Little has been done towards this problem with RQs of multi-task learning from social media being very recent (Benton et al., 2017b). This work, to the best of our knowledge, is the first of its kind to study correlation among diseases in both theory and practice. We examine prior literature to identify correlating themes across illnesses, analyze language data from individuals with different mental health conditions to find practical correlations and trends, and present transfer learning-based classification models to identify undiagnosed illnesses given known features grounded in mental health and psycholinguistic theory and NLP practice.

[3] www.nimh.nih.gov/health/statistics/prevalence/any-mental-illness-ami-among-us-adults.shtml

3 Feature Correlation Across Varying Mental Health Conditions

A natural question that arises when using social media data is its reliability as an information source. Social media is increasingly seen as a popular choice and acceptable platform for healthcare information exchange (Gkotsis et al., 2016a), and its use has been investigated in numerous predictive healthcare tasks. Classical models (e.g., support vector machines) trained on simple text-based features have reliably predicted mental health emergencies (Franco-Penya and Mamani Sanchez, 2016). Audio features have also been found to be excellent markers of mood or other prosodic signals, including for automated detection of depression (Lamers et al., 2014). Language models have demonstrated an ability to learn powerful, quantifiable signals from tweets to predict users' mental states (Coppersmith et al., 2014), and more clinically advanced mental health conditions such as psychosis have also been detected using short appraisals of social media posts (Birnbaum et al., 2017). Predicting depression on social media is a long standing research track (De Choudhury et al., 2021), and social media has also shown that signals to identify suicidal ideation can be traced with high efficacy (Choudhury et al., 2016). Platforms like Reddit[4] can be instrumental in terms of support, resources, and self-disclosure about mental health (Choudhury and De, 2014; Valizadeh et al., 2021).

One of the first traceable thematic identifications of correlated, quantifiable information regarding mental state and wellbeing was by Fleming et al. (1992), suggesting that a lack of social support combined with social isolation was present in patients showing signs of depression or post-partum depression. The same work also identified effects of psychological stress on attitude, emotion, and behavior. The relationship between social isolation, loneliness, and clinical depression was later also validated by MNSc et al. (1996), and the relationship between latent stress and surface depression has since persisted as a recurring theme across mental health literature (Scott et al., 2000).

Homan et al. (2014) found that high levels of stress or distress are related to higher levels of suicidal ideation. Schwartz et al. (2014b) also pointed to trepidation, frustration, annoyance, helplessness, and again stress as major themes correlating with expression of mental illness. Depression and stress

[4] reddit.com

co-exist in latent forms for other surface illnesses such as schizophrenia as well, as demonstrated by Mitchell et al. (2015) who extracted LIWC (Tausczik and Pennebaker, 2010) features from social media data to detect advanced psychosis and schizophrenia in social media.

Perhaps one of the most interesting finds in translational mental health research is the direct relationship between depression, suicidal ideation, and stress (Preoţiuc-Pietro et al., 2015). Preoţiuc-Pietro et al. (2015) provide evidence that depressive language correlates with sustained periods of low sentiment and has similar topical themes to language produced by suicidal or dysphoric individuals.

Although NLP researchers have experimented with a wide range of linguistic features for mental health assessment and analysis, several have emerged as being particularly discriminating. Metadata such as hashtags or the name of a forum (Mills, 2017) can be powerful features to detect mental health conditions such as suicidal ideation (Gkotsis et al., 2016b). Specific words or hashtags can be used to identify personality profiles, as well as stigma or awareness of mental health conditions on social media (Hwang and Hollingshead, 2016). Degrading or negative n-grams (e.g., *crazy*, *mad*, or *nuts*) can distinguish personality types and mental health outlook (Hwang and Hollingshead, 2016), and part-of-speech (POS) tags can also be informative in social media data (Gkotsis et al., 2016b). Tausczik and Pennebaker (2010) characterize speech at a granular level with social and personal profiles, and present LIWC, a powerful tool to extract such features (Malmasi et al., 2016). N-grams have been powerful markers of depression or PTSD (Pedersen, 2015), and can be valuable tools for feature discovery (Tanana et al., 2016). Lexicon-based features, word embedding features, or annotated posts from social media are also informative (Shickel et al., 2016). Across this systematic review of mental health within NLP literature, the following key relations become evident:

- Stressful and emotional events affect measured cognitive complexity (Shen et al., 2009; Vuilleumier, 2006; Isen et al., 1987).
- Depression, stress, and suicide are related with often overlapping diagnoses, and have intersecting themes of general negativity and hopelessness (Fleming et al., 1992; MNSc et al., 1996; Scott et al., 2000; Schwartz et al., 2014b; Homan et al., 2014).
- N-grams, lexicon-based features, word embeddings, and POS features are powerful tools for social media analysis of mental health problems (Gkotsis et al., 2016b; Hwang and Hollingshead, 2016; Pedersen, 2015; Malmasi et al., 2016).

We experiment further with these features in the following subsections.

4 Feature Relationships in Mental Health Data

4.1 Data Sourcing and Ethical Guidelines

To fully understand the relationships among linguistic features in mental health contexts we explore datasets associated with three different MHCs. Gaining access to datasets in this area proved challenging, as also discussed by Harrigian et al. (2021), for numerous reasons including IRB restrictions, personal reluctance, or unresponsiveness to data access requests. We ultimately acquired datasets pertaining to *suicide* (Shing et al., 2018; Zirikly et al., 2019), *stress* (Turcan and McKeown, 2019), and *depression* (Losada and Crestani, 2016; Parapar et al., 2021).

In conducting our exploration, we followed the ethical and privacy guidelines defined by Benton et al. (2017a). No identifiable information is collected, and all data is stored on secured servers and obtained via written agreements from the creators. The institutional review board (IRB) at our institution declared our experiments on these datasets as exempt from further review.

4.2 Data Description

We studied and analyzed each dataset. All datasets were created with a mixed and randomized population of social media users; thus, the selection of participants was not constrained by gender, background, or other factors. Our *suicide* dataset is sourced from Reddit (Shing et al., 2018; Zirikly et al., 2019) and contains posts and labels for users diagnosed as having suicidal ideation or matched controls. Our *stress* dataset is the Dreaddit dataset published by Turcan and McKeown (2019). It is a publicly available dataset with binary labels indicating the presence of stress[5] (*stressed* and *not*

[5]The authors also ask annotators to indicate instances for which the label is unclear; instances for which this is the majority label are later dropped.

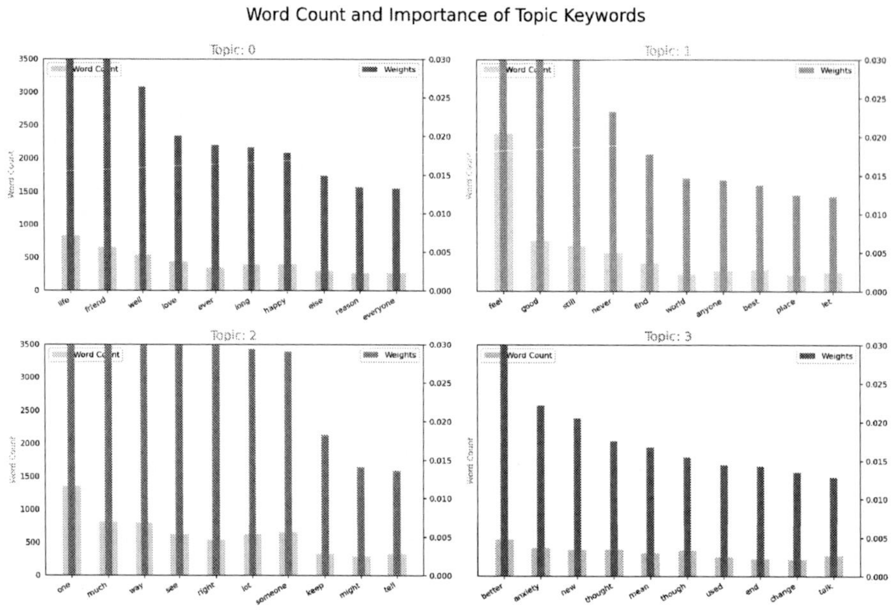

Figure 1: Top four topics identified when applying LDA to *depression*.

stressed) in individuals posting on Reddit. Our *depression* dataset is sourced from Twitter[6] and has binary labels (*depression* and *no depression/control*) and raw and pre-processed tweet text (Losada and Crestani, 2016; Parapar et al., 2021).

Depression contains 531,453 posts from 892 users, and *stress* contains 187,444 posts. Our *suicide* dataset samples 1097 users at random from a pool of 11,129 initial users, and picks 934 from among those to create a four-class dataset with risk assessment classes: *None*, *Low*, *Moderate*, and *Severe*. We aggregate these into binary labels of 0 (*None*, *Low*) and 1 (*Moderate*, *Severe*).

As per our agreements with the creators of these datasets we are unable to share data directly, but we provide a table in the Appendix to summarize dataset statistics. We encourage researchers to examine the data and related private datasets, and thank the respective authors as well as Harrigian et al. (2021) for creating a curated repository of mental health data and pointers facilitating data discovery.

4.3 Data Analysis

As noted in §3, trauma, stress, depression, and mental illness measurably impact reasoning, problem solving, and overall cognitive complexity. Tausczik and Pennebaker (2010) map these effects to psycholinguistic features including sentence complexity, words per sentence, and average word length on a scale of 0-100, where scores less than 50 denote lower cognitive reasoning and analysis.

We perform Latent Dirichlet Allocation (LDA) across the *depression* and *stress* data to identify topical themes. We present a graph in Figure 1 showing the four top themes identified for *depression*, visualized with frequency counts for the thematic terms (a similar graph for *stress* is provided in Figure 3 in the Appendix). To determine thematic titles, we apply Ryan and Bernard (2003)'s Keywords In Context (KWIC) approach, qualitatively examining context and finding the words that adhere to it. We detail our outcomes in Tables 1 and 2, considering the top words identified per theme using LDA and subsequently using KWIC to assign theme names. We find that social support, connections, and familial stress are common topical themes across both illnesses, validating our findings in §3 that similarities in language exist among people suffering from different MHCs. This manifests in our n-gram analyses as well (e.g., with terms such as *feel*, *don't know*, and *life*), further highlighting the intersection of themes across different MHCs.

We further assess the cognitive complexities of a random sample of 380 individuals from *depression* and *suicide*, measured as the average of (a)

[6] `www.twitter.com`

Identified Theme	Keywords in Context
Social Support	Life, Friend, Love, Happy, Everyone, Reason
Feelings & Connections	Feel, Good, Anyone, Never, Find
Action Taken	One, Someone, Might, Tell
Therapeutic	Anxiety, Mean, End, Talk, Better

Table 1: Identified themes applying KWIC to LDA topics for *depression*.

Identified Theme	Keywords in Context
Failed Connections	Relationship, Didn't, Work, Someone, Need
Social and Familial Stress	Doesn't, Feel, Right, Dad, Girl, Kid
Pessimism	Don't, Can't, Family, Know, Good
Chronic Stress	Year, Still, Issue, Hard, Without

Table 2: Identified themes applying KWIC to LDA topics for *stress*.

the ANALYTIC feature extracted by LIWC and (b) the average number of short (length ≤ 6) words per sentence, mapped to a 0-100 scale. We plot the cognitive complexity scores (Y axis) in for each individual in the sample (X axis bars), and observe a slightly lower cognitive complexity for individuals in *suicide* (see Figures 4 and 5 in the Appendix). This is in line with our first finding in §3, and the complementary knowledge that suicidal ideation is often a more extreme expression of depression (Brådvik, 2018).

Finally, to examine the role of sentiment, negativity, and hopelessness (our second finding in §3), we also quantitatively analyze the most frequent trigrams associated with *depression*, *suicide*, and *stress* (see Figures 6, 7, and 8 in the Appendix). We similarly analyze bigrams and unigrams (see Figures 9 and 14 in the Appendix). We find that the top n-grams for all three illnesses are evocative of emotion, confirming substantial overlap across illnesses. N-grams associated with *depression* place additional emphasis on memories (e.g., "campsite tent fire") and specific mental health diagnoses (e.g., "major depressive disorder"), whereas n-grams associated with *suicide* place greater emphasis on confusion (e.g., "basically i'm wondering") and helplessness (e.g., "someone please help"). N-grams associated with *stress* echo many of these themes, with an additional emphasis on uncertainty (e.g., "don't really know").

5 Classification and Transfer Learning

5.1 Task Outline

We model the primary task as a binary classification problem to predict labels at the user level as 1 (*Diagnosed*) or 0 (*Undiagnosed*) for a mental illness or disease D. This can be formulated as:

$$Y_d = M(D)$$

where Y is the label of a classification model M on a domain D. This domain, D, can be defined as:

$$D = \{X, \mathrm{P}(X)\} \quad (1)$$

where X is the feature space and $\mathrm{P}(X)$ is the marginal probability distribution for:

$$X = \{x_1, x_2, ..., x_n\}$$

For our MHC domain, we can define a task, T, as follows:

$$T = \{Y, f(\cdot)\} \quad (2)$$

Here, Y is the label space. This is obtained from a classification function $f(\cdot)$, which learns from our data having features X and labels Y as follows:

$$\{(x_i, y_i) | i \in \{1, 2, ..., n\}, x_i \in X, y_i \in Y\} \quad (3)$$

In Equation 3, each data point in the task is represented by the subscript i, where (x_i, y_i) corresponds to the feature vector and label for point i in a dataset of length n. Represented mathematically, our function predicts a label $y_i = f(x_i)$ using the conditional probability distribution of Y given X:

$$T = \{Y, P(Y|X)\} \quad (4)$$

Thus, given a transfer learning task with source (**S**) and target (**T**), there are four aspects of the task which might differ:

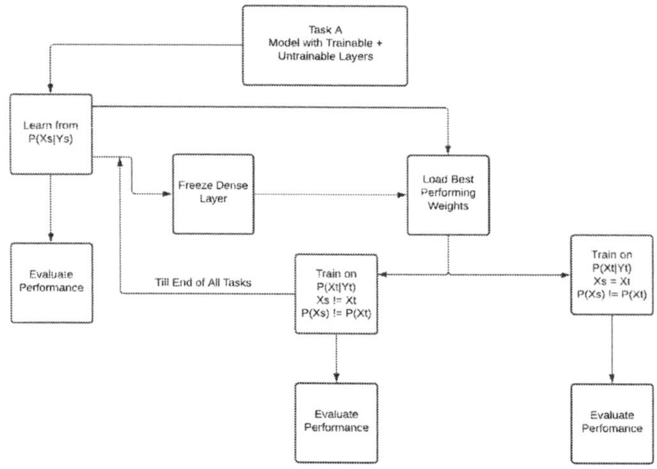

Figure 2: Model Architecture Flow. When Xs = Xt, both datasets use LIWC features. This keeps the feature space the same, with differing marginal distributions owing to separate datasets. When Xs $\neq$ Xt, datasets have LIWC features in the source space and Word2Vec features in the target space.

- The feature space $\mathbf{X}$ of the source and target
- The marginal distribution $\mathbf{P}(\mathbf{X})$
- The label space $\mathbf{Y}$
- The conditional distribution $\mathbf{P}(\mathbf{Y}|\mathbf{X})$

We conduct our experiments under two variable conditions. In the first, we keep the feature space similar across transfer tasks, using LIWC (Tausczik and Pennebaker, 2010) features for both the source and target tasks. In the second we keep the feature space different between the two tasks, using LIWC features for the source task and Word2Vec (Mikolov et al., 2013) features for the target task. The label spaces are also the same, with binary classification labels across all tasks.

5.2 Feature Description

We extract both Word2Vec and LIWC features for each dataset. Word2Vec is a popular vector representation model that learns to predict words given their contexts from millions of online resources (Mikolov et al., 2013). Linguistic Inquiry and Word Count (LIWC) features are common in mental health tasks due to their demonstrated high performance for a wide range of applications including personality modeling, mental state assessment, affective analysis, and language understanding (Ludwig et al., 2013; Park et al., 2014; Schwartz et al., 2013; Pervin and Cervone, 2010; Coviello et al., 2014; Tumasjan et al., 2010; Riffe et al., 2019). They leverage syntactic patterns to provide feature representations that correlate with psycholinguistic characteristics (e.g., measuring cognitive complexity based on word length and words per sentence).

The creators of the *stress*, *depression*, and *suicide* datasets use a variety of features in their own work. We choose Word2Vec features and LIWC features since these exhibit the highest overlap across tasks in prior task-specific work. For instance, Losada and Crestani (2016) use TF-IDF vectorized text, and vectorized embeddings and LIWC features are also used by both Turcan and McKeown (2019) and Shing et al. (2018). The former use Word2Vec embeddings with BERT (Devlin et al., 2019) along with several attributes from LIWC including *clout*, *tone*, and *pronoun* features. The latter use domain-specific word embeddings from a SkipGram model trained on Reddit data, as well as bag-of-words features, topical features, readability scores, and features induced from LIWC, a mental health lexicon (Zirikly et al., 2016), and NRCLex (Mohammad and Turney, 2013).

5.3 Model Architecture and Training

Each model trains on K tasks, where $K \in \{1, 2, .., N\}$, and is comprised of trainable and untrainable layers. Before all of our transfer tasks, each training dataset is padded to the same size (the vocabulary size from the largest training dataset). We consider a convolutional neural network (CNN), as well as to a lesser extent other models such as bidirectional long short-term memory (BiLSTM), LSTM, and RNN models.[7]

Each model in the input layer accepts the training data, consisting of the distribution of the feature space and labels to predict a classification label. Accuracy and F_1 scores are calculated for each task, and during transfer the dense trainable layer is frozen and the weights from the best performing epoch are loaded. Training then proceeds on the next task. This loop continues until all tasks have been learned and evaluation metrics have been calculated. Figure 2 illustrates this process.

Our best performing model is a CNN fine-tuned for transfer learning between datasets and a novel stress→depression→suicide prediction task. This model, as well as a BiLSTM alternative used in preliminary experiments, uses a one-dimensional max pooling layer with a poolsize of 2, flattening, a dropout of 0.5, and a frozen dense layer. The output layer has one node with a sigmoid activation.

6 Results and Discussion

Our experiments offer a first-of-its-kind examination of transfer learning across multiple MHCs. Since there are no directly comparable transfer learning models, we compare individual task performance to the respective benchmarks established by the dataset creators using task-specific models. These models leverage many architectures and feature types, intersecting in their use of vector representations and LIWC features. Specifically, we compare to the following:

- **Depression:** Losada and Crestani (2016) use TF-IDF vectorized embeddings with a logistic regression classifier.
- **Stress:** Turcan and McKeown (2019) use LIWC features and Word2Vec embeddings with a logistic regression classifier.
- **Suicide:** Shing et al. (2018) use LIWC features, Word2Vec embeddings, bag-of-words features, LDA features, and NRCLex features with a CNN classifier.

[7]Preliminary experiments using RNN and LSTM achieved weaker performance than CNN and BiLSTM, so we did not pursue further experimentation with those models.

Model	Depression	Stress	Suicide
Losada and Crestani (2016)	0.66	—	—
Turcan and McKeown (2019)	—	0.79	—
Shing et al. (2018)	—	—	0.42
Ours	**0.75**	**0.80**	**0.76**

Table 3: Performance comparison between existing task-specific models (Losada and Crestani, 2016; Turcan and McKeown, 2019; Shing et al., 2018) and our transfer learning model reported here. Performance is measured using F_1.

For our own transfer CNN model (our highest-performing model), we train on: *stress* when using a target task of **depression**; *depression* when using a target task of **stress**; and *stress* and *depression* when using a target task of **suicide**, based on patterns of MHC expression identified in earlier reviewed literature. We report our findings in Table 3, using F_1 to measure performance. As shown, our model outperforms existing benchmarks with relative performance improvements of 13.64%, 1.27%, and 80.95% for *depression*, *stress*, and *suicide*, respectively and achieving a new state of the art with F_1 scores of 0.75, 0.80, and 0.76. We hope that these results will motivate other researchers to experiment with transfer learning across MHCs.

This answers one of our research questions: It is indeed possible to predict MHCs given information about existing ones, validating findings in mental health literature (Saini and Mandeep, 2020). However, the accuracy with which we can predict unseen mental health conditions depends on the feature space we use. LIWC features, which explicitly encode the psychological meaning of words, work better than Word2Vec features which rely purely on distributional semantics.

We also experiment with an alternative model grounded in psychological evidence that *suicide* may occur as a natural escalation from *stress* and then *depression*. We train our same core CNN model first on *stress*, then on *depression*, and then on *suicide* and achieve an 85% accuracy at the target task of **suicide**. Our BiLSTM model achieves an accuracy of 75% on **depression** when first trained on *stress*, and then an accuracy of 76% on **suicide** when subsequently trained on *depression*, echoing this trend albeit to a lesser degree. The strong performance of this technique further supports our finding that shared language characteristics across MHCs make this a promising and impactful sandbox for experiments with transfer learning.

7 Research Answers

In §1, we asked three important research questions. Following our analyses, we present concrete answers to them in this section.

How do features relate across multiple MHCs? Mental health conditions have similar manifestations in language, and correspondingly in their linguistic signatures. We provide evidence for this in our literature review (§3) and analyses (§4). Although we cannot through linguistic analysis conclusively measure the similarity of two MHCs, we can discern that the language usage and its features have significant overlap across MHCs (see Figure 1 and Tables 1 and 2, and other figures and tables in the Appendix).

Can we represent different MHCs under the same feature spaces and find relations? Yes, using semantically descriptive features such as LIWC it is possible to find relations (§4). We demonstrate that using standard NLP tools such as LDA or n-gram language modeling it is possible to see similar themes and topical relationships (§4).

Can we identify underlying MHCs using the language of known ones? Yes and No! While models trained on one task and transferred efficiently can predict unseen MHCs with a higher accuracy then when predicting them using only target domain data, these are linguistic classifications only (§5). AI models are still far from being able to conclusively identify MHCs, and should not be considered as replacements for professional mental health care.

Given these research answers, we close by discussing how we can carry this forward and what it means for NLP in mental health.

8 Conclusion and Future Directions

In this work, we examine the utility of transfer learning for the identification of three MHCs: depression, stress, and suicidal ideation. These MHCs vary in their clinical classification and severity. Depression is formally defined as a mental illness (Kanter et al., 2008), stress is a process which may ultimately result in mental illness (Salleh, 2008), and suicidal ideation is classified as a disorder (Fehling and Selby, 2021). Although we achieve promising performance in detecting these conditions, nothing—not even actual diagnosis by a human expert—can conclusively identify a mental illness with 100% certainty (Allsopp et al., 2019).

We presented a qualitative exploration of the overlap and interplay between language and mental health across multiple MHCs, and also presented quantitative correlations among words, tokens, themes, topics, and large feature space representations using well-known, established NLP methods. Finally, we introduced a transfer learning model to predict unseen mental health conditions using similar and dissimilar feature spaces, the first of its kind. Our model outperforms the baselines established by benchmark models for detecting depression, stress, and suicide with percent increases in measured performance of 13.64%, 1.27%, and 80.95%, respectively. The model also achieved an 85% accuracy at detecting suicidal ideation in a psychologically informed model that trains on datasets in an order established by clinical evidence, with *stress* followed by *depression*[8] and then ultimately *suicide* (Orsolini et al., 2020).

Although this paper demonstrated preliminary evidence that similarities in feature spaces can be leveraged to better predict unknown MHCs, in the future we wish to explore this further with a larger variety of models. We also plan to further examine the role that transfer learning order has in establishing performance.[9] Other work has found that social media-based models do not always generalize and may incur substantial performance losses (Harrigian et al., 2020), and other factors such as social concerns, self-disclosure bias, and temporal artifacts may also influence model performance (Harrigian et al., 2020). We hope that researchers will use our findings to explore new ways to increase the efficiency and usefulness of AI-supported treatment and diagnosis of MHCs (Allsopp et al., 2019).

[8] www.psychologytoday.com/us/blog/in-practice/201303/why-stress-turns-depression

[9] In some early experiments not reported here, reversing the transfer learning order of our model resulted in performance that peaked at an F_1=0.48.

9 Acknowledgements

This work was funded by a start-up grant from the University of Illinois at Chicago and by the National Institute Of Mental Health of the National Institutes of Health under Award Number R01MH116902. The content is solely the responsibility of the authors and does not necessarily represent the official views of the National Institutes of Health. We thank the anonymous technical and clinical reviewers for their insightful comments. We are also very grateful to all of the teams who shared their valuable datasets with us.

References

Kate Allsopp, John Read, Rhiannon Corcoran, and Peter Kinderman. 2019. Heterogeneity in psychiatric diagnostic classification. *Psychiatry Research*, 279.

Adrian Benton, Glen Coppersmith, and Mark Dredze. 2017a. Ethical research protocols for social media health research. In *Proceedings of the First ACL Workshop on Ethics in Natural Language Processing*, pages 94–102, Valencia, Spain. Association for Computational Linguistics.

Adrian Benton, Margaret Mitchell, and Dirk Hovy. 2017b. Multitask learning for mental health conditions with limited social media data. *Proceedings of the 15th Conference of the European Chapter of the Association for Computational Linguistics: Volume 1, Long Papers*, pages 152–162.

Natalie Berry, Fiona Lobban, Maksim Belousov, Richard Emsley, Goran Nenadic, and Sandra Bucci. 2017. #whywetweetmh: Understanding why people use twitter to discuss mental health problems. *Journal of Medical Internet Research*, 19.

José Bertolote. 2008. The roots of the concept of mental health. *World psychiatry : official journal of the World Psychiatric Association (WPA)*, 7:113–6.

Michael Birnbaum, Sindhu Kiranmai Ernala, Asra Rizvi, Munmun Choudhury, and John Kane. 2017. A collaborative approach to identifying social media markers of schizophrenia by employing machine learning and clinical appraisals. *Journal of Medical Internet Research*, 19:e289.

Su Lin Blodgett, Solon Barocas, Hal Daumé III, and Hanna Wallach. 2020. Language (technology) is power: A critical survey of "bias" in NLP. In *Proceedings of the 58th Annual Meeting of the Association for Computational Linguistics*, pages 5454–5476, Online. Association for Computational Linguistics.

Louise Brådvik. 2018. Suicide risk and mental disorders. *International Journal of Environmental Research and Public Health*, 15:2028.

M.D. Choudhury and S. De. 2014. Mental health discourse on reddit: Self-disclosure, social support, and anonymity. *Proceedings of the 8th International Conference on Weblogs and Social Media, ICWSM 2014*, pages 71–80.

Munmun Choudhury, Emre Kiciman, Mark Dredze, Glen Coppersmith, and Mrinal Kumar. 2016. Discovering shifts to suicidal ideation from mental health content in social media. In *Proceedings of the SIGCHI conference on human factors in computing systems . CHI Conference*, volume 2016, pages 2098–2110.

Glen Coppersmith, Mark Dredze, and Craig Harman. 2014. Quantifying mental health signals in Twitter. In *Proceedings of the Workshop on Computational Linguistics and Clinical Psychology: From Linguistic Signal to Clinical Reality*, pages 51–60, Baltimore, Maryland, USA. Association for Computational Linguistics.

Lorenzo Coviello, Yunkyu Sohn, Adam Kramer, Cameron Marlow, Massimo Franceschetti, Nicholas Christakis, and James Fowler. 2014. Detecting emotional contagion in massive social networks. *PloS one*, 9:e90315.

Munmun De Choudhury, Michael Gamon, Scott Counts, and Eric Horvitz. 2021. Predicting depression via social media. *Proceedings of the International AAAI Conference on Web and Social Media*, 7(1):128–137.

Jacob Devlin, Ming-Wei Chang, Kenton Lee, and Kristina Toutanova. 2019. BERT: Pre-training of deep bidirectional transformers for language understanding. In *Proceedings of the 2019 Conference of the North American Chapter of the Association for Computational Linguistics: Human Language Technologies, Volume 1 (Long and Short Papers)*, pages 4171–4186, Minneapolis, Minnesota. Association for Computational Linguistics.

Kara B. Fehling and Edward A. Selby. 2021. Suicide in dsm-5: Current evidence for the proposed suicide behavior disorder and other possible improvements. *Frontiers in Psychiatry*, 11.

Alison Fleming, E Klein, and C Corter. 1992. The effects of a social support group on depression, maternal attitudes and behavior in new mothers. *Journal of child psychology and psychiatry, and allied disciplines*, 33:685–98.

Hector-Hugo Franco-Penya and Liliana Mamani Sanchez. 2016. Text-based experiments for predicting mental health emergencies in online web forum posts. In *Proceedings of the Third Workshop on Computational Linguistics and Clinical Psychology*, pages 193–197, San Diego, CA, USA. Association for Computational Linguistics.

George Gkotsis, Anika Oellrich, Tim Hubbard, Richard Dobson, Maria Liakata, Sumithra Velupillai, and Rina Dutta. 2016a. The language of mental health problems in social media. In *Proceedings of the Third Workshop on Computational Linguistics and Clinical Psychology*, pages 63–73, San Diego, CA, USA. Association for Computational Linguistics.

George Gkotsis, Sumithra Velupillai, Anika Oellrich, Harry Dean, Maria Liakata, and Rina Dutta. 2016b. Don't let notes be misunderstood: A negation detection method for assessing risk of suicide in mental health records. In *Proceedings of the Third Workshop on Computational Linguistics and Clinical Psychology*, pages 95–105, San Diego, CA, USA. Association for Computational Linguistics.

E. Darío Gutiérrez, Guillermo Cecchi, Cheryl Corcoran, and Philip Corlett. 2017. Using automated metaphor identification to aid in detection and prediction of first-episode schizophrenia. In *Proceedings of the 2017 Conference on Empirical Methods in Natural Language Processing*, pages 2923–2930, Copenhagen, Denmark. Association for Computational Linguistics.

Bonnie Harmer, Sarah Lee, Truc vi H Duong, and Abdolreza Saadabadi. 2022. Suicidal ideation. *StatPearls*.

Keith Harrigian, Carlos Aguirre, and Mark Dredze. 2020. Do models of mental health based on social media data generalize? In *Findings of the Association for Computational Linguistics: EMNLP 2020*, pages 3774–3788, Online. Association for Computational Linguistics.

Keith Harrigian, Carlos Aguirre, and Mark Dredze. 2021. On the state of social media data for mental health research. In *Proceedings of the 7th Workshop on Computational Linguistics and Clinical Psychology: Improving Access*.

Christopher Homan, Ravdeep Johar, Tong Liu, Megan Lytle, Vincent Silenzio, and Cecilia Ovesdotter Alm. 2014. Toward macro-insights for suicide prevention: Analyzing fine-grained distress at scale. In *Proceedings of the Workshop on Computational Linguistics and Clinical Psychology: From Linguistic Signal to Clinical Reality*, pages 107–117, Baltimore, Maryland, USA. Association for Computational Linguistics.

Derek Howard, Marta Maslej, Justin Lee, Jacob Ritchie, Geoffrey Woollard, and Leon French. 2019. Transfer learning for risk classification of social media posts: Model evaluation study (preprint). *Journal of Medical Internet Research*, 22.

Jena D. Hwang and Kristy Hollingshead. 2016. Crazy mad nutters: The language of mental health. In *Proceedings of the Third Workshop on Computational Linguistics and Clinical Psychology*, pages 52–62, San Diego, CA, USA. Association for Computational Linguistics.

Thomas Insel. 2008. Assessing the economic costs of serious mental illness. *The American journal of psychiatry*, 165:663–5.

Alice Isen, Kimberly Daubman, and Gary Nowicki. 1987. Positive affect facilitates creative problem solving. *Journal of personality and social psychology*, 52:1122–31.

Nadine Jung, Christina Wranke, Kai Hamburger, and Markus Knauff. 2014. How emotions affect logical reasoning:evidence from experiments with mood-manipulated participants, spider phobics, and people with exam anxiety. *Frontiers in psychology*, 5:570.

Jonathan Kanter, Andrew Busch, Cristal Weeks, and Sara Landes. 2008. The nature of clinical depression: Symptoms, syndromes, and behavior analysis. *The Behavior analyst / MABA*, 31:1–21.

Sanne M.A. Lamers, Khiet P. Truong, Bas Steunenberg, Franciska de Jong, and Gerben J. Westerhof. 2014. Applying prosodic speech features in mental health care: An exploratory study in a life-review intervention for depression. In *Proceedings of the Workshop on Computational Linguistics and Clinical Psychology: From Linguistic Signal to Clinical Reality*, pages 61–68, Baltimore, Maryland, USA. Association for Computational Linguistics.

Huanhuan Li, Li Wang, Zhanbiao Shi, Yuching Zhang, wu Kankan, and Ping Liu. 2010. Diagnostic utility of the ptsd checklist in detecting ptsd in chinese earthquake victims. *Psychological reports*, 107:733–9.

David Losada and Fabio Crestani. 2016. A test collection for research on depression and language use. In *Experimental IR Meets Multilinguality, Multimodality, and Interaction*, volume 9822, pages 28–39.

Stephan Ludwig, ko de ruyter, Mike Friedman, Elisabeth Brüggen, Martin Wetzels, and Gerard Pfann. 2013. More than words: The influence of affective content and linguistic style matches in online reviews on conversion rates. *Journal of Marketing*, 77:87–103.

Shervin Malmasi, Marcos Zampieri, and Mark Dras. 2016. Predicting post severity in mental health forums. In *Proceedings of the Third Workshop on Computational Linguistics and Clinical Psychology*, pages 133–137, San Diego, CA, USA. Association for Computational Linguistics.

Tomas Mikolov, Kai Chen, G.s Corrado, and Jeffrey Dean. 2013. Efficient estimation of word representations in vector space. *Proceedings of Workshop at ICLR*, 2013.

Max Mills. 2017. Sharing privately: the effect publication on social media has on expectations of privacy. *Journal of Media Law*, 9:1–27.

Margaret Mitchell, Kristy Hollingshead, and Glen Coppersmith. 2015. Quantifying the language of schizophrenia in social media. In *Proceedings of the 2nd Workshop on Computational Linguistics and Clinical Psychology: From Linguistic Signal to Clinical Reality*, pages 11–20, Denver, Colorado. Association for Computational Linguistics.

Marja-Terttu MNSc, Marita PhD, Marja-Terttu Tarkka, and Marita Paunonen. 1996. Social support and its impact on mothers'experiences of childbirth. *Journal of Advanced Nursing*, 23:70 – 75.

Saif M. Mohammad and Peter D. Turney. 2013. Crowdsourcing a word-emotion association lexicon. *Computational Intelligence*, 29(3):436–465.

Aakanksha Naik, Abhilasha Ravichander, Norman Sadeh, Carolyn Rose, and Graham Neubig. 2018. Stress test evaluation for natural language inference. In *Proceedings of the 27th International Conference on Computational Linguistics*, pages 2340–2353, Santa Fe, New Mexico, USA. Association for Computational Linguistics.

Laura Orsolini, Roberto Latini, Maurizio Pompili, Gianluca Serafini, Umberto Volpe, Federica Vellante, Michele Fornaro, Alessandro Valchera, Carmine Tomasetti, Silvia Fraticelli, Marco Alessandrini, Raffaella Rovere, Sabatino Trotta, Giovanni Martinotti, Massimo di Giannantonio, and Domenico De Berardis. 2020. Understanding the complex of suicide in depression: from research to clinics. *Psychiatry investigation*, 17:207–221.

Javier Parapar, Patricia Martín-Rodilla, David Losada, and Fabio Crestani. 2021. *Overview of eRisk 2021: Early Risk Prediction on the Internet*, pages 324–344. Springer Link.

Gregory Park, H. Schwartz, Johannes Eichstaedt, Margaret Kern, Michal Kosinski, David Stillwell, Lyle Ungar, and Martin Seligman. 2014. Automatic personality assessment through social media language. *Journal of personality and social psychology*, 108.

Ted Pedersen. 2015. Screening Twitter users for depression and PTSD with lexical decision lists. In *Proceedings of the 2nd Workshop on Computational Linguistics and Clinical Psychology: From Linguistic Signal to Clinical Reality*, pages

46–53, Denver, Colorado. Association for Computational Linguistics.

Abed-Esfahani Pegah, Howard Derek, Maslej Marta, Sejal Patel, Vamika Mann, Sarah Goegan, and Leon French. 2019. Transfer learning for depression: Early detection and severity prediction from social media postings. *ERisk*, 2380.

L Pervin and D Cervone. 2010. *Personality. Theory and research*. Wiley.

Daniel Preoţiuc-Pietro, Johannes Eichstaedt, Gregory Park, Maarten Sap, Laura Smith, Victoria Tobolsky, H. Andrew Schwartz, and Lyle Ungar. 2015. The role of personality, age, and gender in tweeting about mental illness. In *Proceedings of the 2nd Workshop on Computational Linguistics and Clinical Psychology: From Linguistic Signal to Clinical Reality*, pages 21–30, Denver, Colorado. Association for Computational Linguistics.

Philip Resnik, Rebecca Resnik, and Margaret Mitchell, editors. 2014. *Proceedings of the Workshop on Computational Linguistics and Clinical Psychology: From Linguistic Signal to Clinical Reality*. Association for Computational Linguistics, Baltimore, Maryland, USA.

Daniel Riffe, Stephen Lacy, Brendan Watson, and Frederick Fico. 2019. *Analyzing Media Messages: Using Quantitative Content Analysis in Research*. Routledge.

(James) Rosenquist, J.H. Fowler, and Nicholas Christakis. 2010. Social network determinants of depression. *Molecular psychiatry*, 16:273–81.

Gery Ryan and H. Bernard. 2003. Techniques to identify themes. *Field Methods - FIELD METHOD*, 15:85–109.

Satvinder Saini and Mandeep. 2020. A study of perceived stress and loneliness in older people with depression. *International Journal on Aging Human Development*.

Mohd Salleh. 2008. Life event, stress and illness. *The Malaysian journal of medical sciences : MJMS*, 15:9–18.

H. Schwartz, Johannes Eichstaedt, Margaret Kern, Lukasz Dziurzynski, Stephanie Ramones, Megha Agrawal, Achal Shah, Michal Kosinski, David Stillwell, Martin Seligman, and Lyle Ungar. 2013. Personality, gender, and age in the language of social media: The open-vocabulary approach. *PloS one*, 8:e73791.

H. Andrew Schwartz, Johannes Eichstaedt, Margaret L. Kern, Gregory Park, Maarten Sap, David Stillwell, Michal Kosinski, and Lyle Ungar. 2014a. Towards assessing changes in degree of depression through Facebook. In *Proceedings of the Workshop on Computational Linguistics and Clinical Psychology: From Linguistic Signal to Clinical Reality*, pages 118–125, Baltimore, Maryland, USA. Association for Computational Linguistics.

H. Andrew Schwartz, Johannes Eichstaedt, Margaret L. Kern, Gregory Park, Maarten Sap, David Stillwell, Michal Kosinski, and Lyle Ungar. 2014b. Towards assessing changes in degree of depression through Facebook. In *Proceedings of the Workshop on Computational Linguistics and Clinical Psychology: From Linguistic Signal to Clinical Reality*, pages 118–125, Baltimore, Maryland, USA. Association for Computational Linguistics.

Kathryn Scott, Phyllis Klaus, and Marshall Klaus. 2000. The obstetrical and postpartum benefits of continuous support during childbirth. *Journal of women's health & gender-based medicine*, 8:1257–64.

Judy Hanwen Shen and Frank Rudzicz. 2017. Detecting anxiety through Reddit. In *Proceedings of the Fourth Workshop on Computational Linguistics and Clinical Psychology — From Linguistic Signal to Clinical Reality*, pages 58–65, Vancouver, BC. Association for Computational Linguistics.

Liping Shen, Minjuan Wang, and Ruimin Shen. 2009. Affective e-learning: Using "emotional" data to improve learning in pervasive learning environment. *Educational Technology & Society*, 12:176–189.

Benjamin Shickel, Martin Heesacker, Sherry Benton, Ashkan Ebadi, Paul Nickerson, and Parisa Rashidi. 2016. Self-reflective sentiment analysis. In *Proceedings of the Third Workshop on Computational Linguistics and Clinical Psychology*, pages 23–32, San Diego, CA, USA. Association for Computational Linguistics.

Han-Chin Shing, Suraj Nair, Ayah Zirikly, Meir Friedenberg, Hal Daumé III, and Philip Resnik. 2018. Expert, crowdsourced, and machine assessment of suicide risk via online postings. In *Proceedings of the Fifth Workshop on Computational Linguistics and Clinical Psychology: From Keyboard to Clinic*, pages 25–36.

Michael Tanana, Aaron Dembe, Christina S. Soma, Zac Imel, David Atkins, and Vivek Srikumar. 2016. Is sentiment in movies the same as sentiment in psychotherapy? comparisons using a new psychotherapy sentiment database. In *Proceedings of the Third Workshop on Computational Linguistics and Clinical Psychology*, pages 33–41, San Diego, CA, USA. Association for Computational Linguistics.

Mashrura Tasnim and Eleni Stroulia. 2019. *Detecting Depression from Voice*, pages 472–478. Springer Link.

Yla Tausczik and James Pennebaker. 2010. The psychological meaning of words: Liwc and computerized text analysis methods. *Journal of Language and Social Psychology*, 29:24–54.

Andranik Tumasjan, Timm Sprenger, Philipp Sandner, and Isabell Welpe. 2010. Predicting elections with twitter: What 140 characters reveal about political sentiment. In *Social Science Computer Review*, volume 10. Sage Journals.

Elsbeth Turcan and Kathy McKeown. 2019. Dreaddit: A Reddit dataset for stress analysis in social media. In *Proceedings of the Tenth International Workshop on Health Text Mining and Information Analysis (LOUHI 2019)*, pages 97–107, Hong Kong. Association for Computational Linguistics.

Mina Valizadeh, Pardis Ranjbar-Noiey, Cornelia Caragea, and Natalie Parde. 2021. Identifying medical self-disclosure in online communities. In *Proceedings of the 2021 Conference of the North American Chapter of the Association for Computational Linguistics: Human Language Technologies*, pages 4398–4408, Online. Association for Computational Linguistics.

Patrik Vuilleumier. 2006. Vuilleumier p. how brains beware: neural mechanisms of emotional attention. trends cogn sci 9: 585-594. *Trends in cognitive sciences*, 9:585–94.

U. Yasaswini, Y. Sasidhar, P. Sai, P. Eswar, and V. Swathi. 2021. Detecting depression in tweets using distilbert. *International Journal of Innovative Research in Computer Science & Technology*, 9.

Ayah Zirikly, Varun Kumar, and Philip Resnik. 2016. The GW/UMD CLPsych 2016 shared task system. In *Proceedings of the Third Workshop on Computational Linguistics and Clinical Psychology*, pages 166–170, San Diego, CA, USA. Association for Computational Linguistics.

Ayah Zirikly, Philip Resnik, Özlem Uzuner, and Kristy Hollingshead. 2019. CLPsych 2019 shared task: Predicting the degree of suicide risk in Reddit posts. In *Proceedings of the Sixth Workshop on Computational Linguistics and Clinical Psychology*.

A Dataset Descriptions

In Table 4, we present dataset statistics for *depression*, *stress*, and *suicide*. Further details regarding these datasets can be found in the original papers (Losada and Crestani, 2016; Turcan and McKeown, 2019; Shing et al., 2018). We deeply thank all the authors and creators of these datasets.

B Analytical Figures

In this section we include additional figures produced during data analysis. Figures 4 and 5 show cognitive complexity for individuals with depression and suicidal ideation, and Figure 3 shows graphical representations of LDA analyses on people with stress.

C Extended Qualitative Analysis of N-Gram Frequency

In this section we include figures showing the most frequent n-grams associated with *depression*, *suicide*, and *stress*. Trigrams for *depression*, *suicidal ideation*, and *stress* are shown in Figures 6, 7, and 8, respectively. Bigrams for *depression*, *suicidal ideation*, and *stress* are shown in Figures 9, 10, and 11, and unigrams for the same three MHCs are shown in Figures 12, 13, and 14.

Dataset	Size	Labeling Scheme	Labels Used in Our Experiments	Baseline F_1
Depression	531,453 posts from 892 users	Binary	Binary	0.66
Stress	187,444 posts	Binary	Binary	0.79
Suicide	11,129 initial users, downsampled to 934	Categorical (4 Categories)	Aggregated Binary	0.42

Table 4: Additional descriptive statistics regarding *depression* (Losada and Crestani, 2016), *stress* (Turcan and McKeown, 2019), and *suicide* (Shing et al., 2018).

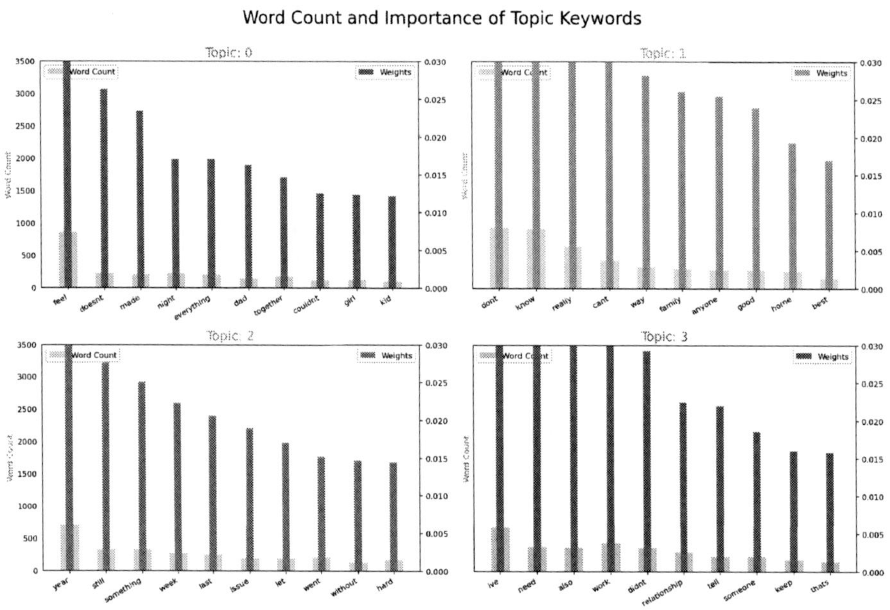

Figure 3: Top four topics identified when applying LDA to *stress*.

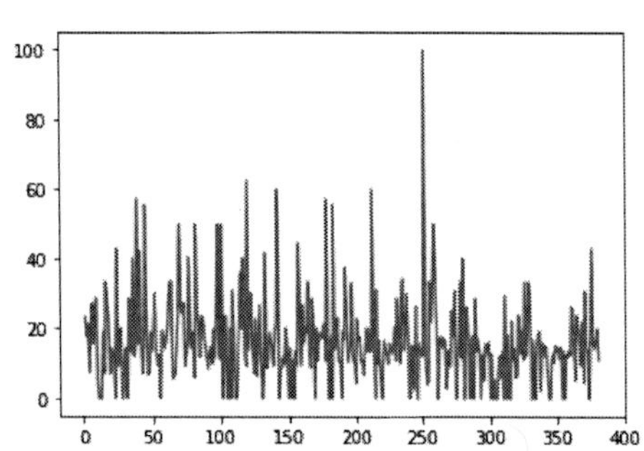

Figure 4: Cognitive complexity of a random subsample with depression.

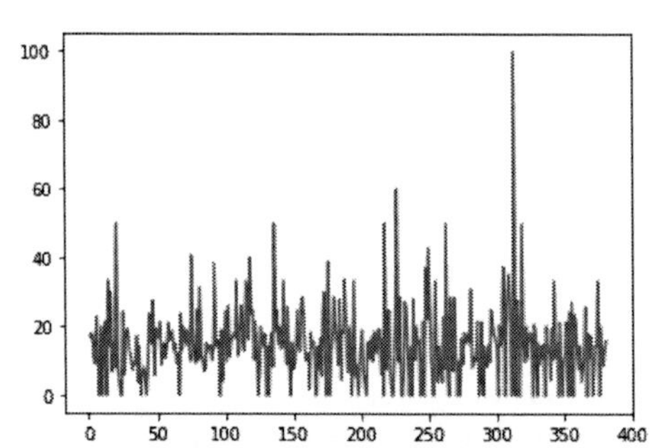

Figure 5: Cognitive complexity of a random subsample with suicidal ideation.

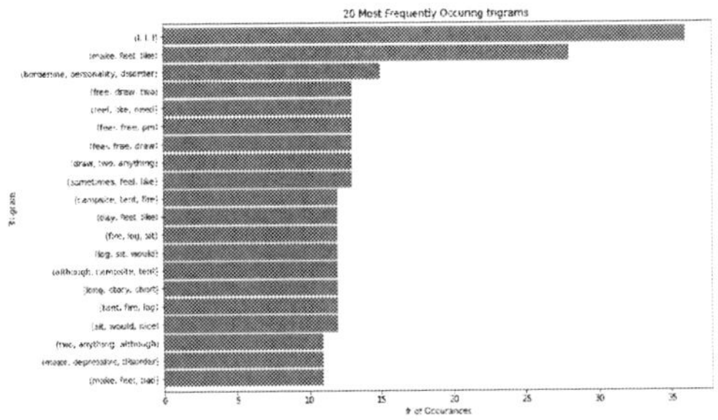

Figure 6: Most frequent trigrams in a random subsample (*depression*).

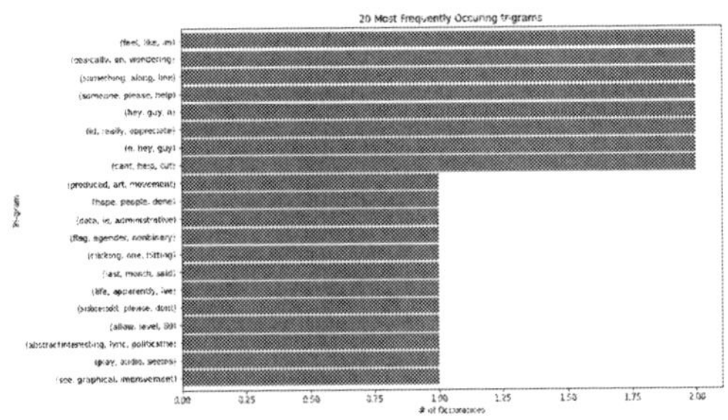

Figure 7: Most frequent trigrams in a random subsample (*suicide*).

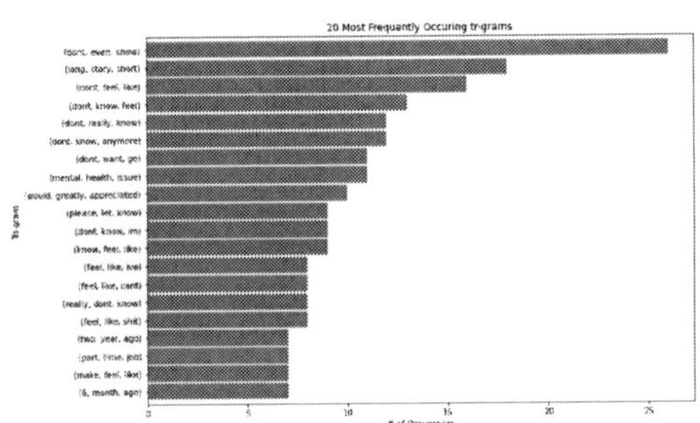

Figure 8: Most frequent trigrams in a random subsample (*stress*).

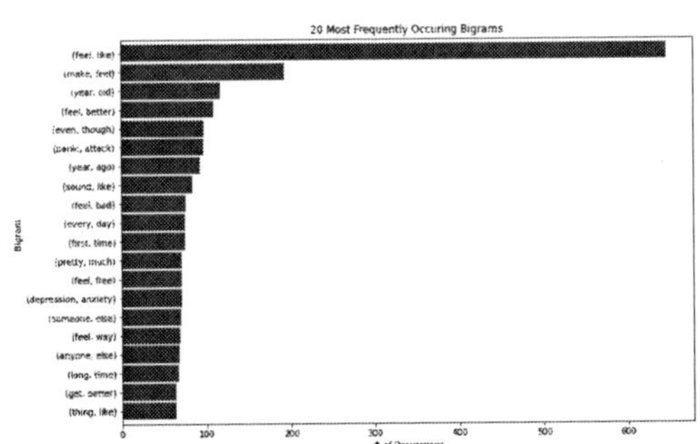

Figure 9: Most frequent bigrams in a random subsample (*depression*).

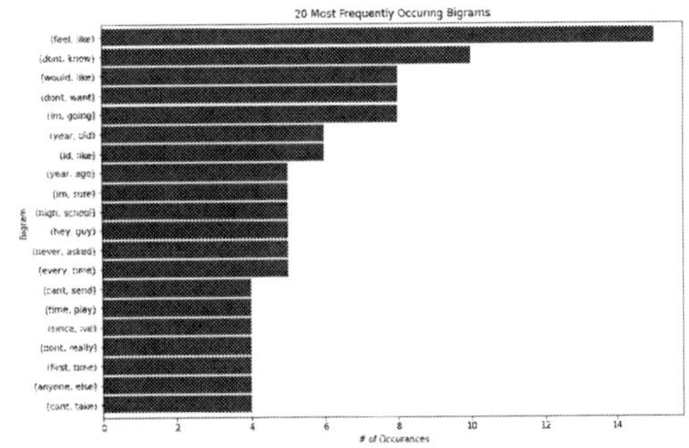

Figure 10: Most frequent bigrams in a random subsample (*suicide*).

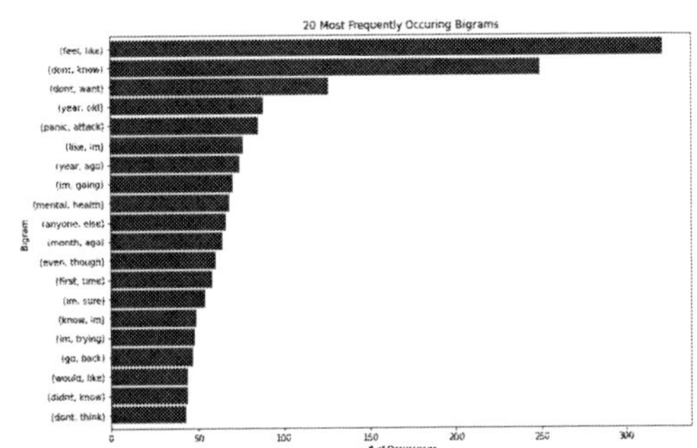

Figure 11: Most frequent bigrams in a random subsample (*stress*).

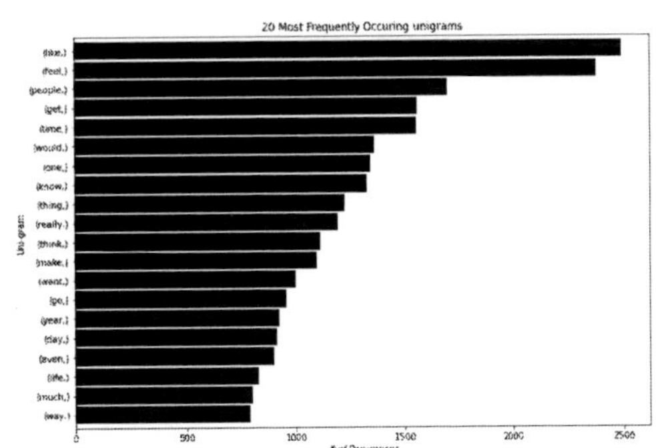

Figure 12: Most frequent unigrams in a random subsample (*depression*).

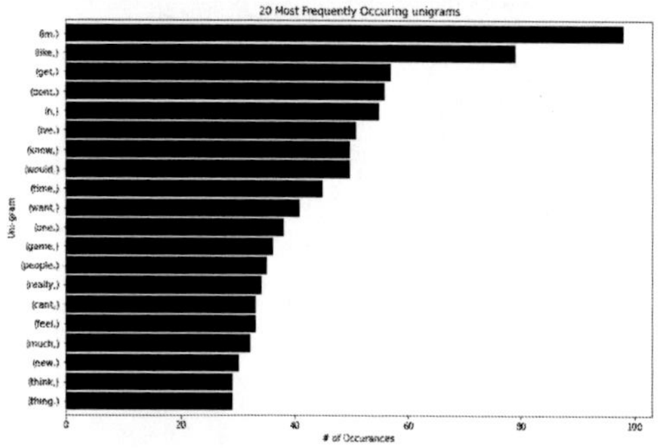

Figure 13: Most frequent unigrams in a random subsample (*suicide*).

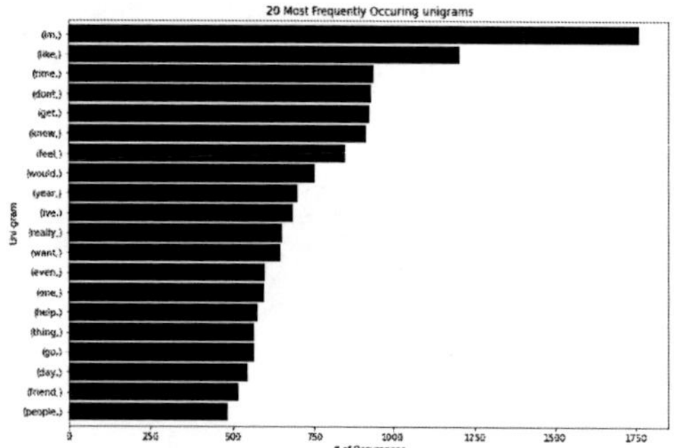

Figure 14: Most frequent unigrams in a random subsample (*stress*).

Comparing emotion feature extraction approaches for predicting depression and anxiety

Hannah A. Burkhardt
University of Washington
haalbu@uw.edu

Michael D. Pullmann
University of Washington
pullmann@uw.edu

Thomas D. Hull
Talkspace
derrick@talkspace.com

Patricia A. Areán
University of Washington
parean@uw.edu

Trevor Cohen
University of Washington
cohenta@uw.edu

Abstract

The increasing adoption of message-based behavioral therapy enables new approaches to assessing mental health using linguistic analysis of patient-generated text. Word counting approaches have demonstrated utility for linguistic feature extraction, but deep learning methods hold additional promise given recent advances in this area. We evaluated the utility of emotion features extracted using a BERT-based model in comparison to emotions extracted using word counts as predictors of symptom severity in a large set of messages from text-based therapy sessions involving over 6,500 unique patients, accompanied by data from repeatedly administered symptom scale measurements. BERT-based emotion features explained more variance in regression models of symptom severity, and improved predictive modeling of scale-derived diagnostic categories. However, LIWC categories that are not directly related to emotions provided valuable and complementary information for modeling of symptom severity, indicating a role for both approaches in inferring the mental states underlying patient-generated language.

1 Introduction

Almost 10% of adults in the United States receive mental health counseling (Zablotsky and Terlizzi, 2020). The principle of measurement-based care dictates that medical treatments should be initiated and evaluated over time based on repeated assessments of patient symptoms and symptom trajectory (Scott and Lewis, 2015). In the context of talk therapy, mental health practitioners estimate treatment progress based on patients' current and historical verbal communications. For evaluating depression and anxiety severity, expressions of emotional state are key aspects of such communications (Beck, 1967; Rottenberg, 2017; Amstadter, 2008).

While prior work predominantly focused on sentiment, i.e. positive/negative polarity, expression of fine-grained emotions (Chancellor and De Choudhury, 2020; Guntuku et al., 2017) may give further insights into depression and anxiety symptomatology. For example, pride may be impacted by depression in a unique way. Gruber et al. (2011) showed that pride, a positive emotion relating to the self, is inversely correlated with depression, which is often associated with a poor self-image. At the same time, they found a smaller effect on joy and amusement, concluding that grouping these emotions into "positive affect" may result in a loss of nuance.

The increasing adoption of digital mental health tools and services, particularly message-based therapy, has afforded new opportunities to assist practitioners in quantifying depression and anxiety severity by assessing emotion in patient-generated text. Linguistic Inquiry and Word Count (LIWC) (Pennebaker et al., 2007; Tausczik and Pennebaker, 2010) is a software package designed to count words belonging to pre-defined categories with an extensive track record of validation for the detection of linguistic indicators of mental state (Tausczik and Pennebaker, 2010). It is commonly used to measure positive and negative affect, a limited set of specific emotions (sadness, anxiety, and anger), and other linguistic dimensions related to style and topic. Several LIWC categories have established relationships with depression, including the affect category sadness (e.g. "sad", "cry", "suffer"), the topic category health (e.g. "alcohol", "rash", "self-care"), and the syntactic category first-person pronouns (e.g. "I", "me", "my"). LIWC has been used to measure depression levels in social media posts (Coppersmith et al., 2014; De Choudhury et al., 2014, 2013a,b), therapy conversations (Burkhardt et al., 2021; Sonnenschein et al., 2018), and other written texts (Rude et al., 2004; Wiltsey Stirman and Pennebaker, 2001). LIWC measurements have also been shown to distinguish between patients with depression and those with anxiety

Proceedings of the Eighth Workshop on Computational Linguistics and Clinical Psychology, pages 105 - 115
July 15, 2022

disorders (Sonnenschein et al., 2018), correlate with self-reported measures of anxiety and worry in written descriptions of emotional responses to COVID-19 (Kleinberg et al., 2020), and predict whether posts emanated from anxiety-related subreddits (Shen and Rudzicz, 2017).

However, word counting methods cannot address linguistic phenomena such as negation ("not bad"), sarcasm, and context-dependence (for example, in the case of polysemy, words have multiple meanings that can only be disambiguated in context), and manually defined dictionaries may omit synonyms for terms they encode. Prior work suggests that neural network (NN)-based natural language processing (NLP) techniques can account for such phenomena and may therefore improve upon this straightforward word-counting method in their ability to identify concepts related to symptom severity. Shen and Rudzicz found that the performance of machine learning models identifying whether or not Reddit posts were drawn from anxiety-related subreddits improved when these models included neural word embeddings rather than LIWC-derived features (2017). However, the distributed representations of posts used in this work do not relate directly to interpretable emotion features. Further, contemporary transformer-based NN language models offer advantages over neural word embeddings in their ability to leverage proximal cues (such as "not") when interpreting the contextual meaning of a word. As noted by the authors, this work suggests a need for further research on automated assessments of linguistic indicators of anxiety disorders, involving larger data sets and explicit diagnostic assessments.

Therefore, using a large set of messages from text-based therapy session, we investigated if emotions extracted using a Bidirectional Encoder Representations from Transformers (BERT) (Devlin et al., 2019) based model trained on GoEmotions, a large dataset of Reddit posts annotated with 27 fine-grained emotions (Demszky et al., 2020), are stronger predictors of depression and anxiety status than counts of emotion-related word categories (LIWC). To this end, we first determined the association of each feature with the outcomes of interest in univariate regression analyses. Further, in order to provide clinical decision support to mental health practitioners, it is paramount to be able to classify previously unseen messages as indicating depression and/or anxiety. We therefore proceeded

	slope	R2
GoEmotions		
sadness	18.84 (16.50 - 21.18)**	0.782
admiration	-16.62 (-18.80 - -14.44)**	0.781
annoyance	12.61 (9.67 - 15.55)**	0.778
disappointment	19.01 (16.87 - 21.14)**	0.778
joy	-16.40 (-19.33 - -13.48)**	0.778
pride	-64.35 (-78.54 - -50.16)**	0.777
excitement	-28.34 (-33.18 - -23.49)**	0.777
disapproval	16.11 (12.99 - 19.23)**	0.776
approval	-7.81 (-9.27 - -6.36)**	0.776
confusion	9.65 (7.30 - 11.99)**	0.775
relief	-24.19 (-30.79 - -17.59)**	0.774
neutral	-0.83 (-1.60 - -0.06)*	0.774
anger	18.67 (14.38 - 22.97)**	0.774
disgust	29.79 (21.72 - 37.86)**	0.774
optimism	-6.15 (-8.28 - -4.03)**	0.773
realization	-1.08 (-2.74 - 0.59)	0.773
amusement	-10.96 (-14.85 - -7.07)**	0.772
fear	10.75 (7.44 - 14.06)**	0.771
nervousness	3.44 (0.84 - 6.05)*	0.771
caring	-2.77 (-6.03 - 0.49)	0.771
gratitude	-2.87 (-9.79 - 4.05)	0.771
embarrassment	11.85 (4.25 - 19.45)*	0.771
curiosity	0.03 (-2.56 - 2.62)	0.771
desire	2.08 (-1.10 - 5.26)	0.771
love	-1.96 (-5.22 - 1.31)	0.771
surprise	-4.00 (-10.18 - 2.18)	0.771
grief	134.76 (104.49 - 165.03)**	0.770
GoEmotions Ekman		
joy	-9.31 (-10.21 - -8.41)**	0.788
anger	18.53 (16.46 - 20.61)**	0.783
sadness	15.81 (14.06 - 17.56)**	0.779
disgust	48.43 (37.93 - 58.93)**	0.778
neutral	-0.11 (-1.17 - 0.96)	0.775
surprise	4.11 (2.47 - 5.75)**	0.774
fear	4.52 (2.25 - 6.80)**	0.772
LIWC		
sad	1.21 (1.02 - 1.40)**	0.781
i	0.25 (0.21 - 0.29)**	0.777
anger	0.84 (0.65 - 1.02)**	0.776
health	0.66 (0.52 - 0.80)**	0.775
anx	0.19 (0.05 - 0.34)*	0.774
we	-0.53 (-0.65 - -0.41)**	0.774
bio	0.41 (0.33 - 0.50)**	0.774

Table 1: PHQ-9 score univariate mixed-effects linear regression models coefficients and variance explained. * $p<0.05$. ** $p<0.001$

to train and evaluate a machine learning classifier using emotion features in conjunction with established depression-related LIWC features to predict depression and anxiety status in a held-out test set.

2 Methods

2.1 Data

We utilized a corpus of messaging therapy sessions from over 6,500 unique patients previously collected via the Talkspace platform (Hull et al., 2020). Talkspace offers a paid service utilizing licensed and credentialed therapists to conduct asynchronous, message-based therapy conversations. All patients and clinicians give written consent to the use of their data in a de-identified, aggregate format as part of the user agreement before they begin using the platform. Over the course of 12 weeks, patients engaged in two-way messaging therapy and completed depression questionnaires (9-item Patient Health Questionnaire, PHQ-9 (Kroenke et al., 2001)) as well as anxiety questionnaires (7-item General Anxiety Disorder questionnaire), every 3 weeks. For each available score, patient messages from the period in question ("(o)ver the last two (2) weeks") were concatenated into a single unit of analysis ("document"), resulting in up to 4 labeled data points per patient (weeks 3, 6, 9, and 12). All messages without a corresponding score were excluded from analysis. Data from baseline assessments were removed, as preliminary analysis suggested that messages before the week 0 mark introduced spurious associations due to differences between typical therapy dialog and the patient-therapist matching process, combined with generally worse symptom severity scores at the beginning of the study period. Participants were young (79% were 35 years old or younger), educated (75% had a Bachelor's degree or higher), and predominantly female (79%). Race and ethnicity were not systematically collected. There were over 13,000 text documents with both PHQ-9 and GAD-7 scores, totaling over 24 million words from over 337,000 messages. The original study was approved as exempt by the local institutional review board. The current study concerned secondary analysis of previously collected de-identified data, which is not considered human subjects research; nonetheless, data were stored on a secure server with study team member access only. All textual data were thoroughly de-identified by an automated algorithm before leaving their

	slope	R2
GoEmotions		
sadness	15.04 (12.96 - 17.12)**	0.728
admiration	-15.02 (-16.97 - -13.07)**	0.727
neutral	-1.00 (-1.72 - -0.29)*	0.725
joy	-16.99 (-19.53 - -14.44)**	0.724
approval	-6.80 (-8.14 - -5.47)**	0.724
fear	18.62 (15.32 - 21.93)**	0.724
annoyance	12.83 (10.20 - 15.46)**	0.724
excitement	-22.74 (-26.98 - -18.49)**	0.723
pride	-56.42 (-69.75 - -43.09)**	0.723
disappointment	14.05 (12.12 - 15.97)**	0.723
disapproval	12.97 (10.18 - 15.76)**	0.723
nervousness	11.91 (9.36 - 14.46)**	0.723
confusion	8.41 (6.33 - 10.48)**	0.721
anger	19.29 (15.38 - 23.19)**	0.721
relief	-22.16 (-28.05 - -16.28)**	0.720
optimism	-6.86 (-8.84 - -4.89)**	0.719
realization	-1.48 (-2.99 - 0.02)	0.718
amusement	-10.34 (-13.73 - -6.96)**	0.717
curiosity	-0.00 (-2.44 - 2.43)	0.717
caring	-1.94 (-5.11 - 1.24)	0.716
gratitude	-3.54 (-7.67 - 0.59)	0.716
desire	1.25 (-1.55 - 4.06)	0.716
love	-3.66 (-7.08 - -0.23)*	0.716
surprise	-6.42 (-11.87 - -0.97)*	0.716
embarrassment	10.33 (3.08 - 17.58)*	0.716
grief	118.01 (90.79 - 145.22)**	0.716
disgust	25.72 (18.68 - 32.76)**	0.715
GoEmotions Ekman		
joy	-8.62 (-9.42 - -7.82)**	0.736
anger	15.92 (14.08 - 17.76)**	0.727
disgust	44.78 (35.38 - 54.18)**	0.726
sadness	12.38 (10.83 - 13.93)**	0.725
neutral	-0.21 (-1.18 - 0.76)	0.722
fear	12.15 (9.97 - 14.33)**	0.722
surprise	3.27 (1.79 - 4.75)**	0.720
LIWC		
anx	0.73 (0.59 - 0.86)**	0.729
i	0.17 (0.13 - 0.21)**	0.726
sad	0.89 (0.72 - 1.05)**	0.726
anger	0.93 (0.76 - 1.10)**	0.724
we	-0.36 (-0.47 - -0.25)**	0.723
health	0.46 (0.33 - 0.59)**	0.717
bio	0.28 (0.21 - 0.36)**	0.716

Table 2: GAD-7 score univariate mixed-effects linear regression models coefficients and variance explained. * $p<0.05$. ** $p<0.001$

source, with all names, places, contact information, social media identifiers, and mentions of specific events removed.

LIWC 2015 was used to obtain the following word-count-based features: first-person singular pronouns ("I"), first-person plural pronouns ("we"), bio, health, sadness, anxiety, anger, positive emotion, and negative emotion. These features were selected on account of their track record of correlation with indicators of depression and anxiety in previous work (Tausczik and Pennebaker, 2010).

A BERT-based GoEmotions classifier pipeline using fine-tuned models available from the Hugging Face transformer library[1] was used to extract emotion features from each document. This model has been shown to approximate published results for performance in extracting emotions from the GoEmotions dataset (macro-average F1 score of $\approx$0.5 to $\approx$0.7, depending on the granularity of the emotions concerned). For further details of the training corpus and procedures used, we refer the reader to Demszky et al. (2020). After splitting documents into sentences and extracting emotions from the first 512 tokens of each sentence, scores were averaged over all sentences in a document to yield one set of emotion scores for the two-week period concerned. Only 38 of $\approx$13,000 documents contained sentences that were truncated due to being over 512 tokens long. The pipeline provides several output settings, resulting in different sets of emotions being extracted. Two sets of emotions were extracted. First, we extracted the set of 6 basic emotions proposed by Ekman (1992), consisting of sadness, joy, surprise, disgust, anger, fear, and a neutral category, which was assigned by annotators when they felt that no particular emotion was expressed. Second, we extracted the full set of 28 categories that were used to annotate the GoEmotions corpus, consisting of 27 fine-grained emotions described by Cowen and Keltner (2017), plus a neutral category. Finally, we calculated positive and negative emotion features by averaging the scores belonging to positive and negative emotions. The negative GoEmotions Ekman emotions are anger, disgust, fear, and sadness; joy is the only positive Ekman emotion. Negative fine-grained GoEmotions (Cowen) emotions encompass anger, annoyance, disappointment, disapproval, disgust, embarrassment, fear, grief, nervousness, remorse, and sadness. Admiration, amusement, approval, caring, desire, excitement, gratitude, joy, love, optimism, pride, and relief are the positive emotions in the fine-grained GoEmotions set. The interested reader is referred to Demszky et al. (2020) for further details on these groupings.

2.2 Comparison of variables

A common approach to identifying associations of individual variables with an outcome of interest is to determine the statistical significance of the association between each candidate variable and the outcome by fitting univariate regressions. Linear regression models, however, require observations to be independent of each other. Because patients contribute between 1 and 4 observations in our dataset, this independence assumption is not met: two observations from the same patient may be expected to be more like each other than two observations from different patients. Mixed-effect linear regressions can be used to account for this. In such models, the within-patient and between-patient effects of the predictor variables on the outcome are separately accounted for. In other words, in addition to the "fixed effect" of the predictor variables on the outcome (the effect of interest), we model a "random effect" that is different for each patient, which is arbitrary but consistent across all observations for a given patient. In essence, the outcome is the linear combination of an emotion's global relationship to PHQ-9/GAD-7 scores and the patient-specific relationship of the emotion on scores (plus an intercept term for each effect as well as a residual error term). The univariate mixed-effect linear regression models for each emotion variable model the patient identity as a random effect and are of the following form:

$$Y_{ij} = \beta_0 + \beta_1 X_{ij} + \gamma_{0i} + \gamma_{1i} X_{ij} + \epsilon_{ij}$$

Where Y_{ij} is the ith outcome (PHQ-9 score, GAD-7 score) for patient i, X_{ij} is the level of emotion in the jth document written by patient i, β_0 and β_1 are the fixed effect parameters (emotion), and γ_{0i} and γ_{1i} are the random effect parameters (patient ID), and ϵ_{ij} is the residual error for patient i's jth document. Models were fitted via Maximum Likelihood Estimation using the Statsmodels package for Python (Seabold and Perktold, 2010). Statsmodels calculates p-values using t-tests. We report the explanatory power of each feature as the amount of variance explained (R2).

Following a similar process, we fitted bivariate

[1] https://github.com/monologg/GoEmotions-pytorch

mixed-effects models using the positive and negative emotion variables from each feature source.

2.3 Prediction

Next, using the Scikit-Learn package for Python (Pedregosa et al., 2011), we trained random forest classifiers to predict binary depression (MDD) and anxiety (GAD) status from 49 features: 7 Ekman emotion categories from GoEmotion, as well as the positive and negative emotion variables calculated from Ekman emotions; 27 fine-grained emotions plus neutral, as well as the positive, and negative emotion variables calculated from the 27 fine-grained emotions; 5 LIWC emotion variables (positive emotion, negative emotion, anxiety, anger, sadness); and 4 LIWC variables with an established relationship to depression (I, we, biology, health) (Rude et al., 2004; De Choudhury et al., 2013b; Eichstaedt et al., 2018; Sonnenschein et al., 2018; Burkhardt et al., 2021). We first trained random forest classifiers using each individual feature set. Then, we trained models using combinations of these feature sets to evaluate their relative contribution (LIWC non-emotion variables combined with each set of emotion variables from the three sources). Then, we trained another random forest classifier on all available features. For this model, relative feature importance was calculated using SHAP (Lundberg and Lee, 2017).

To avoid information leakage due to within-patient effects (Saeb et al., 2017), data were split into training and test sets such that all observations from an individual patient were kept within the same fold. Patients were assigned to the training (80%) and test (20%) populations, resulting in a training set of 4,913 patients (with 10,006 observations) and a test set of 1,638 patients (with 3,321 observations). Average PHQ-9 across all observations did not significantly differ between training and test observations.

Hyperparameters (number of estimators, maximum number of features, maximum tree depth, minimum number of samples for splitting, minimum number of samples per leaf, using or not using bootstrap) were automatically selected (based only on the training data) via 3-fold cross-validation, a process where, for each hyperparameter combination, each of the three folds is held out in turn, while a model is trained on the remaining 2 folds; this way, 3 scores are produced per hyperparameter combination, and their average represents the score for that hyperparameter set. Finally, the hyperparameters that produced the best score are selected, and a final model with those hyperparameters is trained on all training data, then tested on the held-out test set.

A binary prediction target was used to align predictions with the clinical task of classifying a diagnosis as present or absent. A cut-off between 8 and 11 was previously found to have a clinically acceptable tradeoff between sensitivity and specificity when dichotomizing PHQ-9 scores for diagnosis of major depressive disorder (MDD) (Manea et al., 2012). Therefore, we considered a PHQ-9 score of 10 or more (depression severity of moderate, moderately severe, or severe) as indicating MDD for the purposes of this work. A PHQ-9 score of 9 or less (depression severity of mild or none) was considered non-depressed. As the GAD-7 has been found to have acceptable properties for identification of generalized anxiety disorder (GAD) at a cutoff of 7-10 (Plummer et al., 2016; Spitzer et al., 2006), a GAD-7 score of 10 or more was considered an indicator of GAD, and a score of 9 or less was considered an indicator of a negative diagnosis for this condition.

3 Results

3.1 Comparison of variables

The variance in PHQ-9 and GAD-7 scores, respectively, explained by each individual emotion variable and by variable pairs is shown in Figure 1, Table 1, and Table 2. Emotion variables that were obtainable from all three feature sources were anger and sadness as well as the summary dimensions of positive and negative emotion. With BERT-based models, these are composites of individual predictions returned by the model, while LIWC returns a summary value as an individual feature. The variance in PHQ-9 scores explained by these directly comparable variables is shown in Figure 1, along with the variance explained by the combination of positive and negative emotion features. The three feature extraction approaches resulted in features that explained similar portions of the variance; LIWC explained slightly more, except for anger and sadness, where the GoEmotions Ekman and GoEmotions Cowan variables explained more, respectively. The GoEmotions Cowan variable for sadness was more explanatory than the GoEmotions Ekman variable, but the Ekman anger variable outperformed the fine-grained anger vari-

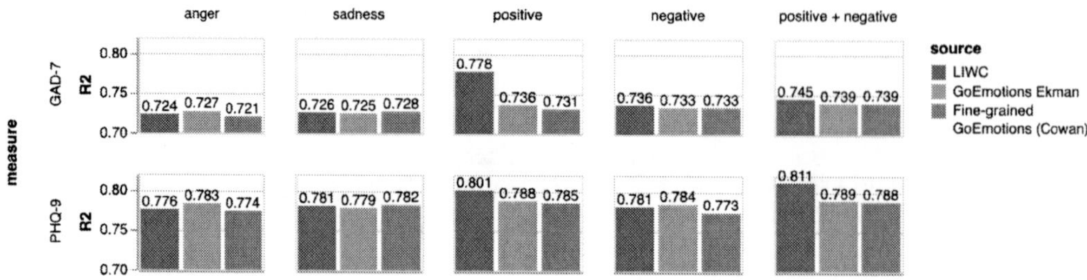

Figure 1: PHQ-9 and GAD-7 score variance explained by comparable features from LIWC, GoEmotions (Ekman set), and GoEmotions (fine-grained set)

able. The combination of positive and negative emotion explained more variance than either positive or negative emotion alone, except when LIWC positive emotion was assessed for GAD-7. Notably, LIWC's positive emotion variable appears to be more explanatory than anger, sadness, and negative emotion for both PHQ-9 and GAD-7, and even the combination of positive and negative emotion for GAD-7.

All individual emotions, as quantified by each of the three feature extraction approaches, are shown in Table 1. Expressions of realization, caring, gratitude, curiosity and desire were not significantly associated with either anxiety or depression. Love and surprise were not predictive of depression, but were associated with anxiety. Both were significantly associated with sadness, fear, and nervousness; however, sadness was more strongly related to depression, and fear and nervousness were more strongly related to anxiety. Joy was roughly equally associated with anxiety and depression, across both GoEmotions feature sets. The emotions with the largest differences between more and less depressed individuals were grief, pride, excitement, relief, and disgust. The emotions with the largest differences with respect to anxiety were grief, disapproval, approval, relief, and disgust.

3.2 Prediction

In contrast to the multivariate models, results from predictive modeling experiments show a clear advantage for deep learning models, with the best overall performance by ROC and F1 score achieved using GoEmotions Cowen features for both MDD and GAD. As shown in Table 3, the models including only the non-emotion LIWC features achieved an area under the receiver-operator characteristic curve (AUROC) of 0.577 for MDD and 0.549 for GAD. When using emotion features only, the fine-grained GoEmotions set performed best. For both MDD and GAD, adding LIWC emotion features to LIWC non-emotion features improved predictive performance less than adding GoEmotions Ekman features, which improved the model less than adding the fine-grained GoEmotions set. Using all emotion features concurrently ("all three") slightly improved performance for both GAD and MDD (by F1 score but not ROC in the latter case).

The relative importance of all features for the MDD and GAD models is shown in Table 4. Fear was ranked higher for predicting GAD than for predicting MDD. Sadness was ranked higher for predicting MDD than for predicting GAD.

4 Discussion

In this work, we showed that neural network models such as the BERT-based GoEmotions classifier can outperform LIWC, a straightforward, broadly adopted word-counting method for extracting emotion features from natural language. We further confirmed that some emotions not traditionally associated with depression and anxiety can be predictive of these diagnoses; specifically, pride. Finally, we showed that using LIWC features together with emotion features derived using GoEmotions predict depression/anxiety status with reasonable accuracy. This finding is important, in that further development of such tools could lead to better detection of emotional change during treatment in a way that could be derived naturally in the client/clinician encounter. NLP applied to such naturalistic data has been used for measuring clinician skills in delivering psychotherapy with some success (Flemotomos et al., 2021); here, rather than using such tools for quality measurement, linguistic analysis of affect could be used to detect depression/anxiety severity and client response to treatment.

Both LIWC variables and GoEmotions variables

	ROC	F1	Pr	Rc
MDD				
LIWC non-emo	0.577	0.413	0.525	0.341
LIWC emo	0.621	0.471	0.561	0.405
GoEmo Ekman	0.643	0.493	0.583	0.427
GoEmo	0.662	0.522	0.613	0.455
LIWC non-emo +				
LIWC emo	0.640	0.484	0.569	0.420
GoEmo Ekman	0.655	0.498	0.585	0.434
GoEmo	0.671	0.514	0.615	0.441
All three	0.671	0.520	0.612	0.453
GAD				
LIWC non-emo	0.549	0.290	0.478	0.209
LIWC emo	0.613	0.405	0.541	0.324
GoEmo Ekman	0.643	0.443	0.550	0.371
GoEmo	0.652	0.444	0.565	0.366
LIWC non-emo +				
LIWC emo	0.617	0.401	0.529	0.324
GoEmo Ekman	0.637	0.441	0.548	0.369
GoEmo	0.654	0.451	0.568	0.374
All three	0.657	0.456	0.567	0.382

Table 3: AUROCs, F1 score (positive class), precision, and recall of random forest model trained with just the non-emotion LIWC features, and trained with the non-emotion LIWC features plus LIWC emotion, GoEmotion Ekman and the full GoEmotion feature sets, for predicting MDD (PHQ-9 score $\geq$ 10) and GAD (GAD-7 score $\geq$ 10).

explained a large portion of the variance in univariate mixed-effect regressions: R2 values ranged from 0.770 to 0.788 when modeling PHQ-9 scores as outcome, and from 0.715 to 0.736 when modeling GAD-7 scores as outcome. Therefore, LIWC and GoEmotions features both capture valuable information. GoEmotions features marginally outperformed 2 out of 4 of the equivalent LIWC features for predicting GAD-7 and 3 ouf of 4 features for predicting PHQ-9. For predicting binary depression (MDD) and anxiety (GAD) status, the emotion set resulting in the best predictive performance when combined with LIWC's non-emotion features was the full GoEmotions set.

However, despite the availability of pre-trained models, neural networks can have high computational demands. Consequently, using BERT-based models may not be justified if the cost of model inference outweighs the potential benefits. Therefore, the decision to include these features should be evaluated for each individual predictive analytics project and dataset, weighing the added predictive performance observed at development time with the costs to include the features in production (e.g. a deployed clinical decision support tool continuously evaluating patient-generated messages in real-time), given the available compute resources. Similarly, on-device processing to preserve data privacy can be accomplished with LIWC (Liu et al., 2022), but doing this with a BERT-based model would challenge some contemporary and most legacy smartphone devices.

Depression affects individuals in many ways and expresses itself in various behavioral and thought patterns that may not be fully captured with the high-level categories of positive and negative affect. GoEmotions' main strength therefore lies in its ability to extract fine-grained features spanning the breadth of human emotion, capturing depressed individuals' emotional experiences comprehensively. The different emotion feature sets appeared to be somewhat complementary, as evidenced by the additive performance metrics shown in Table 3; however, when predicting depression, the combination of non-emotion LIWC features and fine-grained GoEmotions features was as predictive as all features combined, suggesting that all signal is contained within this feature subset. In this work, this breadth enabled us to delineate differences in how different types of emotions are associated with depression and anxiety.

Depression severity was associated with large differences in grief, pride, excitement, relief, and disgust. In agreement with generally lower reactivity (Rottenberg, 2017), less excitement was predictive of depression. Grief manifestations are similar to depression symptoms; though grief in itself is not pathological, it often co-occurs with depression (Aoyama et al., 2018). Additionally, depressed individuals expressing less pride than their non-depressed counterparts might be expected on account of lower self-image, and matches findings presented by Gruber et al. (2011). Caused by a perception of violations of moral and social norms, internally directed disgust, also termed self-disgust or self-loathing, has been reported to be associated with both depression and anxiety symptoms (Ille et al., 2014). We further found that increased disapproval - and conversely, decreased approval - were associated with anxiety symptoms. This may be explained by disturbances in interpersonal sensitivity and an inclination to be self-critical, which have been described as characteristic of anxiety

	MDD	GAD
1	LIWC we	GE negemo
2	GEE posemo	GEE negemo
3	GEE joy	GEE joy
4	GEE sadness	GEE posemo
5	GE negemo	GE posemo
6	LIWC bio	LIWC bio
7	GE disappointment	GE fear
8	GEE negemo	GE sadness
9	LIWC sad	LIWC health
10	LIWC i	LIWC we
11	LIWC health	LIWC posemo
12	GE posemo	GE realization
13	GE excitement	GEE sadness
14	GE admiration	GE nervousness
15	GE sadness	GEE fear
16	GEE anger	LIWC negemo
17	GE confusion	GE pride
18	GE pride	LIWC anx
19	GE disapproval	GE joy
20	GE joy	LIWC i
21	GEE disgust	GE disappointment
22	GE realization	GE admiration
23	LIWC posemo	GEE anger
24	GE relief	GE excitement
25	GE approval	GE disgust
26	GE disgust	GE confusion
27	LIWC negemo	GEE disgust
28	GE grief	GE grief
29	GEE fear	GE neutral
30	GEE neutral	GE relief
31	GE fear	GEE neutral
32	GE desire	GEE surprise
33	GE remorse	LIWC sad
34	GE curiosity	GE desire
35	GE nervousness	GE neutremo
36	GE embarrassment	GE curiosity
37	LIWC anx	GE gratitude
38	GE optimism	GE disapproval
39	GE amusement	GE love
40	GE neutremo	GE embarrassment
41	GE neutral	GE anger
42	GE gratitude	GE approval
43	GE love	GE annoyance
44	GE annoyance	GE amusement
45	GE surprise	GE remorse
46	GEE surprise	GE caring
47	LIWC anger	GE surprise
48	GE caring	GE optimism
49	GE anger	LIWC anger

Table 4: Random forest classifier features in order of importance (most important first) for predicting MDD and GAD, as calculated by SHAP (Lundberg and Lee, 2017). GE = GoEmotions. GEE = GoEmotions Ekman

(Ille et al., 2014).

Non-emotion LIWC features have established utility for predicting depression and anxiety. These features capture aspects of symptomatology outside emotion, such as increased self-focus, social isolation, and usage of health-related words. Non-emotion LIWC features would therefore be expected to be complementary to emotion features, and our work confirms that and leveraging both may achieve the best results. We trained a machine learning model using these features in conjunction with emotion features to predict depression (AUROC 0.671) and anxiety (AUROC 0.657). That these models show similar performance using the same features to predict different outcomes may be explained by the large overlap in symptoms between anxiety and depression, e.g. both are characterized by negative self-talk and hopelessness. Additionally, depression and anxiety are often comorbid; indeed, in this dataset, 74.5% of assessments with a GAD-7 score above the diagnosis threshold also had a positive depression finding, and 70.6% of positive anxiety questionnaires also had a positive anxiety finding.

There are important ethical considerations when analyzing patient-generated natural language to infer mental state. Any passive monitoring of patient-generated data may be considered invasive. Due to the sensitive nature of personal health data, such data are subject to protections that do not apply to non-health data. When health-related insights are derived from data that may be neither private nor health-related (e.g. social media posts), obtaining informed consent and handling inferences with appropriate care is paramount. While academic studies such as the current work are governed by rigorous institutional ethics guidelines regarding consent and data sharing, different rules apply to healthcare organizations and commercial entities. The use of technologies such as the ones presented here may be acceptable if conducted by trusted entities, such as healthcare providers, in order to support care (Areán et al., 2021); on the other hand, consumers may be wary of commercial entities conducting such analyses. Further research, as well as applications of the findings presented here, must take such considerations into account.

This work has several limitations. The data used here stem from predominantly female, young, and well-educated participants, and results may therefore not generalize to populations with a different makeup. If predictive algorithms were to be deployed in practice, fairness may be a concern if predictive performance differs for underrepresented groups. In addition, the GoEmotions dataset used to train the BERT-based models is drawn from Reddit, which has been shown to have a disproportionately high representation of young male users (Duggan and Smith, 2013). Though it is encouraging that models trained on these data produce features that correlate well with symptom severity in the current study, the development of annotated datasets drawn from a more diverse population may lead to models that better address linguistic and cultural differences in the ways in which emotions are expressed.

Several features used in the random forest classification model are expected to be highly redundant (e.g. GoEmotions Cower sadness, GoEmotions Ekman sadness, and LIWC sadness; calculated negative emotion variables which are calculated using sadness). However, interdependent features should not affect the random forest's ability to leverage all features optimally to optimize predictive performance.

This work enables and informs future work. We showed that BERT-based emotion features are associated with depression and anxiety status; however, this work did not assess longitudinally if changes in emotion track with changes in depression and anxiety. While existing work demonstrated this relationship for depression-related LIWC features (Burkhardt et al., 2021), future work may aim to ascertain whether changes in emotion features over time also predict longitudinal patient trajectories. This work also informs feature selection for future work in depression and anxiety prediction. Emotion variables can be obtained with a range of extraction approaches. Our results indicate the GoEmotions variables may be a better choice than LIWC for emotions. Nevertheless, LIWC features have a place in future work. LIWC's syntactic and topic features were shown in prior work to be associated with depression scores as well as longitudinal patient trajectories and continued to demonstrate utility in this work.

We determined that fine-grained emotions measured in the language of individuals are associated with and predict anxiety and depression status. The associations we found reflect previous findings. This work thus contributes evidence of the reliability of such measurement approaches, supporting the use of these methods in future work investigating the nature of depression and anxiety. For example, these features could aid investigations into depression phenotypes through cluster analysis, as well as psychology research investigating the differential expression of similarly-valenced emotions in depression and anxiety, e.g. by aiding data collection.

Additionally, this work has important clinical implications. Measurement-based care is facilitated by periodic progress assessments, but additional data collection incurs additional workload. In text-based therapy, depression and anxiety status may instead be automatically determined from already-available patient messages. In clinical settings, interpretability is essential; thus, models based on interpretable features such as emotions may be preferred over black-box models classifying raw text directly. Future work may therefore investigate opportunities to leverage emotion-based predictive models for clinical decision support.

5 Conclusion

Extraction methods differ in the quality of emotion features extracted. With the data and approaches presented here, emotion features extracted by the GoEmotions BERT-based model not only explained more variance in univariate mixed-effect regressions, but also contributed significantly to predictions of depression and anxiety status by a random forest classifier. Further, while non-emotion variables obtained from LIWC remain valuable in linguistic modeling tasks, GoEmotions' level of granularity offers clinically relevant nuance that prevailing tools cannot capture.

5.1 Acknowledgments

This work was supported by the National Library of Medicine (grant number 67-3780) and by Innovation Grant "Informatics-Supported Authorship for Caring Contacts (ISACC)" from the Garvey Institute for Brain Health Solutions.

5.2 Conflicts of Interest

TDH is an employee of the platform that provided the data.

References

Ananda Amstadter. 2008. Emotion regulation and anxiety disorders. *Journal of anxiety disorders*, 22(2):211–221.

Maho Aoyama, Yukihiro Sakaguchi, Tatsuya Morita, Asao Ogawa, Daisuke Fujisawa, Yoshiyuki Kizawa, Satoru Tsuneto, Yasuo Shima, and Mitsunori Miyashita. 2018. Factors associated with possible complicated grief and major depressive disorders. *Psycho-Oncology*, 27(3):915–921.

Patricia A Areán, Abhishek Pratap, Honor Hsin, Tierney K Huppert, Karin E Hendricks, Patrick J Heagerty, Trevor Cohen, Courtney Bagge, and Katherine Anne Comtois. 2021. Perceived Utility and Characterization of Personal Google Search Histories to Detect Data Patterns Proximal to a Suicide Attempt in Individuals Who Previously Attempted Suicide: Pilot Cohort Study. *Journal of Medical Internet Research*, 23(5):e27918.

Aaron T. Beck. 1967. *Depression: clinical, experimental, and theoretical aspects.* Hoeber Medical Division, Harper & Row, New York.

Hannah A. Burkhardt, George S. Alexopoulos, Michael D. Pullmann, Thomas D. Hull, Patricia A. Areán, and Trevor Cohen. 2021. Behavioral Activation and Depression Symptomatology: Longitudinal Assessment of Linguistic Indicators in Text-Based Therapy Sessions. *Journal of Medical Internet Research*, 23(7):e28244.

Stevie Chancellor and Munmun De Choudhury. 2020. Methods in predictive techniques for mental health status on social media: a critical review. *npj Digital Medicine*, 3(1).

Glen Coppersmith, Mark Dredze, and Craig Harman. 2014. Quantifying Mental Health Signals in Twitter. In *Proceedings of the Workshop on Computational Linguistics and Clinical Psychology: From Linguistic Signal to Clinical Reality*, pages 51–60, Stroudsburg, PA, USA. Association for Computational Linguistics.

Alan S. Cowen and Dacher Keltner. 2017. Self-report captures 27 distinct categories of emotion bridged by continuous gradients. *Proceedings of the National Academy of Sciences of the United States of America*, 114(38):E7900–E7909.

Munmun De Choudhury, Scott Counts, and Eric Horvitz. 2013a. Social media as a measurement tool of depression in populations. In *Proceedings of the 5th Annual ACM Web Science Conference on - WebSci '13*, pages 47–56, New York, New York, USA. ACM Press.

Munmun De Choudhury, Scott Counts, Eric J. Horvitz, and Aaron Hoff. 2014. Characterizing and predicting postpartum depression from shared facebook data. *Proceedings of the ACM Conference on Computer Supported Cooperative Work, CSCW*, pages 625–637.

Munmun De Choudhury, Michael Gamon, Scott Counts, and Eric Horvitz. 2013b. Predicting Depression via Social Media. In *Proceedings of the Seventh International AAAI Conference on Weblogs and Social Media*, pages 128–137.

Dorottya Demszky, Dana Movshovitz-Attias, Jeongwoo Ko, Alan Cowen, Gaurav Nemade, and Sujith Ravi. 2020. GoEmotions: A Dataset of Fine-Grained Emotions.

Jacob Devlin, Ming Wei Chang, Kenton Lee, and Kristina Toutanova. 2019. BERT: Pre-training of deep bidirectional transformers for language understanding. *NAACL HLT 2019 - 2019 Conference of the North American Chapter of the Association for Computational Linguistics: Human Language Technologies - Proceedings of the Conference*, 1(Mlm):4171–4186.

Maeve Duggan and Aaron Smith. 2013. 6% of Online Adults are reddit Users. *Pew Research Center*, pages 1–10.

Johannes C. Eichstaedt, Robert J. Smith, Raina M. Merchant, Lyle H. Ungar, Patrick Crutchley, Daniel Preoţiuc-Pietro, David A. Asch, and H. Andrew Schwartz. 2018. Facebook language predicts depression in medical records. *Proceedings of the National Academy of Sciences of the United States of America*, 115(44):11203–11208.

Paul Ekman. 1992. Are There Basic Emotions? *Psychological Review*, 99(3):550–553.

Nikolaos Flemotomos, Victor R. Martinez, Zhuohao Chen, Karan Singla, Victor Ardulov, Raghuveer Peri, Derek D. Caperton, James Gibson, Michael J. Tanana, Panayiotis Georgiou, Jake Van Epps, Sarah P. Lord, Tad Hirsch, Zac E. Imel, David C. Atkins, and Shrikanth Narayanan. 2021. Automated evaluation of psychotherapy skills using speech and language technologies. *Behavior Research Methods*, pages 690–711.

June Gruber, Christopher Oveis, Dacher Keltner, and Sheri L. Johnson. 2011. A discrete emotions approach to positive emotion disturbance in depression. *Cognition and Emotion*, 25(1):40–52.

Sharath Chandra Guntuku, David B. Yaden, Margaret L. Kern, Lyle H. Ungar, and Johannes C. Eichstaedt. 2017. Detecting depression and mental illness on social media: an integrative review. *Current Opinion in Behavioral Sciences*, 18:43–49.

Thomas D. Hull, Matteo Malgaroli, Philippa S. Connolly, Seth Feuerstein, and Naomi M. Simon. 2020. Two-way messaging therapy for depression and anxiety: longitudinal response trajectories. *BMC Psychiatry*, 20(1):297.

Rottraut Ille, Helmut Schöggl, Hans Peter Kapfhammer, Martin Arendasy, Markus Sommer, and Anne Schienle. 2014. Self-disgust in mental disorders - Symptom-related or disorder-specific? *Comprehensive Psychiatry*, 55(4):938–943.

Bennett Kleinberg, Isabelle van der Vegt, and Maximilian Mozes. 2020. Measuring emotions in the covid-19 real world worry dataset. In *Proceedings of the 1st Workshop on NLP for COVID-19 at ACL 2020*.

Kurt Kroenke, Robert L Spitzer, and Janet B W Williams. 2001. The PHQ-9. *Journal of General Internal Medicine*, 16(9):606–613.

Tony Liu, Jonah Meyerhoff, Johannes C Eichstaedt, Chris J Karr, Susan M Kaiser, Konrad P Kording, David C Mohr, and Lyle H Ungar. 2022. The relationship between text message sentiment and self-reported depression. *Journal of affective disorders*, 302:7–14.

Scott M Lundberg and Su-In Lee. 2017. A unified approach to interpreting model predictions. In I. Guyon, U. V. Luxburg, S. Bengio, H. Wallach, R. Fergus, S. Vishwanathan, and R. Garnett, editors, *Advances in Neural Information Processing Systems 30*, pages 4765–4774. Curran Associates, Inc.

Laura Manea, Simon Gilbody, and Dean McMillan. 2012. Optimal cut-off score for diagnosing depression with the Patient Health Questionnaire (PHQ-9): a meta-analysis. *Canadian Medical Association Journal*, 184(3):E191–E196.

F. Pedregosa, G. Varoquaux, A. Gramfort, V. Michel, B. Thirion, O. Grisel, M. Blondel, P. Prettenhofer, R. Weiss, V. Dubourg, J. Vanderplas, A. Passos, and E. Cournapeau, D. and Brucher, M. and Perrot, M. and Duchesnay. 2011. Scikit-learn: Machine Learning in Python. *Journal of Machine Learning Research*, 12:2825–2830.

James W. Pennebaker, R.J. Booth, and M.E. Francis. 2007. Linguistic Inquiry and Word Count: LIWC.

Faye Plummer, Laura Manea, Dominic Trepel, and Dean McMillan. 2016. Screening for anxiety disorders with the gad-7 and gad-2: a systematic review and diagnostic metaanalysis. *General hospital psychiatry*, 39:24–31.

Jonathan Rottenberg. 2017. Emotions in Depression: What Do We Really Know? *Annual Review of Clinical Psychology*, 13:241–263.

Stephanie S. Rude, Eva Maria Gortner, and James W. Pennebaker. 2004. Language use of depressed and depression-vulnerable college students. *Cognition and Emotion*, 18(8):1121–1133.

Sohrab Saeb, Luca Lonini, Arun Jayaraman, David C. Mohr, and Konrad P. Kording. 2017. The need to approximate the use-case in clinical machine learning. *GigaScience*, 6(5):1–9.

Kelli Scott and Cara C. Lewis. 2015. Using measurement-based care to enhance any treatment. *Cognitive and Behavioral Practice*, 22(1):49–59.

Skipper Seabold and Josef Perktold. 2010. statsmodels: Econometric and statistical modeling with python. In *9th Python in Science Conference*.

Judy Hanwen Shen and Frank Rudzicz. 2017. Detecting anxiety through reddit. In *Proceedings of the Fourth Workshop on Computational Linguistics and Clinical Psychology—From Linguistic Signal to Clinical Reality*, pages 58–65.

Anke R. Sonnenschein, Stefan G. Hofmann, Tobias Ziegelmayer, and Wolfgang Lutz. 2018. Linguistic analysis of patients with mood and anxiety disorders during cognitive behavioral therapy. *Cognitive Behaviour Therapy*, 47(4):315–327.

Robert L. Spitzer, Kurt Kroenke, Janet B.W. Williams, and Bernd Löwe. 2006. A brief measure for assessing generalized anxiety disorder: The GAD-7. *Archives of Internal Medicine*, 166(10):1092–1097.

Yla R. Tausczik and James W. Pennebaker. 2010. The Psychological Meaning of Words: LIWC and Computerized Text Analysis Methods. *Journal of Language and Social Psychology*, 29(1):24–54.

Shannon Wiltsey Stirman and James W. Pennebaker. 2001. Word Use in the Poetry of Suicidal and Nonsuicidal Poets. *Psychosomatic Medicine*, 63(4):517–522.

Benjamin Zablotsky and Emily P. Terlizzi. 2020. Mental Health Treatment Among Adults: United States, 2019. *NCHS data brief*, (380):1–8.

Detecting Suicidality with a Contextual Graph Neural Network

Daeun Lee, Migyeong Kang, Minji Kim, Jinyoung Han*
Sungkyunkwan University
{delee12, gy77, m5512m, jinyounghan}@skku.edu

Abstract

Discovering individuals' suicidality on social media has become increasingly important. Many researchers have studied to detect suicidality by using a suicide dictionary. However, while prior work focused on matching a word in a post with a suicide dictionary without considering contexts, little attention has been paid to how the word can be associated with the suicide-related context. To address this problem, we propose a suicidality detection model based on a graph neural network to grasp the dynamic semantic information of the suicide vocabulary by learning the relations between a given post and words. The extensive evaluation demonstrates that the proposed model achieves higher performance than the state-of-the-art methods. We believe the proposed model has great utility in identifying the suicidality of individuals and hence preventing individuals from potential suicide risks at an early stage.

1 Introduction

Suicide has become a serious problem in society. The OECD (Organization for Economic Cooperation and Development) reported that the suicide rate of South Korea and the USA was 23.0 and 14.5 deaths per 100,000 population in 2017, which ranked 1st and 8th, respectively[1].

The awareness of the severity of suicide has led researchers to develop suicidality detection models using a deluge of user activity data on social media, which can help capture latent warning signs of suicide in an early stage (Sawhney et al., 2020; Lee et al., 2020; Shing et al., 2020). For example, the prior work showed that linguistic characteristics revealed in social media posts could be linked to suicide risks (De Choudhury et al., 2016; Shing et al., 2018). Specifically, applying the lexicon-based methods using suicide dictionaries made by domain experts has been reported as effective in capturing linguistic characteristics to detect suicidality. (Gaur et al., 2019; Lv et al., 2015).

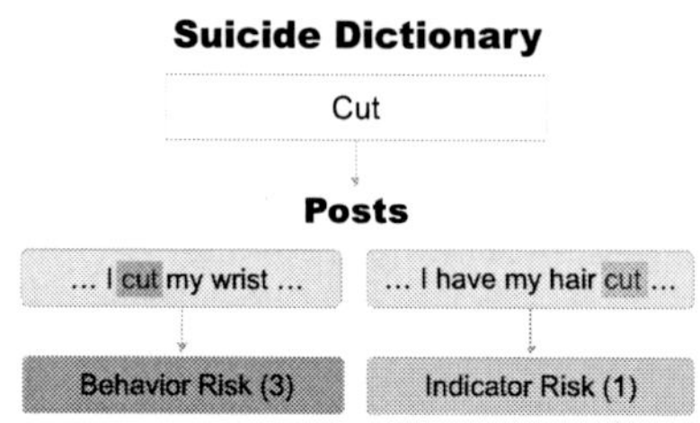

Figure 1: An example of how a word in a suicide dictionary can be misleading in prior work.

While applying the lexicon-based method has been known to be explainable and easy to implement (Kotelnikova et al., 2021; Razova et al., 2021), it may have a limitation: only focusing on whether each word in a post is matched with the suicide lexicon, not considering the context. For example, as illustrated in Figure 1, there are two sentences: "I cut my wrist" and "I have my hair cut". Assuming that the word 'cut' belongs to the suicide dictionary, only the former sentence should be evaluated as having suicidality. However, the latter sentence could also appear to have suicidality if the methods of prior work (Lv et al., 2015; Gaur et al., 2019) are applied. In other words, if the context is incorrectly captured, a model using a suicide lexicon created by experts may not be able to accurately assess the risk of suicidality (Limsopatham and Collier, 2016).

To address this problem, we propose to model the dynamic semantic knowledge between posts and multiple suicide-related words in a suicide dictionary. Capturing the posts' document-word association and word co-occurrence is crucial to un-

*Corresponding author.

[1]https://data.oecd.org/healthstat/suicide-rates.htm

Proceedings of the Eighth Workshop on Computational Linguistics and Clinical Psychology, pages 116 - 125
July 15, 2022 ©2022 Association for Computational Linguistics

derstanding the contextualized suicidality revealed in social media posts. To this end, we apply a graph neural network to jointly learn word and document embeddings over a contextual graph representing the relations between posts and multiple suicide-related words in the dictionary. We build a heterogeneous network describing the relations (i) between social media posts and multiple words in a suicide dictionary and (ii) between suicide words based on the co-occurrence. As node information in the given graph, a post node includes the contextual representation obtained from pretrained BERT (Devlin et al., 2018), and a word node contains the suicide risk level information and contextual representation obtained from the fine-tuned Word2Vec (Mikolov et al., 2013). We learn the proposed heterogeneous graph using the modified GraphSAGE (Hamilton et al., 2017), *Contextual GraphSAGE (C-GraphSAGE)*, to derive a contextualized graph representation.

Instead of using existing suicide dictionaries, we create a word-level suicide dictionary based on social media data using a computational method (Section 3). Since the existing suicide-related lexicon mostly consists of clinical terms (e.g., 'Suicide by self-administered drug') validated by domain experts (Gaur et al., 2019), it may result in a discrepancy with the language used in social media. The created suicide dictionary consists of 279 words and four categories of suicidality levels.

We summarize our contributions as follows.

- We propose a contextualized suicidality detection model *Contextual GraphSAGE (C-GraphSAGE)* using a graph neural network, which can effectively utilize a suicide dictionary. Our evaluation of the real-world dataset demonstrates that the proposed model outperforms the state-of-the-art methods for detecting suicide risk levels using a suicide dictionary.

- We make a word-level English suicide dictionary based on social media data publicly available [2]. We believe the created dictionary can be useful for researchers who want to assess suicidal ideation on social media to prevent potential suicide risks at an early stage.

[2] https://sites.google.com/view/daeun-lee/dataset

2 Related Work

2.1 Suicidality Assessment with Suicide Lexicon

Researchers have investigated that user activity data on social media can provide a cue for analyzing individual suicidality (De Choudhury et al., 2016; Shing et al., 2018). Specifically, prior research showed that linguistic characteristics revealed in social media posts (Sawhney et al., 2020, 2021a) could be linked to suicidal ideation. In particular, utilizing suicide dictionaries made by domain experts has been demonstrated as effective (Lv et al., 2015; Cao et al., 2019; Gaur et al., 2019; Lee et al., 2020), and such lexicon-based methods are known to be fast, explainable, and easy to implement (Kotelnikova et al., 2021; Razova et al., 2021). For example, Lv et al. (2015) developed and validated that a Chinese suicide dictionary made by domain experts helps predict suicidality. Similarly, Gaur et al. (2019) demonstrated the predictive power of suicide dictionaries with domain knowledge.

With the recent advancement of deep learning technologies, high-performing deep learning models have been proposed for accurately assessing suicidality (Sawhney et al., 2021a,b; Cao et al., 2020). In this way, incorporating a suicide dictionary into a deep learning model has received great attention (Cao et al., 2019; Lee et al., 2020). For example, Cao et al. (2019) built suicide-oriented word embeddings to intensify the sensibility of suicide-related lexicons and employed a two-layered attention mechanism. Lee et al. (2020) proposed a deep learning method to utilize existing suicide dictionaries for the low-resource language where a knowledge-based suicide dictionary has not yet been developed. However, the prior work focused on how each word in a post is associated with the words/phrases in a suicide dictionary, e.g., via lexical matching (Lv et al., 2015; Gaur et al., 2019) or fixed word embeddings (Cao et al., 2019; Lee et al., 2020), which may fail to capture the semantic information of suicide lexicons in the suicide-related context.

2.2 Suicidality Assessment with Graph Neural Networks

Among the recent deep learning technologies, graph neural networks (GNNs) have received growing attention in the suicidality assessment task. In particular, GNNs were adopted to extract social information from a user's neighborhood in a social network formed between different users posting about suicidality (Sinha et al., 2019; Sawhney et al., 2021b). Furthermore, Cao et al. (2020) built personal knowledge graphs on Sina Weibo to utilize rich social interaction data in suicidal ideation detection. Since capturing the posts' document-word association and word co-occurrence is crucial to understanding the contextualized suicide intent revealed in social media posts using the suicide dictionary, we apply a GNN to jointly learn word and document embeddings over a textual graph representing the relations between posts and multiple suicide-related words in the dictionary. Note that GNN has been explored to be useful in jointly learning word and document embeddings over a textual graph representation from the perspective of using lexicon for many NLP tasks (Yao et al., 2019; Tang et al., 2020).

3 Suicide Dictionary

A suicide-related word list can help build a simple detector that automatically responds with helpline links to suicidal content. However, the existing English suicide-related lexicon[3] mainly was made of clinical terms validated by domain experts (Gaur et al., 2019), which results in the discrepancy with the language used in social media. Hence, the authors (Gaur et al., 2019) just used the suicide lexicon as a criterion for checking the presence of a concept in the user's posts. Instead of using the existing English suicide lexicon mostly consisting of clinical terms, we propose to create a word-level English suicide dictionary based on social media data. The proposed computational method can be easily applied to other languages that do not have their own suicide lexicons.

Creating a Suicide Dictionary. We create a word-level English suicide dictionary in a computational way using the UMD Reddit Suicidality Dataset (Shing et al., 2018; Zirikly et al., 2019).

The dataset contains 79,569 posts uploaded to 37,083 subreddits of 866 Reddit users posted on the r/SuicideWatch subreddit from 2008 to 2015. In addition, each post is labeled the suicidality severity conducted by crowdsourcing and domain experts (i.e., No risk, Low risk, Moderate risk, and Severe risk). We only use the posts uploaded to the r/SuicideWatch and 15 mental-health-related subreddits (e.g., r/depression, r/anxiety, r/selfharm, etc.) (Gaur et al., 2018) as a target group and use the posts of users who had not posted on either r/SuicideWatch or mental-health related subreddits as a control group.

[3]https://github.com/AmanuelF/Suicide-Risk-Assessment-using-Reddit

Risk Level	# of Words	Examples
No Risk	55	mother, friend, hope, hug, talk
Low Risk	48	emptiness, overthink, stress, desperate
Moderate Risk	83	scared, lonely, psychiatric, pain
Severe Risk	111	cutting, die, hallucination, dread

Table 1: Example words of the generated suicide dictionary.

Before constructing a dictionary, we anonymize the dataset by removing personally identifiable information such as names, email addresses, and URLs. After removing stopwords and lemmatizing the text using spaCy (Honnibal and Montani, 2017), we extract keywords for each post using KeyBERT (Grootendorst, 2020), and then apply the sparse additive generative model (SAGE) (Eisenstein et al., 2011) to determine the words specialized for each label compared to the entire lexicon. Finally, the constructed dictionary includes 297 suicide-related words. Note that the words belonging to the control group are excluded from the corpus set of each label.

Validation and Correction. We recruited two clinical psychotherapists and a psychiatrist to validate and correct the computationally generated suicide dictionary. All annotators verify how well each label of the suicide word complies with the existing sharing task guideline (Shing et al., 2018; Zirikly et al., 2019), and correct it if it does not meet the criteria. Each annotator performs the validation process independently. The final risk label of each suicide word is set to the label agreed by more than or equal to two annotators. As a result of removing 18 differently validated words from all three annotators, there are 279 words in the final dictionary. Table 1 describes the example of words for each class in the generated suicide dictionary.

4 The Model

We propose a suicidality detection model *C-GraphSAGE* that can capture the severity of suicidality of a post on social media. Figure 2 il-

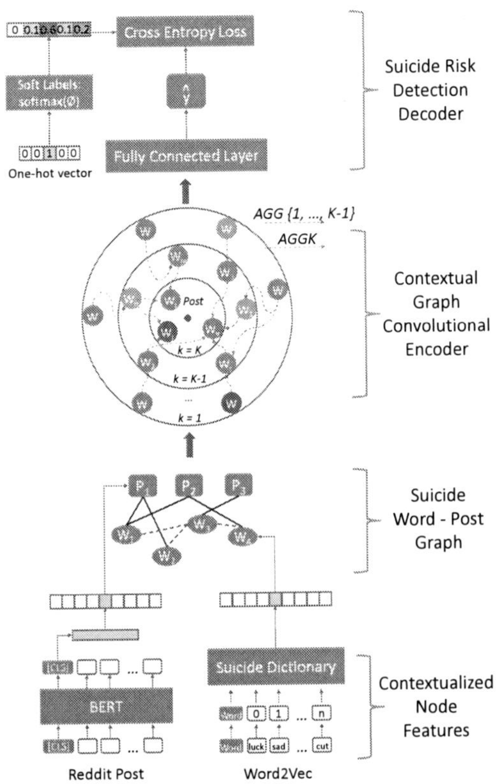

Figure 2: The overall architecture of the model.

lustrates the overall architecture of the proposed model. The model first takes a heterogeneous network that includes posts and suicide words as input. We then apply GraphSAGE (Hamilton et al., 2017) to the given graph to learn the informative representation of suicide-related context by capturing (i) post-words associations and (ii) relations between suicide-related words. Finally, the extracted node presentation from the network is fed into the classification layer. The given post is classified into one of five risk categories: Support (*SU*), Indicator (*IN*), Ideation (*ID*), Behavior (*BR*), and Attempt (*AT*).

4.1 Heterogeneous Network

We build a heterogeneous graph $G = (V_P \cup V_W, E_{PW} \cup E_{WW})$ to represent the relations between social media posts $\{p_i\}_{i=1}^{m} \in P$ and multiple words in a suicide dictionary $\{w_i\}_{i=1}^{n} \in W$, where m and n indicate the number of posts and suicide words, respectively. A graph G consists of two types of nodes, post V_P and suicide word V_W nodes, and two types of edges, post-word E_{PW} and word-word E_{WW} edges. An edge in E_{PW} is linked between a post and its corresponding word if a post contains a specific word in the dictionary. Note that no weight is attached on E_{PW}. An edge in E_{WW} is linked if two words in the suicide dictionary occur together in a post in the UMD Reddit Suicidality Dataset (Shing et al., 2018; Zirikly et al., 2019), which is utilized in constructing a suicide dictionary (in Section 3). A weight on an edge in E_{WW} can be computed by the positive Point-wise Mutual Information (PMI) score that can capture collocations and relations between two terms (Yao et al., 2019; Tang et al., 2020) as follows:

$$PMI(w_i, w_j) = log\frac{p(w_i, w_j)}{p(w_i)p(w_j)} \quad (1)$$

Note that we only attach the edge weight on a suicide word pair with the positive PMI value, which indicates a high semantic correlation of two words in a document.

Contextualized Node Features. In order to generate node features of posts X_P and suicide words X_W, we employ the pre-trained BERT for posts and pre-trained Word2Vec for suicide words, respectively, to capture the contextual representation of text features. Specifically, to obtain X_P, a post p is fed into the BERT model and obtain the `[CLS]` token as a sentence-level representation of the claim as follows:

$$X_{p_i} = BERT(p_i) \in \mathbb{R}^{1 \times d_{cls}} \quad (2)$$

where d_{cls} is the dimension size of a contextualized embedding of `[CLS]` and p_i is i^{th} post. For representing each suicide word w_i, we apply the word-embedding from the pre-processed texts using the Word2Vec model, Gensim (Rehurek and Sojka, 2010). The word vectors are pre-trained with the Skip-Gram representation model using the UMD Reddit Suicidality Dataset (Shing et al., 2018; Zirikly et al., 2019), while the size of the window and the dimension are set to 5 and 200, respectively. Finally, X_W is (i) the suicide risk level (i.e., 0, 1, 2, 3) of each word RL_W and (ii) word embeddings WV_W from pre-trained Word2Vec as follows:

$$RL_{w_i} = \begin{cases} 3, & \text{Severe Risk} \\ 2, & \text{Moderate Risk} \\ 1, & \text{Low Risk} \\ 0, & \text{No Risk} \end{cases} \quad (3)$$

$$WV_{w_i} = Word2Vec(w_i) \in \mathbb{R}^{1\times d_{wv}} \quad (4)$$

$$X_{w_i} = RL_{w_i} \oplus WV_{w_i} \in \mathbb{R}^{1\times (d_{wv}+1)} \quad (5)$$

where d_{wv} is the dimension size of a Word2Vec and w_i is i^{th} word in the suicide dictionary.

4.2 Contextualized Graph Convolutional Encoder

To generate node embedding from the given heterogeneous graph model, we apply the GraphSAGE (Hamilton et al., 2017), a well-known model for a graph neural network (GNN) that supports batch-training without updating states over the whole graph and has shown experimental success compared to other graph representation learning models (Tang et al., 2020). The model first recursively updates embedding for each node v from V_P and V_W by aggregating information from node v's immediate neighbors $N(v), u \in N(v)$, through the aggregation function at each search depth k. After that, h_v^k, node v's representation at step k, is updated by combining h_v^{k-1} and the information obtained from $h_{\mathcal{N}(v)}^{(k)}$, which is the representation of v's neighboring nodes at step k. As suggested in Hamilton et al. (2017), the neighboring nodes are uniformly sampled with a fixed-size set for each search depth. The initial output is $h_v^0 = X_v$. The series of updating processes is defined as follows.

$$h_{\mathcal{N}(v)}^{(k)} = \text{aggregate}_k\left(\{h_u^{k-1}, \forall u \in \mathcal{N}(v)\}\right) \quad (6)$$

$$h_v^{(k)} = \sigma\left(W^k \cdot \text{concat}(h_v^{k-1}, h_{\mathcal{N}(v)}^k)\right) \quad (7)$$

As shown in Figure 3, we propose to use an aggregation function (Eq. 6) based on a convolutional neural network (CNN) instead of existing aggregators such as pool, LSTM, and mean, used in Hamilton et al. (2017). A CNN is proven to be effective in detecting local patterns (Minaee et al., 2021), hence it generates a feature map over the neighbor node embeddings that can explicitly capture relations of words in the suicide dictionary.

Given the target node v's neighboring nodes $\{u_i\}_{i=1}^{j} \in \mathcal{N}(v)$' embedding $\left\{h_{u_1}^{k-1}, h_{u_2}^{k-1}, \cdots, h_{u_j}^{k-1}\right\} \in \mathbb{R}^{j\times d}$, where d is the dimension of node feature, a convolution operation involving a filter $q \in \mathbb{R}^{l*d}$ generates a feature c_i from a window of nodes $u_{i:i+l-1}$ as follows.

$$c_i = \sigma(q \cdot u_{i:i+l-1} + b) \quad (8)$$

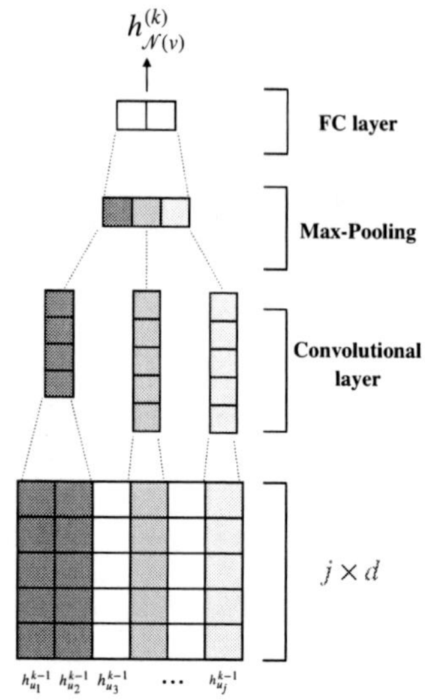

Figure 3: The example of aggregating information from neighborhood of the target node by CNN.

where b is a bias term and $ReLu$ (Nair and Hinton, 2010) is adopted as the non-linear function σ. The filter is employed to each possible window of neighboring nodes to produce a feature map as follows.

$$c = [c_1, c_2, \cdots, c_{j-l+1}] \in \mathbb{R}^{j-l+1} \quad (9)$$

To capture the diverse local structure, we adopt multiple filters with different sizes. For example, the set of kernel sizes used in this paper is [1,2,3]. In this way, the filter can create up to 3 neighbor nodes' combinations. We then apply a max-pooling operation (Collobert et al., 2011) over the feature map and take the maximum value $\hat{c} = max\,\{c\}$ as the feature corresponding to the filter. Finally, we derive a node v's neighbor nodes' representation as follows.

$$h_{\mathcal{N}(v)}^{(k)} = \mathcal{F}_c(\hat{c}) \in \mathbb{R}^{1\times d} \quad (10)$$

Note that, if node v has neighbors with different node types, we sum representations of neighbor nodes. Since we predict the suicidality level of the post, we only consider the node V_p's representation.

4.3 Suicidality Detection Decoder

To predict the suicidality level of a post, the proposed decoder identifies suicidal severity for each node by learning the graph representation as follows.

$$\hat{y} = \mathcal{F}_c(h_v^{(k)}) \quad (11)$$

Like Sawhney et al. (2021a), we adopt the ordinal regression loss (Diaz and Marathe, 2019) as an objective function. Instead of using an one-hot vector representation of the true labels, they used a soft encoded vector representation by considering the ordinal nature between suicidality levels. While ground truth labels are denoted as $\mathcal{Y} = \{SU = 0, IN = 1, ID = 2, BR = 3, AT = 4\} = \{r_{i i=0}^{4}\}$, soft labels as probability distributions of ground truth labels is denoted by y = $[y_0, y_1, y_2, y_3, y_4]$. The probability y_i of each risk-level r_i is

$$y_i = \frac{e^{-\phi(r_t,r_i)}}{\sum_{k=1}^{\lambda} e^{-\phi(r_t,r_i)}} \forall r_i \in \mathcal{Y} \quad (12)$$

where $e^{-\phi(r_t,r_i)}$ is a cost function that penalizes how far the true risk-level r_t is from a risk-level $r_i \in \mathcal{Y}$, which is formulated as $e^{-\phi(r_t,r_i)} = \alpha |r_t - r_i|$, where α is a penalty parameter for incorrect prediction.

Finally, the cross-entropy loss is calculated using the probability distribution y and classification score $\hat{y}$ obtained in Eq(11) as follows:

$$\mathcal{L} = -\frac{1}{n} \sum_{j=1}^{n} \sum_{i=1}^{\lambda} y_{ij} \log \hat{y}_{ij} \quad (13)$$

where n is the batch size and λ is the number of risk-levels.

5 Experiments

We evaluate the our proposed model by answering the following research questions:

- RQ1: Is the proposed suicide dictionary made by a computational method effective in detecting suicidality risk?
- RQ2: Can using the suicide dictionary help improve the model performance?
- RQ3: Is the C-GraphSAGE efficient in utilizing the suicide dictionary?

5.1 Dataset

To learn our proposed model, we utilize *The Golden Standard Dataset* introduced by (Gaur et al., 2019), which consists of Reddit posts collected from the 9 suicide-related subreddits (e.g., r/SuicideWatch and r/depression). The dataset is within the time frame from 2005 to 2016 and annotated with 5 suicidality levels (i.e., Supportive, Indicator, Ideation, Behavior, and Attempt) by mental health experts[4]. While the dataset contains both user-level and post-level data, we utilize the post-level data in this paper since our model aims to detect suicidality levels for a given social media post, and a post-level prediction can be useful for immediate or early intervention on suicidality risks. Finally, the dataset includes 1346, 420, 337, 77, and 49 posts for the Supportive, Indicator, Ideation, Behavior, and Attempt levels, respectively. In addition, we implement a stratified 60:20:20 split such that the train, validation, and test sets consist of 1,427, 356, and 446 posts, respectively.

5.2 Evaluation Metrics

To consider the ordinal nature of suicidality risk levels, we adopt the modified definitions of False Positive (FP), False Negative (FN) (Gaur et al., 2019) in our experiments as follows.

$$FP = \frac{\sum_{i=1}^{N_T} I(\hat{y_i} > y_i)}{N_T} \quad (14)$$

$$FN = \frac{\sum_{i=1}^{N_T} I(y_i > \hat{y_i})}{N_T} \quad (15)$$

where $\hat{y_i}$ is the predicted level, y_i is the actual level for i^{th} test data, and N_T is the size of the test data. $\Delta(y_i, \hat{y_i})$ is the difference between y_i and $\hat{y_i}$. The evaluation metric terms for precision and recall are renamed as graded precision and graded recall, respectively.

5.3 Baselines and Experiment Settings

We compare the proposed model against the following three types of models: (1) Lexicon-based approaches; Rule-based (Gaur et al., 2019), SVM (Lv et al., 2015), and Random Forest (RF) (Amini et al., 2016), (2) Deep learning approaches w/o lexicon; Contextual CNN (Gaur et al., 2019), SISMO (Sawhney et al., 2021a), and BERT (Devlin et al., 2018), and (3) Lexicon + deep learning; Cao et al. (2019) and Reformed BERT. Detailed experimental settings for reproducibility are summarized in the Appendix **??**.

We tune hyperparameters based on the highest FScore obtained from the validation set for all the models. We use the grid search to explore (i) the number of kernel output size in aggregate function $\tilde{q}$, (ii) the number of post features in hidden state $\tilde{H}^D$, (iii) the initial learning rate lr, and (iv) the dropout rate σ. The optimal hyperparameters were

[4]https://github.com/AmanuelF/Suicide-Risk-Assessment-using-Reddit

Type of Model	Model	Loss	G-Precision	G-Recall	G-F1
Suicide lexicon only	Rule-based (Gaur et al., 2019)	/	0.33	0.74	0.46
	SVM (Lv et al., 2015)	Hinge Loss	0.51	0.66	0.58
	RF (Amini et al., 2016)	Gini Impurity	0.65	0.67	0.66
Deep learning only	Contextual CNN (Gaur et al., 2019)	Cross Entropy	0.78	0.57	0.66
	SISMO (Sawhney et al., 2021a)	Soft Label	0.77	0.77	0.77
	SDM w/o Lexicon (Cao et al., 2019)	Cross Entropy	0.73	0.75	0.74
	BERT w/o Lexicon (Devlin et al., 2018)	Soft Label	0.81	0.80	0.80
Suicide lexicon + Deep learning	SDM w/ Lexicon (Cao et al., 2019)	Cross Entropy	0.75	0.78	0.77
	BERT w/ Lexicon (Devlin et al., 2018)	Soft Label	0.82	0.79	0.81
	C-GraphSAGE (Ours)	Soft Label	**0.85**	**0.82**	**0.84**

Table 2: Performance comparisons of the proposed model and baselines.

found to be: $\tilde{q} = 50, \tilde{H^D} = 512, lr = 3e-5$, and $\sigma = 0.1$.

6 Results

In this section, we present our experiment results to answer the three above research questions. Table 2 summarizes the overall performance results of the proposed model (C-GraphSAGE) and the baselines.

6.1 RQ1: Is the proposed suicide dictionary made by a computational method effective in detecting suicidality risks?

Model	Lexicon	Precision	Recall	FScore
Rule-Based (Gaur et al., 2019)	Gaur et al. (2019)	0.26	0.70	0.38
	Ours	**0.33**	**0.74**	**0.46**
RF	Gaur et al. (2019)	0.51	0.65	0.57
	Ours	**0.65**	**0.67**	**0.66**

Table 3: Performance Comparisons between the existing suicide dictionary made by domain experts and the proposed computationally created dictionary (Ours).

To answer the first question, we evaluate the suicidality detection models (Rule-based (Gaur et al., 2019) and Random Forest (RF)) with two different suicide dictionaries: (1) the domain knowledge-based one made by experts (Gaur et al., 2019), and (2) a computationally created one (Ours). As shown in Table 3, the performance with the suicide dictionary created by a computation method (Ours) outperforms the domain knowledge-based lexicon. Furthermore, it indicates that a word-level English suicide dictionary based on social media data is helpful to be mapped with social media posts for detecting suicidality. In other words, the proposed computational method to create a suicide dictionary effectively detects suicidality.

6.2 RQ2: Can using the suicide dictionary help improve the model performance?

Overall, deep learning models with a suicide dictionary (i.e., C-GraphSAGE, 'SDM w/ lexicon', and 'BERT w/ lexicon') perform better than the models that use only text information such as C-CNN, SISMO, 'SDM w/o lexicon', and 'BERT w/o lexicon'. This shows that a model using a suicide dictionary can present the suicide-related context of posts, resulting in high performance. Note that 'SDM w/ lexicon' uses the fine-tuned word embedding model to capture domain knowledge from a pre-built suicide dictionary (Cao et al., 2019), whereas 'SDM w/o lexicon' adopts pre-trained FastText embeddings (Bojanowski et al., 2017) for encoding posts. Also, 'the BERT w/ lexicon' adds the suicide words on the BERT-Tokenizer.

6.3 RQ3: Is the C-GraphSAGE efficient in utilizing the suicide dictionary?

C-GraphSAGE outperforms the other model using a suicide dictionary, the Reformed BERT, offering an insight that capturing dynamic semantic information from a suicide dictionary is beneficial rather than considering only the presence of suicide words. We attribute this to the strength of the graph neural network model that can learn better representations from the relations between posts and words in the suicide dictionary and the associations between suicide words in the suicide-related context. As a result, C-GraphSAGE is helpful in accurately identifying suicidality levels, which shows outstanding utility in preventing suicide risks.

6.4 Ablation Study

We perform an ablation study to examine the effectiveness of different aggregation functions over the proposed C-GraphSAGE, as shown in Table 4. We compare the proposed CNN-based aggregation

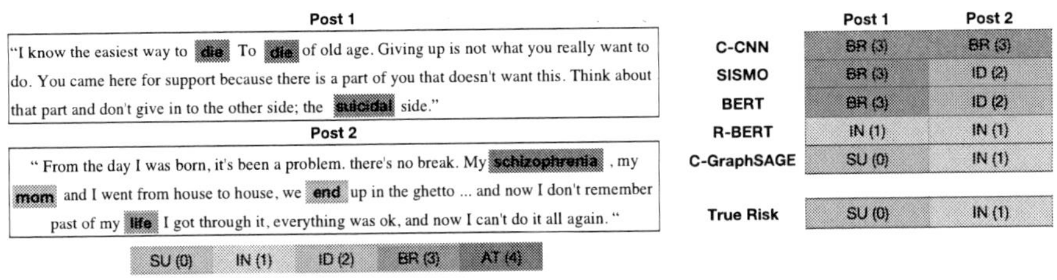

Figure 4: A qualitative analysis on the two cases shows the C-GraphSAGE can capture the risk levels accurately.

Aggregation Function	G-Precision	G-Recall	G-F1
C-GraphSAGE + Pool	0.81	0.79	0.80
+ LSTM	0.81	0.79	0.80
+ MEAN	0.87	0.78	0.82
+ biLSTM	0.88	0.78	0.83
+ CNN (Ours)	0.85	**0.82**	**0.84**

Table 4: An ablation study on different aggregation functions over C-GraphSAGE.

function with the three popular aggregation functions $\in \{LSTM, Pool, Mean\}$ (Hamilton et al., 2017) as well as $bi-LSTM$ (Tang et al., 2020). As shown in Table 4, the model performance significantly improves when we use the aggregation function based on a CNN than other aggregators. Notably, the CNN aggregator outperforms the biLSTM (Tang et al., 2020). This is because an RNN works well in capturing long-term dependencies, whereas a CNN can effectively identify structural patterns. In other words, it is crucial to capture local relations between words than the order of words in our case. We believe that the proposed aggregator can effectively capture neighboring node information, thereby enhancing the robustness of the model for unseen data.

6.5 Qualitative Analysis

To provide detailed insight and interpretability, we qualitatively analyze two cases where C-GraphSAGE performs better than other models in Figure 4. We compare how to predict suicidality by each model given the input that contains the same suicide words. Both posts contain high-level suicide words, but the actual suicidality is relatively low. The proposed model C-GraphSAGE predicts the corresponding risk accurately, whereas other models that assess risk only by the presence of suicide words are likely to classify suicidality levels more highly than actual levels.

7 Concluding Discussion

This paper proposed a suicidality detection model, C-GraphSAGE, which can capture the context of suicidality by learning the relations between social media posts and suicide-related words. Using a word-level English suicide dictionary validated by domain experts, the proposed model achieved higher performance than the state-of-the-art methods in detecting suicidality levels. We believe the proposed model has great utility in identifying potential suicidality levels of individuals with social media data, preventing individuals from potential suicide risks at an early stage.

Ethical Concerns. This study is reviewed and approved by the Institutional Review Board (SKKU2020-10-021). All datasets are anonymized. Hence no personal information can be identifiable.

Limitation. Assessing suicidality using social media data is subjective (Keilp et al., 2012), and the analysis of this paper can be interpreted in diverse ways across the researchers. The experiment data may be sensitive to demographic, annotator, and media-specific biases (Hovy and Spruit, 2016). The analytical patterns learned by C-GraphSAGE may fail to generalize to other social media due to the relatively small data and/or short time window appeared in Reddit. Nevertheless, an interpretable model can help to follow and improve other targets with different statistical patterns and biases (Jacobson et al., 2020).

There is an overlap in data collection periods between the data used to create the suicide dictionary (2008 – 2015) and the data used in the experiment (2005 – 2016). Since all the datasets are anonymized, a Jaccard similarity analysis (Jaccard, 1908) is performed in a grid manner to determine a similarity between all post pairs in two datasets. The result shows that the Jaccard coefficient is quite low (max = 0.5 , mean = 0.1, std = 0.05), meaning that both groups are unrelated.

Practical Applicability. The proposed suicidality detection model can be used for screening or identifying individuals at risk on social media to prioritize early intervention for clinical support.

Acknowledgments

This research was supported by the Ministry of Education of the Republic of Korea and the National Research Foundation of Korea (NRF-2022S1A5A8054322), and the MSIT (Ministry of Science and ICT), Korea, under the ICAN (ICT Challenge and Advanced Network of HRD) program (IITP-2021-2020-0-01816) supervised by the IITP (Institute of Information & Communications Technology Planning & Evaluation).

References

Payam Amini, Hasan Ahmadinia, Jalal Poorolajal, and Mohammad Moqaddasi Amiri. 2016. Evaluating the high risk groups for suicide: a comparison of logistic regression, support vector machine, decision tree and artificial neural network. *Iranian journal of public health*, 45(9):1179.

Piotr Bojanowski, Edouard Grave, Armand Joulin, and Tomas Mikolov. 2017. Enriching word vectors with subword information. *Transactions of the Association for Computational Linguistics*, 5:135–146.

Lei Cao, Huijun Zhang, and Ling Feng. 2020. Building and using personal knowledge graph to improve suicidal ideation detection on social media. *IEEE Transactions on Multimedia*.

Lei Cao, Huijun Zhang, Ling Feng, Zihan Wei, Xin Wang, Ningyun Li, and Xiaohao He. 2019. Latent suicide risk detection on microblog via suicide-oriented word embeddings and layered attention. In *Proceedings of the 2019 Conference on Empirical Methods in Natural Language Processing and the 9th International Joint Conference on Natural Language Processing (EMNLP-IJCNLP)*, pages 1718–1728.

Ronan Collobert, Jason Weston, Léon Bottou, Michael Karlen, Koray Kavukcuoglu, and Pavel Kuksa. 2011. Natural language processing (almost) from scratch. *Journal of machine learning research*, 12(ARTICLE):2493–2537.

Munmun De Choudhury, Emre Kiciman, Mark Dredze, Glen Coppersmith, and Mrinal Kumar. 2016. Discovering shifts to suicidal ideation from mental health content in social media. In *Proceedings of the 2016 CHI conference on human factors in computing systems*, pages 2098–2110.

Jacob Devlin, Ming-Wei Chang, Kenton Lee, and Kristina Toutanova. 2018. Bert: Pre-training of deep bidirectional transformers for language understanding. *arXiv preprint arXiv:1810.04805*.

Raul Diaz and Amit Marathe. 2019. Soft labels for ordinal regression. In *Proceedings of the IEEE/CVF Conference on Computer Vision and Pattern Recognition*, pages 4738–4747.

Jacob Eisenstein, Amr Ahmed, and Eric P Xing. 2011. Sparse additive generative models of text. In *Proceedings of the 28th international conference on machine learning (ICML-11)*, pages 1041–1048. Citeseer.

Manas Gaur, Amanuel Alambo, Joy Prakash Sain, Ugur Kursuncu, Krishnaprasad Thirunarayan, Ramakanth Kavuluru, Amit Sheth, Randy Welton, and Jyotishman Pathak. 2019. Knowledge-aware assessment of severity of suicide risk for early intervention. In *proceedings of the 2019 World Wide Web Conference*, pages 514–525.

Manas Gaur, Ugur Kursuncu, Amanuel Alambo, Amit Sheth, Raminta Daniulaityte, Krishnaprasad Thirunarayan, and Jyotishman Pathak. 2018. " let me tell you about your mental health!" contextualized classification of reddit posts to dsm-5 for web-based intervention. In *Proceedings of the 27th ACM International Conference on Information and Knowledge Management*, pages 753–762.

Maarten Grootendorst. 2020. Keybert: Minimal keyword extraction with bert.

William L Hamilton, Rex Ying, and Jure Leskovec. 2017. Inductive representation learning on large graphs. In *Proceedings of the 31st International Conference on Neural Information Processing Systems*, pages 1025–1035.

Matthew Honnibal and Ines Montani. 2017. spaCy 2: Natural language understanding with Bloom embeddings, convolutional neural networks and incremental parsing. To appear.

Dirk Hovy and Shannon L Spruit. 2016. The social impact of natural language processing. In *Proceedings of the 54th Annual Meeting of the Association for Computational Linguistics (Volume 2: Short Papers)*, pages 591–598.

Paul Jaccard. 1908. Nouvelles recherches sur la distribution florale. *Bull. Soc. Vaud. Sci. Nat.*, 44:223–270.

Nicholas C Jacobson, Kate H Bentley, Ashley Walton, Shirley B Wang, Rebecca G Fortgang, Alexander J Millner, Garth Coombs III, Alexandra M Rodman, and Daniel DL Coppersmith. 2020. Ethical dilemmas posed by mobile health and machine learning in psychiatry research. *Bulletin of the World Health Organization*, 98(4):270.

John G Keilp, Michael F Grunebaum, Marianne Gorlyn, Simone LeBlanc, Ainsley K Burke, Hanga Galfalvy, Maria A Oquendo, and J John Mann. 2012.

Suicidal ideation and the subjective aspects of depression. *Journal of affective disorders*, 140(1):75–81.

Anastasia Kotelnikova, Danil Paschenko, Klavdiya Bochenina, and Evgeny Kotelnikov. 2021. Lexicon-based methods vs. bert for text sentiment analysis. *arXiv preprint arXiv:2111.10097*.

Daeun Lee, Soyoung Park, Jiwon Kang, Daejin Choi, and Jinyoung Han. 2020. Cross-lingual suicidal-oriented word embedding toward suicide prevention. In *Proceedings of the 2020 Conference on Empirical Methods in Natural Language Processing: Findings*, pages 2208–2217.

Nut Limsopatham and Nigel Collier. 2016. Normalising medical concepts in social media texts by learning semantic representation. In *Proceedings of the 54th Annual Meeting of the Association for Computational Linguistics (Volume 1: Long Papers)*, pages 1014–1023.

Meizhen Lv, Ang Li, Tianli Liu, and Tingshao Zhu. 2015. Creating a chinese suicide dictionary for identifying suicide risk on social media. *PeerJ*, 3:e1455.

Tomas Mikolov, Kai Chen, Greg Corrado, and Jeffrey Dean. 2013. Efficient estimation of word representations in vector space. *arXiv preprint arXiv:1301.3781*.

Shervin Minaee, Nal Kalchbrenner, Erik Cambria, Narjes Nikzad, Meysam Chenaghlu, and Jianfeng Gao. 2021. Deep learning–based text classification: A comprehensive review. *ACM Computing Surveys (CSUR)*, 54(3):1–40.

Vinod Nair and Geoffrey E Hinton. 2010. Rectified linear units improve restricted boltzmann machines. In *Icml*.

Elena Razova, Sergey Vychegzhanin, and Evgeny Kotelnikov. 2021. Does bert look at sentiment lexicon? *arXiv preprint arXiv:2111.10100*.

Radim Rehurek and Petr Sojka. 2010. Software framework for topic modelling with large corpora. In *In Proceedings of the LREC 2010 workshop on new challenges for NLP frameworks*. Citeseer.

Ramit Sawhney, Harshit Joshi, Saumya Gandhi, and Rajiv Shah. 2020. A time-aware transformer based model for suicide ideation detection on social media. In *Proceedings of the 2020 Conference on Empirical Methods in Natural Language Processing (EMNLP)*, pages 7685–7697.

Ramit Sawhney, Harshit Joshi, Saumya Gandhi, and Rajiv Ratn Shah. 2021a. Towards ordinal suicide ideation detection on social media. In *Proceedings of the 14th ACM International Conference on Web Search and Data Mining*, pages 22–30.

Ramit Sawhney, Harshit Joshi, Rajiv Shah, and Lucie Flek. 2021b. Suicide ideation detection via social and temporal user representations using hyperbolic learning. In *Proceedings of the 2021 Conference of the North American Chapter of the Association for Computational Linguistics: Human Language Technologies*, pages 2176–2190.

Han-Chin Shing, Suraj Nair, Ayah Zirikly, Meir Friedenberg, Hal Daumé III, and Philip Resnik. 2018. Expert, crowdsourced, and machine assessment of suicide risk via online postings. In *Proceedings of the Fifth Workshop on Computational Linguistics and Clinical Psychology: From Keyboard to Clinic*, pages 25–36.

Han-Chin Shing, Philip Resnik, and Douglas W Oard. 2020. A prioritization model for suicidality risk assessment. In *Proceedings of the 58th Annual Meeting of the Association for Computational Linguistics*, pages 8124–8137.

Pradyumna Prakhar Sinha, Rohan Mishra, Ramit Sawhney, Debanjan Mahata, Rajiv Ratn Shah, and Huan Liu. 2019. # suicidal-a multipronged approach to identify and explore suicidal ideation in twitter. In *Proceedings of the 28th ACM International Conference on Information and Knowledge Management*, pages 941–950.

Pingjie Tang, Meng Jiang, Bryan Ning Xia, Jed W Pitera, Jeffrey Welser, and Nitesh V Chawla. 2020. Multi-label patent categorization with non-local attention-based graph convolutional network. In *Proceedings of the AAAI Conference on Artificial Intelligence*, volume 34, pages 9024–9031.

Liang Yao, Chengsheng Mao, and Yuan Luo. 2019. Graph convolutional networks for text classification. In *Proceedings of the AAAI Conference on Artificial Intelligence*, volume 33, pages 7370–7377.

Ayah Zirikly, Philip Resnik, Ozlem Uzuner, and Kristy Hollingshead. 2019. Clpsych 2019 shared task: Predicting the degree of suicide risk in reddit posts. In *Proceedings of the sixth workshop on computational linguistics and clinical psychology*, pages 24–33.

Identifying Distorted Thinking in Patient-Therapist Text Message Exchanges by Leveraging Dynamic Multi-Turn Context

Kevin Lybarger **Justin S. Tauscher** **Xiruo Ding**
Dror Ben-Zeev **Trevor Cohen**
{lybarger, jtausch, xiruod, dbenzeev, cohenta}@uw.edu
University of Washington, Seattle, WA

Abstract

There is growing evidence that mobile text message exchanges between patients and therapists can augment traditional cognitive behavioral therapy. The automatic characterization of patient thinking patterns in this asynchronous text communication may guide treatment and assist in therapist training. In this work, we automatically identify distorted thinking in text-based patient-therapist exchanges, investigating the role of conversation history (context) in distortion prediction. We identify six unique types of cognitive distortions and utilize BERT-based architectures to represent text messages within the context of the conversation. We propose two approaches for leveraging dynamic conversation context in model training. By representing the text messages within the context of the broader patient-therapist conversation, the models better emulate the therapist's task of recognizing distorted thoughts. This multi-turn classification approach also leverages the clustering of distorted thinking in the conversation timeline. We demonstrate that including conversation context, including the proposed dynamic context methods, improves distortion prediction performance. The proposed architectures and conversation encoding approaches achieve performance comparable to inter-rater agreement. The presence of any distorted thinking is identified with relatively high performance at 0.73 F1, significantly outperforming the best context-agnostic models (0.68 F1).

1 Introduction

Cognitive behavioral therapy (CBT) is an evidence based treatment applicable to a wide range of mental health conditions including depression, anxiety, addiction, bipolar disorder, and schizophrenia spectrum disorders (Yurica and DiTomasso, 2005; Hofmann et al., 2012). One primary clinical activity of CBT is the identification and re-framing of systematic errors in thinking, termed *cognitive distortions*, that create a skewed perception of reality (Beck, 1963). Cognitive distortions are known to exacerbate psychiatric symptoms without intervention (Dudley et al., 2016); however, there are many types of cognitive distortions (e.g., overgeneralization or catastrophizing), which can make identification and appropriate intervention by clinicians more complicated (Burns, 1980).

CBT has traditionally been administered through in-person office visits; however, there is increasing need for remote therapy options, to extend provider reach and increase access (Lin and Espay, 2021). Remote therapy options include internet-delivered therapy, application-based therapy, teletherapy, and text messaging (Lin and Espay, 2021; D'Arcey et al., 2020). There is growing evidence that asynchronous text-message-based exchanges between patients and therapists can augment conventional synchronous therapy and improve patient outcomes (D'Arcey et al., 2020). The expansion of text-message-based CBT provides an opportunity to develop clinician supports via novel natural language processing (NLP) methods that can guide patient treatment and assist in therapist training.

In this work, we explore the automatic identification and categorization of cognitive distortions in a corpus of text-message conversations between patients with serious mental illness and their therapists. Prior work identifying cognitive distortions in text treats each text sample (e.g. sentence or message) as an independent event without context. However, in this conversational paradigm, the preceding turns in the conversation may provide important contextual cues for recognizing distorted thinking. Here, we utilize state-of-the-art deep learning NLP methods to explore the role of conversation history in identifying cognitive distortions in patient-therapist text message exchanges. By identifying distorted thinking in text messages within the broader context of the dialogue, the dialogue-based prediction architectures emulate the real-world process of mental health clinicians who

Proceedings of the Eighth Workshop on Computational Linguistics and Clinical Psychology, pages 126 - 136
July 15, 2022 ©2022 Association for Computational Linguistics

account for conversation context when assessing for distortions. The dialogue-based architectures also mirror the cognitive distortion annotation process associated with the data set used in this work. We present multiple BERT-based architectures for identifying distortions in multi-turn conversations and propose methods for dynamically representing the conversation context. We demonstrate that leveraging the dialogue context and incorporating the proposed dynamic conversation context yields statistically significant performance improvement, reaching performance levels comparable to inter-rater agreement. Distorted thinking is identified in the text messages at 0.73 F1.

2 Related Work

There is a relatively small body of work exploring the automatic identification and categorization of cognitive distortions in user-generated text. Wiemer-Hastings et al. (2004) explored the identification of dysfunctional thoughts in 188 text examples from the cognitive distortion literature. The authors manually curated linguistic features that were used in a decision tree. Simms et al. (2017) annotated 459 Tumblr blogs for the presence of cognitive distortions. Features were extracted using the Linguistic Inquiry and Word Count (LIWC) tool (Tausczik and Pennebaker, 2010)[1], and several classifiers were explored with logistic regression (LR) achieving the best performance. Shickel et al. (2020) investigated the identification of cognitive distortions in online journal entries from college students and samples from crowdsourced participants prompted to give examples of defined distortion types. The authors investigated many classification architectures, including LR, Support Vector Machines (SVM), recurrent neural networks (RNN), convolutional neural networks (CNN), and Bidirectional Encoder Representations from Transformers (BERT) (Devlin et al., 2019). The authors reported the highest performance using LR with with term frequency-inverse document frequency (TF-IDF) features. Shreevastava and Foltz (2021) explored the classification of 10 distinct cognitive distortions in 3,000 therapist question-answer samples. Several classifiers and feature encoding approaches were explored, and the best performance was achieved by an SVM operating on the SentenceBERT encoding, without fine-tuning BERT.

We explored the identification of cognitive distortions in patient-therapist text message exchanges and implemented the best performing models from Shickel et al. (2020) (LR with TF-IDF) and Shreevastava and Foltz (2021) (SVM with SentenceBERT without fine-tuning) as baselines. We found that fine-tuning BERT for multi-label classification achieves state of the art performance in our cognitive distortion prediction task, so we focus the experimentation in this work on BERT architectures. We are not aware of any cognitive distortion prediction work that leverages conversation history as context for identifying distorted thinking.

In this work, we identify cognitive distortions in text-based conversations, exploring the role of conversation history. This distortion prediction task shares similarities with other multi-turn conversational tasks, including retrieval-based dialogue response generation and question answering. Dialogue response and question answering are often approached using hierarchical architectures that first encode each turn, then aggregate the turn embeddings to create a conversation embedding, and lastly generate predictions using the conversation embedding. Conversation turns are frequently mapped to a vector embedding using CNN, RNN, and transformers (e.g. BERT), and conversation embeddings are derived from the turn embeddings using approaches like self-attention, RNN, Markov models, and graphical models (Mensio et al., 2018; Zayats and Ostendorf, 2018; Vickneswaran et al., 2020; Aliannejadi et al., 2020; Zeng et al., 2021). Drawing inspiration from these hierarchical approaches, we experiment with an approach where each turn is encoded using BERT and then the sequence of turn encodings is mapped to a fixed length vector using a uni-directional RNN. There is also conversation modeling work that encodes multiple turns as a single input sequence to BERT, separating the turns with the $[SEP]$ token (Huang et al., 2019), which we also explore here.

Lu et al. (2020) explored a retrieval-based response generation task and proposed a data augmentation technique for model training. The authors created additional positive samples by sampling contiguous multi-turn excerpts from conversations and assuming the last turn is a correct response. Additional negative samples were created by sampling contiguous multi-turn excerpts, randomly removing intermediate turns, and assuming the last turn is an incorrect response. We adapt this turn masking approach to our cognitive distor-

[1] https://www.liwc.app/

tion task to create dynamic conversation context in training, as described in Section 3.2.

3 Methods

3.1 Data

This work utilized a corpus of text message exchanges between patients and therapists that was created as part of a randomized controlled trial that augmented routine care for people with serious mental illness using a text-message-based intervention (Ben-Zeev et al., 2020). The trial was conducted in the Midwest and Pacific Northwest regions of the United States between December 2017 and October 2109. In the intervention, patients participating in standard care engaged with trained clinicians in back-and-forth text-message conversations for 12-weeks. Patients attended an in-person baseline visit to establish rapport and initial goals. Subsequently, clinicians attempted to contact patients up to three times a day to provide support strategies, including reminders, psycho-education, cognitive challenges, self-monitoring prompts, and relaxation techniques. Interactions could be initiated by either patient or clinician each day, and messages could be sent consecutively by a single party in cases where no response was given. The text-message exchanges represent a new model of care that is asynchronous and continuous. The trial demonstrated that augmenting care with mobile texting is logistically feasible, acceptable to patients, safe for patients, and clinically promising. A full description of the trial, including intervention feasibility, acceptability, engagement, and clinical outcomes is available (Ben-Zeev et al., 2020). The randomized controlled trial was approved by the University of Washington's Institutional Review Board (IRB), and study participants provided informed consent. Here, we utilize the text message data for secondary analysis with patient and therapist identifiers removed. All data were stored on a secure server, with patient and clinician identifiers removed prior to annotation and analysis.

The corpus created by the text-message intervention includes messages from 39 patients and 9 therapists with 7,436 patient and 6,959 therapist text messages. The patients who contributed data to the current analysis all had diagnoses of either schizophrenia, schizoaffective disorder, major depressive disorder, or bipolar disorder. The patient demographics were 56% male (N=22), 49% White (N= 17), 29% Black (N=10), 17% multiracial (N=6), and 8% Hispanic/Latinx (N=3). The patients had a mean age of 45.4 (SD=11.1), 12.8 years of education (SD=2.4), and 2.8 lifetime psychiatric hospitalizations (SD=3.4). Patients had variable levels of engagement in the text-message intervention with the average number of client messages per day ranging from 0.3 messages/day to 12.5 messages/day. The average length for the patient and therapist text messages is 15.9 and 22.0 tokens, respectively.

The text message conversations were annotated by a doctoral-level licensed mental health counselor and a masters-level psychologist experienced in working with people with serious mental illness. The corpus is annotated for six cognitive distortion types:

- *Catastrophizing (C)* - Exaggerating or discounting the importance of an event.
- *Jumping to conclusions (J)* - Interpreting a situation without facts or evidence, including mind reading and fortune telling.
- *Mental filtering (M)* - Focusing on one detail of a situation exclusively while ignoring other relevant information.
- *Should statements (S)* - Motivating oneself with absolute expectations, for example should, must, or ought.
- *Overgeneralization (O)* - Extending a single occurrence or isolated incident as evidence of an ongoing or never-ending pattern.
- *Unspecified (U)* - Message included a type of distortion not included in the five categories above or was too incoherent to code specifically.

Table 1 presents example text messages for each distortion type. Annotators reviewed text-messages in the context of a full patient-clinician transcript before applying cognitive distortion annotations at the individual message level. The therapist messages were used to interpret the patient messages; however, no cognitive distortion labels were assigned to therapist messages.

A patient text message may be annotated for multiple cognitive distortions. An *any distortion* (A) label was assigned to each patient text message, indicating whether there is at least one distortion type (logical "or" of distortion types at the message-level). Table 2 presents the distribution of the distortion types. Almost a third of the patient messages include distorted thinking; however, most

Distortion	Example
Catastrophizing (C)	"I just feel so emotional right now right now everything going wrong."
Jumping to conclusions (J)	"My family thinks I have no talents."
Mental filter (M)	"I can't say I have anything to be grateful for"
Should statements (S)	"The team is stopping by so I feel like I have to have my shit together."
Overgeneralizing (O)	"Its always hard to depend on people."
Unspecified (U)	"I felt like bugs were crawling on me and thought I saw some but didn't"

Table 1: Example text messages for each cognitive distortion type.

of the individual distortion types are relatively infrequent, resulting in an imbalanced label distribution. Approximately 20% of the annotated corpus was doubly annotated to assess inter-rater agreement. The Kappa values for the distortion types are: A=0.53, C=0.44, J=0.53, M=0.33, S=0.39, O=0.46, and U=0.01. To facilitate comparison with prediction performance, the inter-rater agreement was assessed as an F1 score, where one of two annotators was assumed to be the ground truth. Table 4 presents the inter-rater agreements as F1 scores. Notably, the agreement for the *unspecified* (U) category is considerably lower than for other categories.

Distortion	Count	Frequency
A	2,145	29%
C	1,113	15%
J	610	8%
M	656	9%
O	268	4%
S	198	3%
U	420	6%

Table 2: Label distribution.

3.2 Distortion Classification

3.2.1 Classification Task

We interpret this cognitive distortion prediction task as a multi-label binary text classification task, where the distortion label set is $\mathcal{V} = \{A, C, J, M, O, S, U\}$. For a given distortion type v in $\mathcal{V}$, a value of 1 indicates the presence of the cognitive distortion type in the target message, m_i. We explore the role of conversation history (context) in assessing the presence of distorted thinking by including preceding messages ($m_{i-n}, \dots m_{i-2}$, m_{i-1}) in modeling, where n indicates the number of context messages or preceding turns used.

3.2.2 Classifier Architectures

We identify cognitive distortions in patient messages using two BERT architectures, which are presented in Figure 1. The first architecture, *BERT-only*, consists of BERT with a linear output layer operating on the pooled output vector. *BERT-only* encodes each target message, m_i, and the context messages, $m_{i-n}, \dots m_{i-1}$, as a single input sequence, where the messages (turns) are delineated by the $[SEP]$ token. The input messages are ordered chronologically, so the last message is the target message ($m_{i-n}, \dots, m_{i-1}, m_i$). The linear output layer projects the pooled output vector for the multi-turn conversation to the number of distortion types (7). In the second architecture, *BERT+LSTM*, each message is separately encoded by BERT, and the pooled output vectors for the messages are sequentially encoded using a uni-directional Long Short-Term Memory (LSTM) RNN. A linear output layer operating on the last hidden state of the LSTM generates the distortion type predictions. For both architectures, a sigmoid activation function converts the label scores to probabilities.

We experimented with including speaker role information to differentiate patients and therapists, for example, "[CLS] [therapist] After seeing her how is you anxiety? [SEP] [patient] It's ok ..." We also experimented with including patient and therapist identifiers, for example, "[CLS] [fe2k] After seeing her how is you anxiety? [SEP] [l2kd] It's ok ...," where "fe2k" and "l2kd" are unique anonymized identifiers for patients and therapists. These approaches did not yield a meaningful performance improvement and are omitted.

3.2.3 Message Context

We explore the introduction of additional randomness in the context messages ($m_{i-n}, \dots, m_{i-1}$) to create dynamic context during model training. We investigate four context representation approaches: *none*, *fixed*, *random length*, and *random mask*. The

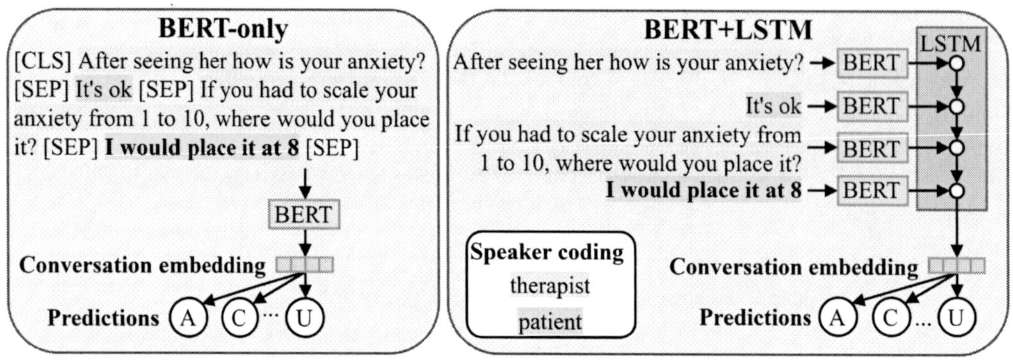

Figure 1: Conversation models. In the text-message examples, **boldface text** indicates the target message, and non-boldface text indicates the context messages.

none context approach does not incorporate any preceding messages as context ($n = 0$), and only the target message is used in training and inference. For the *fixed* context, n context messages preceding the target message are used in both training and inference. For the *random length* context, the number of context messages used in training is randomly selected from a uniform distribution ($uniform(0, n)$, inclusive) for each training sample. The *random length* approach provides contexts of varied lengths during training, and all context messages are sequential with the target message. For the *random mask* context, context messages are randomly masked (removed) for each training sample with probability, p_{mask}. Similar to *random length*, *random mask* provides target messages with varied context lengths; however, with *random mask* the context and target messages will not necessarily be contiguous, as some context messages are randomly removed. For the *random length* and *random mask* context approaches, n context messages are used in inference, similar to the *fixed* approach to utilize all available information. The context length, n, was treated as a tuneable hyperparameter, and context lengths from 0 to 4 were explored. Early experimentation demonstrated that prediction performance improves as the context length increases until $n = 3$, at which point the performance plateaus. All the presented results either include no context ($n = 0$ for *none*) or context of $n = 3$ for *fixed*, *random length*, and *random mask*.

3.2.4 Experimental Paradigm

Model performance was evaluated using a nested cross-validation procedure, to reduce error estimation bias (Varma and Simon, 2006). The annotated data set ($\mathcal{D}$) was split into five folds $(1, 2, ...5)$. To ensure each fold contains sequential messages, each patient-therapist conversation for the entirety of the study was arranged chronologically and split into five folds of approximately equal length ($\approx 20\%$ of each patient-therapist conversation in each fold). There was no overlap between the folds, such that a given message was only included as a target or context in a single fold. These folds were used to create train ($\mathcal{D}_{train}$), validation ($\mathcal{D}_{val}$), and test ($\mathcal{D}_{test}$) splits. Hyperparameters were tuned by training on $\mathcal{D}_{train}$ and evaluating on $\mathcal{D}_{val}$. Final model performance was assessed by training on $\mathcal{D}_{train} \cup \mathcal{D}_{val}$ and evaluating on $\mathcal{D}_{test}$. As a form of repeated holdout testing, we iterated over folds assigned to $\mathcal{D}_{train}$, $\mathcal{D}_{val}$, and $\mathcal{D}_{test}$, re-tuning the hyperparameters for each iteration. For example, fold assignments for iteration $\#1$ were $\mathcal{D}_{train} = \{1, 2, 3\}$, $\mathcal{D}_{val} = 4$, $\mathcal{D}_{test} = 5$, iteration $\#2$ were $\mathcal{D}_{train} = \{2, 3, 4\}$, $\mathcal{D}_{val} = 5$, $\mathcal{D}_{test} = 1$, and so forth. Several of the distortions are very infrequent, and this nested cross validation procedure is intended to better characterize performance across the distortion types. Performance was averaged across the fold iterations and was assessed using F1-score. Hyperparameters were optimized to maximize average F1 across the fold iterations for the *any distortion* label. To assess final performance with significance testing, each fold iteration was repeated 10 times, to generate a distribution of 10 averaged F1 scores for each distortion type. Significance was evaluated using a two-sided T-test with unequal variance and a significance threshold of $p < 0.05$.

All presented results utilized the pretrained BERT model, *MentalBERT* (Ji et al., 2021), which was further pretrained on a Reddit corpus derived

Model	Context	Epochs by fold
BERT-only	none	[4, 4, 4, 6, 4]
	all	[6, 4, 4, 4, 4]
BERT-only	random length	[4, 4, 4, 4, 6]
	random mask	[6, 10, 4, 4, 8]
	all	[4, 8, 4, 4, 4]
BERT+LSTM	random length	[4, 4, 4, 6, 4]
	random mask	[4, 4, 6, 6, 4]

Table 3: Tuned hyperparameters by fold ($[1, 2, 3, 4, 5]$)

from mental health-related subreddits. Other pretrained models may offer performance gains over MentalBERT (Naseem et al., 2022); however, we leave this experimentation to future work. The following configuration and parameters were common to all experimentation: optimizer = AdamW, maximum gradient norm = 1.0, learning rate = 5e-5, batch size = 20, BERT dropout = 0.2, and maximum message length = 120 word pieces. For *BERT-only*, the maximum conversation length was 512 word pieces. For *BERT+LSTM*, the LSTM hidden size = 768. We experimented with context message counts, n, ranging from 0 to 4. We found that performance plateaus around $n = 3$. In the *random mask* experimentation, $p_{mask} = 0.2$. The number of training epochs was tuned for each fold and model configuration, and Table 3 presents the selected epochs for each configuration. To account for the class imbalance associated with label infrequency, a balanced loss function was used in all experimentation, where the loss weights for each label are inversely proportional to positive class frequency.

3.2.5 Distortion clustering

To explore the clustering of distortions in time and the role of conversation context, we calculated the pointwise mutual information (PMI) and conditional probability of distortions in the target and context messages. PMI assesses the association between events. To understand the relationship between distortions in the target message and preceding context messages, PMI is defined here as,

$$PMI(x = v, y = A) = log\frac{p(x = v, y = A)}{p(x = v)p(y = A)},$$

where x is the occurrence of distortion type $v \in \mathcal{V}$ in the target message, and y is the occurrence of *any distortion* (A) in the preceding context messages. We also assessed the association between distortions in target and context messages using the conditional probability, $P(y = A|x = v)$, where x and y are defined similarly to the PMI calculation.

4 Results

4.1 Prediction Performance

Table 4 presents the average cognitive distortion classification performance, as F1, averaged across 10 runs for each of the five fold iterations (each F1 score in the table is the average of 50 values). Each fold iteration involves training on the training and validation folds and evaluating on the withheld test fold. The *BERT-only* model with *none* context is the baseline model for evaluating the role of conversation history on prediction performance.

The inclusion of conversation context in the *BERT-only* and *BERT+LSTM* architectures yields an improvement over *BERT-only* without conversation context for a majority of the distortion labels. The *BERT-only* model with *random length* context achieved the best performance, with significance, for *any distortion* (A) and *catastrophizing* (C). The *BERT-only* model with *random mask* context achieved the best performance, with significance, for *jumping to conclusions* (J). The *BERT+LSTM* model with *fixed* context achieved the best performance, with significance, for *unspecified* (U). For the remaining distortion types (*mental filter* (M), *overgeneralizing* (O), and *should statements* (S)) there is not a statistically significant difference between the top performing model configurations. The dynamic context approaches, *random length* and *random mask*, yield a modest but statistically significant improvement over the *fixed* context for the more frequent and higher performing distortions (*any distortion*, *catastrophizing*, and *jumping to conclusions*).

4.2 Error Analysis

The results in Table 4 demonstrate the inclusion of preceding messages as context improves cognitive distortion prediction performance for the most frequently occurring distortions. We assessed the relationship between distortions in the target message and distortions in the context messages using the PMI, $PMI(x = v, y = A)$, and conditional probability, $P(y = A|x = v)$, defined in Section 3.2.5. Table 5 presents the PMI and conditional probabilities for the two data partitions, *All* and *Improved*. The *All* partition include all 7,436 pa-

Model	Context	F1						
		A (mean±STD)	C	J	M	O	S	U
BERT-only	none	0.68 ± 0.005	0.43	0.46	0.37	0.29	**0.20**	0.32
BERT-only	fixed	0.72 ± 0.003	0.47	0.47	**0.38**	0.29	0.19	0.31
	random length	**0.73** ± 0.003†	**0.48**†	0.46	0.37	0.29	**0.20**	0.33
	random mask	0.72 ± 0.003	0.46	**0.48**†	**0.38**	**0.30**	**0.20**	0.34
BERT+LSTM	fixed	0.72 ± 0.004	0.46	0.46	0.37	0.28	0.16	**0.38**†
	random length	0.72 ± 0.003	0.45	0.44	0.36	0.27	0.15	0.34
	random mask	0.72 ± 0.004	0.46	0.45	0.36	0.28	0.14	0.35
Inter-rater agreement		0.65	0.52	0.56	0.39	0.41	0.48	0.02

Table 4: Cognitive distortion prediction performance, averaged across 10 runs for each fold (1-5). The highest performance for each distortion is **bolded**, and † indicates the best performance with significance ($p < 0.05$). Performance for *any distortion* (A) is presented as mean ± standard deviation. Performance for the remaining distortion types is only presented as the mean, due to space constraints.
Inter-rater agreement is also presented for the doubly annotated subset of the corpus.

Distortion (v)	$PMI(x = v, y = A)$			$P(y = A \mid x = v)$		
	All	Improved	Δ	All	Improved	Δ
A	0.90	3.25	2.35	0.77	0.87	0.10
C	0.98	3.28	2.30	0.83	0.90	0.07
J	0.89	3.22	2.33	0.76	0.84	0.08
M	0.71	3.14	2.43	0.64	0.78	0.14
O	0.79	3.13	2.34	0.69	0.77	0.09
S	0.86	3.16	2.30	0.74	0.79	0.06
U	1.05	3.32	2.27	0.90	0.93	0.04

Table 5: PMI, $PMI(x = v, y = A)$ and conditional probability, $P(y = A|x = v)$, where x is the occurrence of distortion type v in the target message, and y is the occurrence of *any distortion* in the context messages.

tient messages in the annotated corpus. The PMI and conditional probability for *All* messages indicates that distortions cluster in time, specifically that distortions are more likely to occur in the context messages, if there are distortions in the target message (the reverse is also true).

We hypothesized that some of the improved distortion prediction performance associated with the inclusion of context is related with the model implicitly identifying distortions in the context messages. For *BERT-only* with *none* context and *BERT-only* with *random length* context, we identified the models that achieved median *any distortion* F1 performance amongst the 10 runs. We then identified all the samples for which the model without context (*BERT-only* with *none*) was incorrect in assigning the *any distortion* label and the model with context (*BERT-only* with *random length*) was correct is assigning the *any distortion* label. The *Improved* subset in Table 5 includes only the target messages where the model without context was incorrect and the model with context was correct in assigning the *any distortion* label. The *Improved* subset includes 535 target messages. In Table 5, the Δ columns indicates the change from *All* to *Improved.* The PMI and conditional probability are higher for the *Improved* partition across all distortion types, suggesting that at least a portion of the performance improvement associated with the inclusion of context is associated with the presence of distorted thinking in the context. The distortion types with the highest conditional probability in the *Improved* subset in Table 5 (*A*, *C*, *J*, and *U*) are also the distortion types for which the inclusion of context yielded a statistically significant improvement in prediction performance in Table 4.

#	Index	Speaker	Message
1	m_{i-3}	patient	my dad just recently has been trying to get to know me
	m_{i-2}	patient	I'm gonna call [NAME] but the voices r saying no
	m_{i-1}	therapist	Have the voices ever turned out wrong on what they said or ... told you to do?
	m_i	patient	**Some times they are**
2	m_{i-3}	therapist	I'd like to talk about what makes you nervous about leaving your house alone
	m_{i-2}	patient	I guess it started when I never left the house for all those years
	m_{i-1}	therapist	right. and what prevented you from leaving your house back then?
	m_i	patient	**I've never lived here before**

Table 6: Examples where the inclusion of context improves the performance for *any distortion*. In the text-message examples, **boldface text** indicates the target message, and non-boldface text indicates the context messages.

The *Improved* subset in Table 5 includes messages that were labeled incorrectly without the inclusion of context messages but labeled correctly when preceding messages were included as context. We manually reviewed the messages in this *Improved* subset to identify themes in the target and context messages. Table 6 presents example conversations that highlight two of the common themes identified during the manual review of the *Improved* subset. The examples in Table 6 were false negatives for the model without context and true positives for the model with context. In example #1, the target message (m_i) is ambiguous and has no discernible meaning without context. With the inclusion of the context messages ($m_{i-3}, ...m_{i-1}$), we can infer that "they" refers to auditory hallucinations (voices) and "are" affirms that the voices are sometimes incorrect. There are many messages in the *Improved* subset, where the context messages confer meaning to otherwise ambiguous target messages. In example #2, the target message has interpretable meaning without the preceding messages as context and does not necessarily convey distorted thinking. However, the preceding context messages include distorted thinking by the patient and a description of anxiety by the therapist. This context informs the interpretation of the target message and indicates the target message is a continuation of this distorted thinking. There are many examples where an individual message does not necessarily convey distorted thinking when viewed in isolation, but the broader context of the conversation indicates distorted thinking.

5 Discussion

We explored the automatic identification of cognitive distortions in text-based exchanges between patients and therapists, focusing on the role of conversation context. We utilized multiple transformer-based classification architectures and proposed two methods for dynamically utilizing conversation context in training, *random length* and *random mask*. Our results demonstrate that the inclusion of context improves cognitive distortion prediction performance for several distortion types, with the best performing architecture encoding the target message and context messages as a single input sequence to BERT (*BERT-only*). Results also demonstrate that using *random length* for the context during training improves performance over using a *fixed* length context, for several distortion types. The performance of the context-aware models approaches the inter-rater agreement for a majority of the distortion types. *BERT-only* with *random length* context identifies *any distortion* with relatively high performance at 0.73 F1; however, lower performance (F1 $<$ 0.5) is achieved in resolving specific distortion types (e.g. *catastrophizing* or *jumping to conclusions*). The error analysis suggests that at least a portion of the performance gains associated with the inclusion of context messages is attributable to the tendency for messages expressing cognitive distortions to cluster in time.

This work presents context-aware classification approaches that improve performance in identifying cognitive distortions in text messages. The improved performance associated with the inclusion of context will benefit downstream clinical applications, including clinical decision-support systems, therapist training, and clinical research. In the community health setting, the adoption of new treatment modalities and technology for serious mental illness is hindered by the availability of training and expertise among community-based clinicians (Perry et al., 2020). The adoption of new interventions is resource intensive, and training and supervision for novel interventions may improve the adoption of new interventions, like texting (Moyers et al., 2005). Our work exploring the automatic identification of cognitive distortions could mediate the development of clinician training and support tools that improve the uniformity and quality of care and reduce required human resources, by flagging patient content that requires intervention. In terms of clinical research, this work may support the implementation of interventions that target cognitive distortions, assess the extent to which such interventions are effective in reducing distortion frequency, and improve understanding of the relationships between distorted thinking, symptom severity and mental status.

This study is limited by the number of participating patients and therapists. Text-based therapy conversations are likely heterogeneous and vary by patient-therapist dyad, patient clinical condition, and other factors. Due to the size of the annotated corpus, the data set was split such that each patient appears both in the train and test partitions, although there is no overlap between the messages in the train and test partitions. Additional work with an expanded data set is needed to assess the generalizability of the classifiers to a diverse patient population, including patients not represented in the training data.

Similar to prior cognitive distortion work (Shickel et al., 2020), classification performance is limited by the challenge of manually annotating distortions, including the soft boundaries between distortion types. We are currently adding additional cognitive distortion type labels to the text-message corpus to include more fine-grained distortion categories that can be condensed into functionally related higher-level categories. The inclusion of additional cognitive distortion types and aggregation of individual distortion types into higher-level thought patterns may improve annotation consistency. As part of this annotation effort, we are expanding the annotation guidelines and providing additional annotator training to improve annotation detail and quality.

This work investigates the use of preceding conversational turns as context for prediction. There are many other forms of context, and mechanisms for representing it, that may be considered in future work. With a sufficiently large corpus of text conversations, it may be feasible to learn patient representations that capture important linguistic patterns, thinking styles, and other information relevant to characterizing thought patterns and mental state. The patient representations could take the form of learned patient embeddings, for example special patient-specific BERT tokens. Additional contextual information could include message metadata (e.g. time of day or time between responses) or patient demographics/attributes (e.g. age, gender, tech literacy, or diagnoses). Models incorporating such information may add to our understanding of the contexts in which distortions occur and further improve automated methods to detect them.

6 Conclusions

The improvements in performance shown in this work demonstrate that modeling conversational context is important for identifying cognitive distortions in text-based exchanges between patients and therapists. By identifying cognitive distortions in patient messages within the larger context of the conversation, the modeling better emulates the process mental health clinicians use to assess for distortions. Distorted thinking in the patient messages tends to cluster in time, such that distortions are more likely to occur in context messages, if there are distortions in the target message (and vice-versa). Some of the improved performance associated with the inclusion of context is likely attributable to the model implicitly identifying distortions in the context messages. Additionally, the inclusion of context also captures important cues in therapist messages for the presence of distorted thinking in patient messages. Conversational context is likely to improve performance in identifying cognitive distortions, with implications for the development of decision support tools, and quantification of distortions in observational data.

Acknowledgements

This work was supported by a UW Medicine Garvey Institute for Brain Health Solutions Innovation Grant, a grant from the National Institute of Mental Health (R56MH109554), and the National Institutes of Health, National Library of Medicine (NLM) Biomedical and Health Informatics Training Program at UW (Grant Nr. T15LM007442). The content is solely the responsibility of the authors and does not necessarily represent the official views of the National Institutes of Health.

References

Mohammad Aliannejadi, Manajit Chakraborty, Esteban Andrés Ríssola, and Fabio Crestani. 2020. Harnessing evolution of multi-turn conversations for effective answer retrieval. In *Proceedings of the Conference on Human Information Interaction and Retrieval*, page 33–42.

Aaron T Beck. 1963. Thinking and depression: I. Idiosyncratic content and cognitive distortions. *Archives of General Psychiatry*, 9(4):324–333.

Dror Ben-Zeev, Benjamin Buck, Suzanne Meller, William J Hudenko, and Kevin A Hallgren. 2020. Augmenting evidence-based care with a texting mobile interventionist: a pilot randomized controlled trial. *Psychiatric Services*, 71(12):1218–1224.

David D Burns. 1980. *Feeling Good: The New Mood Therapy*. William Morrow and Company.

Jessica D'Arcey, Joanna Collaton, Nicole Kozloff, Aristotle N Voineskos, Sean A Kidd, George Foussias, et al. 2020. The use of text messaging to improve clinical engagement for individuals with psychosis: systematic review. *Journal of Medical Internet Research - Mental Health*, 7(4):e16993.

Jacob Devlin, Ming-Wei Chang, Kenton Lee, and Kristina Toutanova. 2019. BERT: Pre-training of deep bidirectional transformers for language understanding. In *North American Chapter of the Association for Computational Linguistics*, pages 4171–4186.

Robert Dudley, Peter Taylor, Sophie Wickham, and Paul Hutton. 2016. Psychosis, delusions and the "jumping to conclusions" reasoning bias: a systematic review and meta-analysis. *Schizophrenia Bulletin*, 42(3):652–665.

Stefan G Hofmann, Anu Asnaani, Imke JJ Vonk, Alice T Sawyer, and Angela Fang. 2012. The efficacy of cognitive behavioral therapy: A review of meta-analyses. *Cognitive Therapy and Research*, 36(5):427–440.

Yen-Hao Huang, Ssu-Rui Lee, Mau-Yun Ma, Yi-Hsin Chen, Ya-Wen Yu, and Yi-Shin Chen. 2019. EmotionX-IDEA: Emotion BERT–an affectional model for conversation. *arXiv preprint*.

Shaoxiong Ji, Tianlin Zhang, Luna Ansari, Jie Fu, Prayag Tiwari, and Erik Cambria. 2021. MentalBERT: Publicly available pretrained language models for mental healthcare. *arXiv preprint*.

Amanda Lin and Alberto J Espay. 2021. Remote delivery of cognitive behavioral therapy to patients with functional neurological disorders: Promise and challenges. *Epilepsy & Behavior Reports*, 16:100469.

Junyu Lu, Xiancong Ren, Yazhou Ren, Ao Liu, and Zenglin Xu. 2020. Improving contextual language models for response retrieval in multi-turn conversation. In *Proceedings of ACM SIGIR Conference on Research and Development in Information Retrieval*, pages 1805–1808.

Martino Mensio, Giuseppe Rizzo, and Maurizio Morisio. 2018. Multi-turn QA: A RNN contextual approach to intent classification for goal-oriented systems. In *International World Wide Web Conference - Companion Proceedings*, pages 1075–1080.

Theresa B. Moyers, Tim Martin, Jennifer K. Manuel, Stacey M.L. Hendrickson, and William R. Miller. 2005. Assessing competence in the use of motivational interviewing. *Journal of Substance Abuse Treatment*, 28.

Usman Naseem, Byoung Chan Lee, Matloob Khushi, Jinman Kim, and Adam Dunn. 2022. Benchmarking for public health surveillance tasks on social media with a domain-specific pretrained language model. In *Proceedings of NLP Power! The First Workshop on Efficient Benchmarking in NLP*, pages 22–31. Association for Computational Linguistics.

Kristen Perry, Sari Gold, and Erika M. Shearer. 2020. Identifying and addressing mental health providers' perceived barriers to clinical video telehealth utilization. *Journal of Clinical Psychology*, 76(6):1125–1134.

Benjamin Shickel, Scott Siegel, Martin Heesacker, Sherry Benton, and Parisa Rashidi. 2020. Automatic detection and classification of cognitive distortions in mental health text. In *IEEE International Conference on Bioinformatics and Bioengineering*, pages 275–280.

Sagarika Shreevastava and Peter Foltz. 2021. Detecting cognitive distortions from patient-therapist interactions. In *Proceedings of the Workshop on Computational Linguistics and Clinical Psychology: Improving Access*, pages 151–158.

Taetem Simms, Clayton Ramstedt, Megan Rich, Michael Richards, Tony Martinez, and Christophe Giraud-Carrier. 2017. Detecting cognitive distortions through machine learning text analytics. In *IEEE International Conference on Healthcare Informatics*, pages 508–512.

Yla R Tausczik and James W Pennebaker. 2010. The psychological meaning of words: LIWC and computerized text analysis methods. *Journal of Language and Social Psychology*, 29(1):24–54.

Sudhir Varma and Richard Simon. 2006. Bias in error estimation when using cross-validation for model selection. *BMC Bioinformatics*, 7(1):1–8.

Jarsigan Vickneswaran, Piruntha Navanesan, Vahesan Vijayaratnam, and Uthayasanker Thayasivam. 2020. Simplified approach for predicting emotions of multi-turn textual utterances. In *International Conference on Advances in ICT for Emerging Regions*, pages 71–76.

Katja Wiemer-Hastings, Adrian S Janit, Peter M Wiemer-Hastings, Steve Cromer, and Jennifer Kinser. 2004. Automatic classification of dysfunctional thoughts: a feasibility test. *Behavior Research Methods, Instruments, & Computers*, 36(2):203–212.

Carrie L. Yurica and Robert A. DiTomasso. 2005. *Cognitive distortions*, pages 117–122. Springer.

Victoria Zayats and Mari Ostendorf. 2018. Conversation modeling on Reddit using a graph-structured LSTM. *Transactions of the Association for Computational Linguistics*, 6:121–132.

Xingshan Zeng, Jing Li, Lingzhi Wang, and Kam-Fai Wong. 2021. Modeling global and local interactions for online conversation recommendation. *ACM Transactions on Information Systems*, 40(3):1–33.

Learning to Automate Follow-up Question Generation using Process Knowledge for Depression Triage on Reddit Posts

Shrey Gupta[1*], **Anmol Agarwal**[1*], **Manas Gaur**[2],
Kaushik Roy[2], **Vignesh Narayanan**[2], **Ponnurangam Kumaraguru**[1], **Amit Sheth**[2]
[1]International Institute of Information Technology, Hyderabad, India
{shrey.gupta, anmol.agarwal}@students.iiit.ac.in,
pk.guru@iiit.ac.in
[2]AI Institute, University of South Carolina, SC, USA
{mgaur, kaushikr}@email.sc.edu, {vignar, amit}@sc.edu

Abstract

Conversational Agents (CAs) powered with deep language models (DLMs) have shown tremendous promise in the domain of mental health. Prominently, the CAs have been used to provide informational or therapeutic services (e.g., cognitive behavioral therapy) to patients. However, the utility of CAs to assist in mental health triaging has not been explored in the existing work as it requires a controlled generation of follow-up questions (FQs), which are often initiated and guided by the mental health professionals (MHPs) in clinical settings. In the context of 'depression', our experiments show that DLMs coupled with process knowledge in a mental health questionnaire generate 12.54% and 9.37% better FQs based on similarity and longest common subsequence matches to questions in the PHQ-9 dataset respectively, when compared with DLMs without process knowledge support. Despite coupling with process knowledge, we find that DLMs are still prone to hallucination, i.e., generating redundant, irrelevant, and unsafe FQs. We demonstrate the challenge of using existing datasets to train a DLM for generating FQs that adhere to clinical process knowledge. To address this limitation, we prepared an extended PHQ-9 based dataset, PRIMATE, in collaboration with MHPs. PRIMATE contains annotations regarding whether a particular question in the PHQ-9 dataset has already been answered in the user's initial description of the mental health condition. We used PRIMATE to train a DLM in a supervised setting to identify which of the PHQ-9 questions can be answered directly from the user's post and which ones would require more information from the user. Using performance analysis based on MCC scores, we show that PRIMATE is appropriate for identifying questions in PHQ-9 that could guide generative DLMs towards controlled FQ generation (with minimal hallucination) suitable for aiding triaging. The dataset created as a part of this research can be obtained from here.

1 Introduction

Conversational agents (CAs) powered by DLMs are software designed to interact with human users for specific tasks. For mental health purposes, particularly depression, CAs have been studied extensively in prior work for helping patients follow generic mental health guidelines, typically by providing reminders to assist patients in adhering to the medication and therapy strategy outlined by a mental health professional (MHP)[1][2]. However, previous work on depression have not examined the use of CAs for triaging. For the purpose of triaging, CAs should learn to generate controlled and clinical process knowledge-guided discourse that can assist MHPs in diagnosis. Our research suggests a clinically grounded and explainable methodology to develop conversational information-seeking tools, first to learn "what symptoms the user is suffering" and "what extra information is needed for triaging."

CAs are susceptible to irrelevant and sometimes harmful questions when generating FQs or responses to a patient suffering from depression (Miner et al., 2016). The primary reason for irrelevant and harmful questions is that CAs cannot incorporate contextual information in generating appropriate follow-up questions (FQs) (see Figure 1). Further, the sensitivity of the conversation and a controlled generation process are essential characteristics of patient-clinician interactions, which are difficult to embed in DLM-based CAs. Therefore, question generation (QG) in mental health is challenging, and research to develop CAs for automating triage has not been explored.

[1]https://tinyurl.com/yfp3bhr2
[2]https://woebothealth.com/

*Authors contributed equally

Proceedings of the Eighth Workshop on Computational Linguistics and Clinical Psychology, pages 137 - 147
July 15, 2022 ©2022 Association for Computational Linguistics

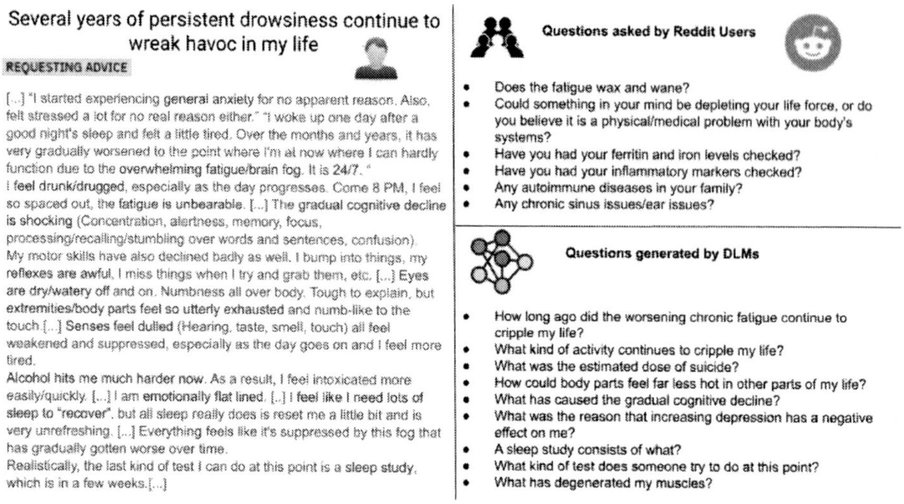

Figure 1: Reddit is a rich source for bringing crowd perspective in training DLMs over conversational data. On the **left** is a sample post from r/depression_help which sees inquisitive interaction from other Reddit users. At the **top-right** are the FQs asked by the Reddit users in the comments. These FQs are aimed at understanding the severity of the mental health situation of the user and are hence, diagnostically relevant. At the **bottom-right** are the questions generated by DLMs. It can be seen that these are not suitable FQs.

Procedures for generating semantically related and logically ordered questions in the mental health domain are a form of process knowledge manifested in various clinical instruments for mental health triage. For example, the severity of depression is measured using Patient Health Questionnaire (PHQ-9). Enforcing DLMs to follow process knowledge, like in PHQ-9, would make CAs generate FQs similar to an MHP when they are seeking information from the patient (Karasz et al., 2012). Unfortunately, datasets that meet this criterion are currently unavailable. Though clinical diagnostic interviews exist, they are not rich, sufficiently dense, and varied to train DLMs (Manas et al., 2021; Gratch et al., 2014). Further, we require dataset(s) that includes *support seeking queries* and *natural questions* that show help providing behavior. For this purpose, anonymized user-generated conversational data in Mental Health support communities on Reddit provides a rich source of fine-grained, contextual, and diverse information suitable for fine-tuning DLMs. Specific to depression, we explored posts and comments in r/depression_help.

In the current research, we emphasize the limitations of T5, a state-of-the-art DLM[3] to generate process knowledge-like FQs using the data from r/depression_help (Raffel et al., 2019). We filtered the dataset by retaining only posts with at least one comment that seeks additional information from the user seeking support. Further filtering of comments was performed using PHQ-9 to assist T5 in generating relevant FQs (see Figure 2). We found that the outcome is substantial for the single turn question answering model; however, not suitable for mental health triage, which is a discourse. We conducted a series of experiments keeping our focus on 'depression' and leveraged its associated process knowledge for mental health triage: the PHQ-9 (Kroenke et al., 2001). To the best of our knowledge, FQ generation relating to *depression* has never been studied using PHQ-9 for *discourse modeling* and *generation*.

We make the following key contributions: (a) **Extending PHQ-9:** PHQ-9 questions are limited in scope for common NLP tasks like finetuning. In collaboration with MHPs, we prepared a list of 134 sub-questions for nine PHQ-9 questions for better fine-tuning of T5. (b) We analyzed the performance of three variants of T5 using BLEURT (Sellam et al., 2020) and ROUGE-L scores that measure semantic relatedness and exact match similarity of generated question to sub-questions of PHQ-9. (c) **PRIMATE Dataset:** Lessons learned during our experiments suggested that T5 must be trained in a supervised setting to capture 'what

[3]Current DLMs are either variants of T5 or built from T5

the user has already mentioned about his/her depression condition in the post-text' and then generate FQs. Along with MHPs, we constructed a novel **PRIMATE** (**PR**ocess knowledge **I**ntegrated **M**ental he**A**lth da**T**as**E**t) dataset that would train DLMs to capture PHQ-9-answerable information from user text. In this research, we restrict our experiments and discussion on whether **PRIMATE** can help capture context from the user post relevant to some PHQ-9 questions and pointing out which other PHQ-9 questions would form candidates to direct FQ generation. Our approach and insights have applications to Anxiety (GAD-7), Suicide (C-SSRS), and other mental health disorders as well.

2 Related Work

Recently, DLMs have attracted much attention for question answering, thanks to their successes in NLP applications (Thoppilan et al., 2022; Borgeaud et al., 2021). Research on question generation has focused on improving the legibility and relevance of questions. This is because DLMs continue to hallucinate while generating questions in general-purpose domains, which can lead to factually incorrect responses. This can have severe consequences in the mental health domain (Thoppilan et al., 2022). Recently, inappropriate and toxic behaviors of language models have been extensively studied and reported in the literature (Dinan et al., 2021; Weidinger et al., 2021). Solutions around fine-tuning, augmenting a neural retriever to support generation, and rules on generation quality have been defined as possible remedies (Manas et al., 2021). These have been effective for the general-purpose domain; however, the research surrounding DLMs is yet to unfold in mental health. ELIZA (Weizenbaum, 1983) could transform users' statements into questions but employs labor-intensive templates to generate safe and relevant questions. Models like RAG and REALM were developed to include external knowledge to support question generation (Lewis et al., 2020; Guu et al., 2020). However, these models are still susceptible to incoherent and irrelevant FQ generation . Further, their end-to-end learning approach is rigid to support process-guided question generation and discourse, often followed in a clinical setting for triage (Gaur et al., 2021).

In theory, DLMs should be capable enough of extracting pieces of information from user description that portrays the understanding of the user and leverage it for generating the next FQ. For such a task, supervised training of DLMs with process knowledge and coupling it with information retrieval over domain-specific mental health knowledge is a viable solution. This is because mental health knowledge sources (e.g., SCID (Structured Clinical Interviews for DSM-5) have structured/semi-structured information on how interviews are performed (Brodey et al., 2018). Our research substantiates that DLMs (e.g., T5) generate low quality follow-up questions in the context of depression for triage, and granting external knowledge through PHQ-9 reduces the rate at which models generate meaningless FQs (Thoppilan et al., 2022; Komeili et al., 2021). In the current research, we define an approach for supervised training of DLMs on a specific dataset that would yield probability distribution over PHQ-9 (with support from Extended PHQ-9). These probabilities will confirm whether the DLM can identify cues from user text that can inform a set of PHQ-9 questions. Remaining PHQ-9 questions are potential FQs.

Datasets: Prior datasets such as Counsel Chat (CounselChat), Counseling Conversations (Huang, 2015), Role Play (Demasi et al., 2019), Crisis Text Line (Althoff et al., 2016) and Reddit C-SSRS (Gaur et al., 2019) have been created to train CA for mental health counseling. Trained CAs can engage in a single turn question answering; however, conducting a conversation requires capturing user context and leveraging clinical instruments to guide the generation of FQs.

3 Question Generation (QG)

Dataset for QG: Our approach to data collection involves scraping posts and comments from r/depression_help, a subreddit on Reddit, which is meant to provide advice and support to help individuals suffering from depression. The posts on this subreddit contain flair tags such as *SEEKING HELP*, *SEEKING ADVICE*, and *REQUESTING SUPPORT*. We filter down the data curated from this subreddit based on the flair tag attribute to retain only *advice, help* or *support* seeking posts and their comments. After filtering, our dataset had approximately 21,000 posts. Each post contains a title, description, and comments. On average, each post has 5 comments. Next, we chunked the main text of each post into smaller groups of sentences (chunks) of less than 512 tokens while making sure

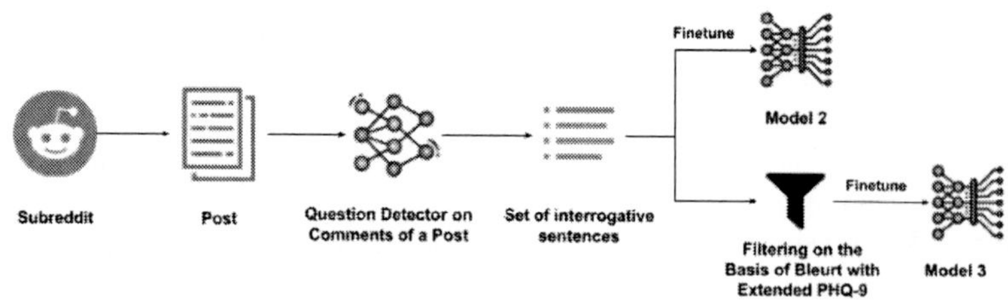

Figure 2: An illustration of our pipeline for developing Model 2 and Model 3 using T5 as the deep language models. Starting with posts (including comments) from r/depression_help, we filter out comments that are neither interrogative nor information seeking in nature to yield a posts-questions dataset for fine-tuning T5. This dataset was further filtered using extended PHQ-9 before using it to fine-tune T5 (Model 3).

Model 1 QGs	Model 2 QGs	Model 3 QGs
• How long ago did the worsening chronic fatigue continue to cripple my life? *[*User already mentions in the title of the post he has been facing the fatigue for years]* • What was the estimated dose of suicide? *[*Unsafe incoherent question]* • How could body parts feel far less hot in other parts of my life? *[*Incoherent question]* • What has caused the gradual cognitive decline? *[*It's the physician who needs to answer this question after the conversation]* • What effect is that I'm less optimistic, and it's embarrassing ? *[*Incoherent attempt to generate a question from the sentence in the post which mentions "I am still pretty optimistic , but I 'm just too out of it , it 's embarrassing"]* • What was the reason that increasing depression has a negative effect on me? *[*Irrelevant and incoherent]* • A sleep study consists of what? *[*Irrelevant to the diagnostic procedure]* • What did Caffeine feel in my body? *[*Wrong pronoun used and irrelevant to the diagnosis]*	• Have you ever tried many depressants or mental health things? *[*Redundant question as user has already mentioned in the post text that he has tried antidepressants]* • Are you seeing a therapist? *[*Redundant question as user had already mentioned in the post text that he has seen a therapist in the past]* • Do you even ever try to focus on negative things and/or pain, relief, and patience? *[*Incoherent question]* • And what if you forgot to fucking realize that, then it will be fucking so much that you'll be fucking fucking can we ask what they want. *[*Unsafe objectionable question]* • Do you mean heavy bipolar disorder? *[*Irrelevant to the context of the post]* • Are you aware about the qualities of antidepressants? *[*Inquisitive in nature but irrelevant from a diagnostic perspective]* • What does antidepressant work at? *[*Irrelevant from a diagnostic perspective]* • Do you have a close friend you can go to who you can talk with, that way you can get out of the house? *[*FQ]*	Are you on any antidepressants? *[*User mentions in the post he has tried them before but no mention has been made if he is still on them]* • Do you have any positive thoughts? *[*FQ]* • Are you in any danger of hurting yourself? *[*Slightly unsafe]* • Is it that you aren't happy with your feelings? *[*Irrelevant question]* • Have you tried some exercise? *[*Redundant question as user already mentions he has tried it]* • Do you wake up frequently? *[*FQ]* • How is your sleep quality? *[*FQ]* • When you wake up, what do you do? *[*FQ]* • Is there anything that helps you calm the symptoms for now? *[*FQ]* • What are your hobbies? *[*Generic FQ]* • What are your interests? *[*Generic FQ]*

Table 1: Examples of questions generated by T5 when tasked to generate FQs when the user query for the **post** in Figure 1 was provided as input. **Model 1**, which is a pre-trained T5 (Raffel et al., 2019), often generates questions which are irrelevant, unsafe, incoherent, and redundant. **Model 2**, which is T5 fine-tuned on r/depression_help seems to be relatively coherent and inquisitive compared to **Model 1**. However, both models generate questions about the topic that user has discussed in their query. As a result, we see that pre-trained and fine-tuned DLMs fail to generate FQs. By enforcing FQ generation using using a dataset curated using extended PHQ-9, generated questions have been mostly inquisitive. This is shown by **Model 3**. Still, a lot of generations are around the problem the user mentioned.

no sentence is segmented in between. The motivation for chunking is to ensure no context is lost from the post due to the limitation of T5 to process 512 tokens as input (DLMs in general suffer from such representation limits). We also appended the post title to each chunk to ensure that main idea of each post was captured in it's chunks. This curated dataset tests T5's capability to generate FQs similar to any of the questions in the extended PHQ-9 questionnaire.

Extending PHQ-9 to support FQ generation: PHQ-9 questions are subject to different interpreta-

Post text	Model 1 QGs	Model 2 QGs	Model 3 QGs
I feel depressed . I have felt depressed for a while. I have done a lot of things to try and help myself *through therapy* ,[...] , staying busy , going to university . all I have found is that each time I complete any of these things that I still feel like I am empty . When I do not do those things for more than 10 mins by myself that my thoughts just get dark and I be little everything I have ever done . [....] Also I get depressed because I have dreams where I 'm constantly bombarded by women who are not my wife who want to have relationships with me and it 's a nightly battle to not give in to it when I 'm dreaming . I know it sounds silly but I fear that if I give into the dreams that I might do it in real life . *I do n't want to tell my wife* because she does n't understand other weird dreams that I have had in the past .	• What if I do not do it for more than 10 mins by myself? *[*Closed book question already answered in the post and irrelevant to the diagnostic procedure]* • What might happen to any women that want to have relationships with me? *[*Irrelevant to the diagnostic process]* • What do the women at work do? *[*Irrelevant to the diagnostic process)]* • What kind of dreams are I was depressed? *[*Closed book question generated directly from a statement in the post]*	Would you be happy with them? *[*Irrelevant to the diagnostic process)]* • What's your job? *[*Irrelevant]* • Are you able to reach out to your doctor or psychiatrist? *[*User has already mentioned in the original post that he has talked to his therapist]* • what are your dreams? *[*Closed book question already answered in the post]* • What do you enjoy about your dreams? *[*FQ]* • Do you want to talk to people about what dreams are ? *[*FQ]* • Do you feel as satisfied as yourself when you are dreams? *[*FQ]*	Have you talked to a therapist? *[*Already answered by the user in the post itself]* • Is there anything that helps you calm your feeling for now? *[*FQ]* • "Have you ever gotten any help?" *[*Already answered in the query]* • Do you feel isolated? *[*FQ]* • What are your hobbies? *[*FQ]* • What are your interests? *[*FQ]* • How long have you been waiting for your wife to talk about these dreams? *[*FQ]* • Have you told your wife you're depressed or not? *[*Inquisitive in nature but already answered by the user in original post]*

Table 2: In this example, the generated questions from both Model 2 and Model 3 seem to be relevant FQs, but they are not assessing the severity of the mental health condition, despite Model 3 being fine-tuned on a dataset filtered by PHQ-9 questions. In comparison to the qualitative outcome in Table 1, this showcases the inability of T5 to support mental health triage.

tions depending on patient-MHP interaction. Additionally, nine questions are limited in scope for use in tasks like fine-tuning and similarity-based performance evaluations. Therefore, to increase the strength of PHQ-9, we collaborated with MHPs to create sub-questions for each question in PHQ-9. First, we used Google SERP API[4] and Microsoft Bing Search API[5] to retrieve "People-Also-Ask" questions. For each question, we retrieved 40 questions by manually searching and assessing their relevance to PHQ-9 questions. Next, we provided the set of 360 questions to three MHPs for assessment. MHPs evaluated the questions on two grounds:(a) Whether they would ask such a question to a patient? (relevance) (b) If yes, when should such a question be asked? (rank). Based on their ratings, we created a final set of 134 sub-questions for the nine questions in PHQ-9[6] resulting in a total of 143 questions.

Models for FQ Generation: We used an off-the-bench T5-base QG model that was fine-tuned on the SQuAD 2.0 question generation dataset (Rajpurkar et al., 2018) **[Model 1]**. Next, we fine-tuned Model 1 on r/depression_help posts and comments. To align with our task of making T5 generate relevant FQs, we filtered out comments which were non-interrogative. We kept only the interrogative statements asked by Reddit users in the comments **[Model 2]**. Not all interrogative comments by Reddit users are *diagnostically relevant* FQs (Eg: "Can you use MS Excel?", "Were you interactions on FaceTime?"). To remove such questions, we further filtered the dataset by calculating the maximum BLEURT score between the question (present in the comments) and the questions in extended PHQ-9. We applied a threshold of 0.60 to this score[7]. This removed harmful and diagnostically irrelevant questions while preserving contextual, semantically relevant, and legible questions **[Model 3]**. See Fig 1 for examples of diagnostically relevant questions.

[4]https://serpapi.com/

[5]https://www.microsoft.com/en-us/bing/apis/bing-web-search-api

[6]Questions in extended PHQ-9 : link

[7]empirically judged

$\|\hat{Q}\|(\downarrow)$	Hit Rate on BLEURT			Hit Rate on Rouge-L		
$\delta(\rightarrow)$	0.4	0.5	0.7	0.4	0.5	0.7
Model 1: Pre-trained T5						
5	0.5417	0.1233	0.0020	0.1241	0.0386	0.0005
10	0.5400	0.1203	0.0010	0.1290	0.0400	0.0010
15	0.5368	0.1250	0.0013	0.1266	0.0384	0.0009
Model 2: Fine-Tuned T5 on r/depression_help						
5	0.6657	0.2804	0.0097	0.3445	0.1560	0.0100
10	0.6691	0.2792	0.0104	0.3481	0.1590	0.0098
15	0.6726	0.2787	0.0104	0.3476	0.1588	0.0094
Model 3: T5 Fine-tuned on r/depression_help filtered by PHQ-9						
5	0.9489	0.7088	0.1261	0.7457	0.4937	0.0903
10	**0.9542**	**0.7126**	0.1272	0.7460	**0.5002**	**0.0947**
15	0.9514	0.7098	**0.1274**	**0.7484**	0.4945	0.0916

Table 3: Experimental results comparing different models in generating questions that match the sub-questions in PHQ-9. $\hat{Q}$ is the set of generated questions in each chunk. The performance is recorded over all the generated questions ($\hat{\mathbf{Q}}$). δ was used as the threshold on the similarity between generated question and PHQ-9 sub-questions while calculating hit rate. BLEURT records semantic similarity, whereas Rouge-L records the longest common subsequence exact match between generated question and PHQ-9 sub-questions. The highest performance on semantic and string similarity is bolded. Acceptable performance in Model 3 achieved using PHQ-9 motivated us to prepare **PRIMATE**.

A User's Post

Should I use the psychological help service that my university provides for free ?.

Lately I have been [feeling really low **(Q2, Q3)**]. [I can't make myself leave the bed **(Q3, Q9)**], [I start crying out of the blue and everything is just so heavy **(Q1, Q4)**]. I think I have [always suffered from some kind of depression **(Q2)**] but I have never been to therapy because [I could not afford it **(Q1)**] on my own and [my family did not ever suspect anything **(Q1)**]. Now I live on my own in another city .

[...] my university provides psychological help for students for free . Do you think I should give it a go ? [......] I have nothing to lose because it's free . Did you ever try anything like that ?

Process Knowledge Annotation using PHQ-9

Q1: Feeling bad about yourself or that you are a failure or have let yourself or your family down, YES

Q2: Feeling down depressed or hopeless, YES

Q3: Feeling tired or having little energy, YES

Q4: Little interest or pleasure in doing things, YES

Q5: Moving or speaking so slowly that other people could have noticed Or the Opposite being so fidgety or restless that you have been moving around a lot more than usual, **NO**

Q6: Poor appetite or overeating, **NO**

Q7: Thoughts that you would be better off dead or of hurting yourself in some way, **NO**

Q8:Trouble concentrating on things such as reading the newspaper or watching television, **NO**

Q9: Trouble falling or staying asleep or sleeping too much, YES

Figure 3: A post in **PRIMATE** which is annotated with PHQ-9.The questions marked "YES" are answerable by DLMs using the mental health specific cues from user text. The questions marked "NO" are the questions a DLM should consider asking as FQs. Sentences within [] were taken as signals that the "YES" marked questions had already been answered in the post .

Analysis of Models for Question Generation: Out of the 21k questions, performance of Models 1, 2, and 3 were examined on those 2003 posts that had at least one interrogative comment. Each of the three models was made to generate FQs in sets of 5, 10, and 15 through nucleus sampling (Holtzman et al., 2019). For a generated question, BLEURT score was computed with each question in Extended PHQ-9 and the maximum among those scores was taken as the score for the generated question. A clear distinction between models 1, 2, and 3 is the nature of the questions asked. Model 1 generated closed book questions, whereas Model 2 and 3 seem to show some inquisitive nature and seem more focused on the mental health domain, which can be attributed to the after effect of fine-tuning on Reddit (see Table 1 and 2). We captured the performance of the models quantitatively using 'hit rate' as a metric. For a generated question ($\hat{q}$), we denote :

$$score(\hat{q}) = max(bleurt\_score(\hat{q}, q_1), bleurt\_score(\hat{q}, q_2), ..., bleurt\_score(\hat{q}, q_{143})),$$

where $q_1, q_2...q_{143} \in$ Extended-PHQ-9. Across all 2003 posts, we had C = 2575 chunks[8]. Let total number of questions generated by a model be $|\hat{\mathbf{Q}}|$ and $|\hat{Q}|$ denote the number of question generated by the model for a given chunk. For experimentation, we set $|\hat{Q}|$ to have values {5, 10, 15}. Thus, $|\hat{\mathbf{Q}}| = |\hat{Q}| * C$. Then the **Hit Rate** for a model was computed as:

$$\text{Hit Rate(model, } |\hat{Q}|) = \frac{\sum_{\hat{q} \in \hat{\mathbf{Q}}} \mathbf{I}(score(\hat{q}) > \delta)}{|\hat{\mathbf{Q}}|},$$

where δ is the threshold on the similarity between generated question in a chunk and sub-questions in PHQ-9 and $I[\varphi]$ is the indicator function taking values 0 or 1 for a predicate φ (Table 3 has the scores).

Inference: **(1)** Regardless of fine-tuning and filtering based on PHQ-9 questions, inherently, T5 does not capture the meaning and usage of the words in the mental health context. Moreover, T5 fails to generate legible and relevant FQs as safe as PHQ-9 questions. Therefore, we scrutinize the generated FQs by mapping them to most similar questions in extended PHQ-9. Examples of irrelevant generations by T5 that it thought were relevant are: (a) "Wtf?" (generated FQ) was found most similar to "Do you have hope?" (PHQ-9) (b) "What did Boyfriend suffocate me with during his break up a week after I got a diagnosis?" (generated FQ) was found most similar to "What do you think makes you a failure" (PHQ-9). The previous generated question is redundant as the answer to it was already present in the original post. **(2)** Many generated questions contain extreme language due to the informal nature of the Reddit platform, which is very sensitive issue, especially in the mental health domain. Examples are: "Did you f***ing realize that f***ing people are f***ing too?" (generated FQ) was found to be the most similar to "What do you think makes you a failure?". Thus, T5 and its variants need to capture "what the user knows and has already mentioned in his post" by checking which PHQ-9 questions are already answerable using the user's post before generating the next probable FQs in order to avoid redundancy.

[8]Chunking was done as DLM accepts a maximum input length of 512 tokens.

4 PRIMATE for FQ Generation

We conceptualize our approach on the duality of data and the process knowledge contained in PHQ-9 (see Figure 4). First, a BERT Answerability Evaluator identifies which questions in PHQ-9 are already answerable (using the user's initial description of his/her condition in the post) and which ones need more information to be answerable. The latter type of questions form candidates for training a generative DLM for FQ generation. We present **PRIMATE**, a dataset consisting of Reddit posts containing user situations describing their health conditions and whether the questions in PHQ-9 are answerable using the content in the posts. Each question is attributed with a binary "yes" or "no" label stating whether the user's description already contains the answer to that question (see Table 4). **PRIMATE** was created from a month long annotation-evaluation cycle between MHPs and crowd workers. A total of five crowd workers performed this task, achieving an initial annotator agreement of 67% using Fleiss kappa. Subsequently, the MHPs assessed the quality of annotations and provided their suggestion for improvement, leading to an acceptable agreement score of 85%. A sample annotated post in **PRIMATE** is shown in Figure 3.

BERT as Answerability Evaluator: While Model 3 shows respectable performance (Table 3), even the FQs generated by Model 3 may not yield the most efficient capture of the PHQ-9 related questions (evident from the low hit rate at a higher threshold) (δ). The MHPs would probably have a more streamlined, focused questioning strategy. For efficient MHPs and AI collaboration, we propose to guide the questioning in a more systematic way by predicting if the user post already has answers to the PHQ-9 questions. This is first posed as a binary classification problem over nine PHQ-9 questions. Thereafter, the approach is to generate questions similar to the PHQ-9 questions that do not have answers in the post. Thus, we train

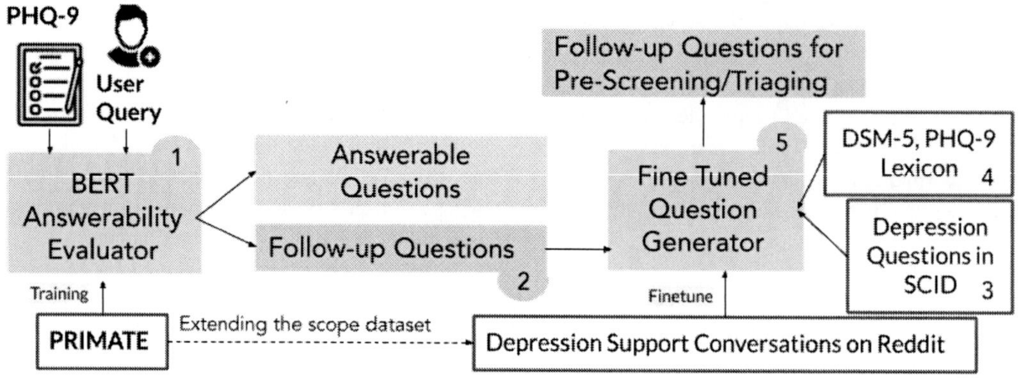

Figure 4: 1. Answerability evaluator: A BERT model is trained in a supervised setting to be an evaluator of whether a PHQ-9 question can be answered in a given user post (binary) using PRIMATE. For nine PHQ-9 questions, we require nine such evaluators. 2. Follow up questions: PHQ-9 questions that are not already answerable using the user post form candidates for follow up. 3. SCID: Corresponding to each PHQ-9 question, the SCID describes a clinician approved sub-sequence of questions to obtain the answer to the follow up question. 4. Use existing PHQ-9 and DSM-5 lexicons (Yazdavar et al., 2017) to filter the question to be generated. 5. Generate FQs using T5 fine-tuned on external domain-specific knowledge and the large-scale depression support conversation dataset created from Reddit and **PRIMATE**.

PHQ-9	**Number of Posts**	
Questions	**With Answer (Yes)**	**W/o Answer (No)**
Q1	1679	324
Q2	1664	339
Q3	686	1317
Q4	949	1054
Q5	530	1473
Q6	195	1808
Q7	741	1262
Q8	196	1807
Q9	374	1629

Table 4: Distribution of 2003 posts in **PRIMATE** according to whether the text in the post answers a particular PHQ-9 question. Through this imbalance, **PRIMATE** presents its importance in training DLM(s) to identify potential FQs in PHQ-9 that would guide a generative DLM(s) to conduct a discourse with a patient with a vision to assist MHPs in triage. Q1-Q9 are described in Figure 3

BERT[9] (a transformer-based DLM) as a classifier on the **PRIMATE** dataset. We plan to further use the classification outcome from the BERT model to drive the direction of further questioning with the patient in a more controlled manner. This process can lead to high efficiency and completion of the mental health triaging in as few questions as possible.

δ ($\rightarrow$)	0.5	0.7	0.9	Class-
PHQ-9($\downarrow$)	MCC	MCC	MCC	Type
Q1	0.0	0.17	0.17	W
Q2	0.43	0.45	0.52	S
Q3	0.41	0.46	0.33	M
Q4	0.14	0.19	0.13	W
Q5	0.63	0.65	0.66	S
Q6	0.47	0.43	0.27	W
Q7	0.66	0.68	0.7	S
Q8	0.1	0.0	0.0	W
Q9	0.62	0.56	0.39	M

Table 5: We record the Matthews Correlation Coefficient (MCC) to measure the performance of the Evaluator (see Figure 4). The MCC score for all 9 questions across different thresholds is in the range 0 to +1 (low to high positive relationships). The MCC for some configurations runs into a divide by zero error, and we replace this value with 0.0. **W**: model is unable to learn cues to determine answerability in a post. **M**: model is uncertain whether a particular PHQ-9 question is answerable or not. **S**: answerability can be determined by the model with high reliability. Class-Type: Classification Type when $\delta = 0.9$

[9]BERT end-to-end training perform well compared to baselines Electra(Clark et al., 2019), and MedBERT(Gu et al., 2021)

Performance Analysis: We report the Matthews Correlation Coefficient (MCC) scores in table 5. MCC is a reliable metric to assess a model's classification over an imbalanced dataset, particularly useful when we are interested in all four categories of confusion matrix: true positives (answerable questions (AQ)), true negatives (FQ candidates), and false alarms (false negatives and positives). As **PRIMATE** shows a disproportional distribution of AQs (yes) and FQs (no), MCC is an appropriate metric (Chicco and Jurman, 2020). We base our analysis on the consistency of BERT classifier on varying threshold (δ) in table 5. A score between 0.0 to 0.30 (Type **W: Weak**) on MCC means the model is only able to find a negligible to weak positive relationship between input and output. In our context, a score in this range for a particular PHQ-9 question means that model is unable to effectively learn the cues needed to judge the answerability of that question in user posts. A score between 0.30 and 0.40 (Type **M: Maybe**) means that the model is able to learn a moderately positive relationship, interpreted as ambiguity in the model to judge whether a particular PHQ-9 question is answerable from user posts. MCC scores between 0.40 to 0.70 (Type **S: Strong**) for a question in PHQ-9 means that the model can effectively judge whether that question is answerable in user posts . Any score above 0.70 makes the model's judgements even more reliable. This experiment completes steps 1 and 2 in Figure 4. Steps 3, 4 and 5 are concerned with the task of FQ generation by fine-tuning the T5 DLM as a generator over r/depression_help and other depression support communities on Reddit. The FQ generations will be controlled using the process knowledge in SCID which is consulted for interviewing by MHPs. Further, PHQ-9 lexicons are leveraged for promoting diversity and filtering irrelevant FQ generations. We leave this process of FQ generations to shape discourse as future work.

5 Conclusion

This paper demonstrated the importance of data and process knowledge to adapt DLMs for generating FQs that would assist MHPs in triaging depression. Our experiments show that without process knowledge, DLMs hallucinate by generating unsafe, incoherent, and irrelevant questions that are not helpful for MHPs in pre-screening or triaging. The challenge lies in the inability of the DLMs to judge from the set of generated questions, which is a potential effective FQ to ask based on the user information. The improved question generation performance of DLMs fine-tuned on conversational data filtered by process knowledge encouraged us to prepare PRIMATE. **PRIMATE** can train DLMs to judge 'whether a user's description of their mental health condition already contains an answer to a particular question in PHQ-9', which would eventually guide coherent FQ generations. We leave our approach for FQ generation as future work, but provide sufficient details on the broader forms of knowledge needed in realizing such a pipeline.

Limitations: We are yet to scale our understanding to other mental health disorders, such as anxiety using GAD-7 and Suicidality using C-SSRS (Jiang et al., 2020). Further, we are yet to investigate whether **PRIMATE**, along with the knowledge in SCID can make DLMs transferable across multiple mental health disorders, especially the ones comorbid with depression. Also, there is a need for a clinically explainable safety metric for our task.

Ethical Considerations: Mental health communities on Reddit offer a crowd perspective on various disorders wherein the FQs in the comments highlight the good intentions of Reddit users to help users with conditions, such as depression. We take such interactions as a proxy for improving patient-MHP interactions. (Benton et al., 2017) described that studies involving user-generated content are exempted from the IRB requirement as long as the data source is public and the user's identity is not recognizable. Apart from being publicly available, Reddit users are anonymous, and we further work with random user IDs. Since we make **PRIMATE** public for research use, we use a Data Use Agreement (Losada and Crestani, 2016) for responsible dissemination of the dataset.

6 Acknowledgment

We acknowledge partial support from the National Science Foundation (NSF) award #2133842 "EAGER: Advancing Neuro-symbolic AI with Deep Knowledge-infused Learning," with PI Dr. Amit Sheth. Any opinions, findings, and conclusions or recommendations expressed in this material are those of the author(s) and do not necessarily reflect the views of the NSF.

References

Tim Althoff, Kevin Clark, and Jure Leskovec. 2016. Large-scale analysis of counseling conversations: An application of natural language processing to mental health. *Transactions of the Association for Computational Linguistics*, 4:463–476.

Adrian Benton, Glen Coppersmith, and Mark Dredze. 2017. Ethical research protocols for social media health research. In *Proceedings of the First ACL Workshop on Ethics in Natural Language Processing*, pages 94–102, Valencia, Spain. Association for Computational Linguistics.

Sebastian Borgeaud, Arthur Mensch, Jordan Hoffmann, Trevor Cai, Eliza Rutherford, Katie Millican, George van den Driessche, Jean-Baptiste Lespiau, Bogdan Damoc, Aidan Clark, Diego de Las Casas, Aurelia Guy, Jacob Menick, Roman Ring, Tom Hennigan, Saffron Huang, Loren Maggiore, Chris Jones, Albin Cassirer, Andy Brock, Michela Paganini, Geoffrey Irving, Oriol Vinyals, Simon Osindero, Karen Simonyan, Jack W. Rae, Erich Elsen, and Laurent Sifre. 2021. Improving language models by retrieving from trillions of tokens. *CoRR*, abs/2112.04426.

Benjamin Brodey, Susan E Purcell, Karen Rhea, Philip Maier, Michael First, Lisa Zweede, Manuela Sinisterra, M Brad Nunn, Marie-Paule Austin, and Inger S Brodey. 2018. Rapid and accurate behavioral health diagnostic screening: initial validation study of a web-based, self-report tool (the sage-sr). *Journal of Medical Internet Research*, 20(3):e9428.

Davide Chicco and Giuseppe Jurman. 2020. The advantages of the matthews correlation coefficient (mcc) over f1 score and accuracy in binary classification evaluation. *BMC genomics*, 21(1):1–13.

Kevin Clark, Minh-Thang Luong, Quoc V Le, and Christopher D Manning. 2019. Electra: Pre-training text encoders as discriminators rather than generators. In *International Conference on Learning Representations*.

CounselChat. Mental health answers from counselors.

Orianna Demasi, Marti A Hearst, and Benjamin Recht. 2019. Towards augmenting crisis counselor training by improving message retrieval. In *Proceedings of the Sixth Workshop on Computational Linguistics and Clinical Psychology*, pages 1–11.

Emily Dinan, Gavin Abercrombie, A. Stevie Bergman, Shannon L. Spruit, Dirk Hovy, Y-Lan Boureau, and Verena Rieser. 2021. Anticipating safety issues in E2E conversational AI: framework and tooling. *CoRR*, abs/2107.03451.

Manas Gaur, Amanuel Alambo, Joy Prakash Sain, Ugur Kursuncu, Krishnaprasad Thirunarayan, Ramakanth Kavuluru, Amit Sheth, Randy Welton, and Jyotishman Pathak. 2019. Knowledge-aware assessment of severity of suicide risk for early intervention. In *The World Wide Web Conference*, pages 514–525.

Manas Gaur, Kalpa Gunaratna, Vijay Srinivasan, and Hongxia Jin. 2021. Iseeq: Information seeking question generation using dynamic meta-information retrieval and knowledge graphs. *arXiv preprint arXiv:2112.07622*.

Jonathan Gratch, Ron Artstein, Gale Lucas, Giota Stratou, Stefan Scherer, Angela Nazarian, Rachel Wood, Jill Boberg, David DeVault, Stacy Marsella, et al. 2014. The distress analysis interview corpus of human and computer interviews. In *Proceedings of the Ninth International Conference on Language Resources and Evaluation (LREC'14)*, pages 3123–3128.

Yu Gu, Robert Tinn, Hao Cheng, Michael Lucas, Naoto Usuyama, Xiaodong Liu, Tristan Naumann, Jianfeng Gao, and Hoifung Poon. 2021. Domain-specific language model pretraining for biomedical natural language processing. *ACM Transactions on Computing for Healthcare (HEALTH)*, 3(1):1–23.

Kelvin Guu, Kenton Lee, Zora Tung, Panupong Pasupat, and Ming-Wei Chang. 2020. Realm: Retrieval-augmented language model pre-training. *arXiv preprint arXiv:2002.08909*.

Ari Holtzman, Jan Buys, Li Du, Maxwell Forbes, and Yejin Choi. 2019. The curious case of neural text degeneration. *arXiv preprint arXiv:1904.09751*.

Rongyao Huang. 2015. *Language use in teenage crisis intervention and the immediate outcome: A machine automated analysis of large scale text data*. Ph.D. thesis, Master's thesis, Columbia University.

Zheng Ping Jiang, Sarah Ita Levitan, Jonathan Zomick, and Julia Hirschberg. 2020. Detection of mental health from reddit via deep contextualized representations. In *Proceedings of the 11th International Workshop on Health Text Mining and Information Analysis*, pages 147–156.

Alison Karasz, Christopher Dowrick, Richard Byng, Marta Buszewicz, Lucia Ferri, Tim C Olde Hartman, Sandra Van Dulmen, Evelyn van Weel-Baumgarten, and Joanne Reeve. 2012. What we talk about when we talk about depression: doctor-patient conversations and treatment decision outcomes. *British Journal of General Practice*, 62(594):e55–e63.

Mojtaba Komeili, Kurt Shuster, and Jason Weston. 2021. Internet-augmented dialogue generation. *CoRR*, abs/2107.07566.

Kurt Kroenke, Robert L Spitzer, and Janet BW Williams. 2001. The phq-9: validity of a brief depression severity measure. *Journal of general internal medicine*, 16(9):606–613.

Patrick Lewis, Ethan Perez, Aleksandra Piktus, Fabio Petroni, Vladimir Karpukhin, Naman Goyal, Heinrich Küttler, Mike Lewis, Wen-tau Yih, Tim Rocktäschel, et al. 2020. Retrieval-augmented generation for knowledge-intensive nlp tasks. *Advances in Neural Information Processing Systems*, 33:9459–9474.

D. Losada and F. Crestani. 2016. A test collection for research on depression and language use. In *Proc. of Experimental IR Meets Multilinguality, Multimodality, and Interaction, 7th International Conference of the CLEF Association, CLEF 2016*, pages 28–39, Evora, Portugal.

Gaur Manas, Vamsi Aribandi, Ugur Kursuncu, Amanuel Alambo, Valerie L Shalin, Krishnaprasad Thirunarayan, Jonathan Beich, Meera Narasimhan, Amit Sheth, et al. 2021. Knowledge-infused abstractive summarization of clinical diagnostic interviews: Framework development study. *JMIR Mental Health*, 8(5):e20865.

Adam S Miner, Arnold Milstein, Stephen Schueller, Roshini Hegde, Christina Mangurian, and Eleni Linos. 2016. Smartphone-based conversational agents and responses to questions about mental health, interpersonal violence, and physical health. *JAMA internal medicine*, 176(5):619–625.

Colin Raffel, Noam Shazeer, Adam Roberts, Katherine Lee, Sharan Narang, Michael Matena, Yanqi Zhou, Wei Li, and Peter J Liu. 2019. Exploring the limits of transfer learning with a unified text-to-text transformer. *arXiv preprint arXiv:1910.10683*.

Pranav Rajpurkar, Robin Jia, and Percy Liang. 2018. Know what you don't know: Unanswerable questions for squad. *CoRR*, abs/1806.03822.

Thibault Sellam, Dipanjan Das, and Ankur Parikh. 2020. Bleurt: Learning robust metrics for text generation. In *Proceedings of the 58th Annual Meeting of the Association for Computational Linguistics*, pages 7881–7892.

Romal Thoppilan, Daniel De Freitas, Jamie Hall, Noam Shazeer, Apoorv Kulshreshtha, Heng-Tze Cheng, Alicia Jin, Taylor Bos, Leslie Baker, Yu Du, YaGuang Li, Hongrae Lee, Huaixiu Steven Zheng, Amin Ghafouri, Marcelo Menegali, Yanping Huang, Maxim Krikun, Dmitry Lepikhin, James Qin, Dehao Chen, Yuanzhong Xu, Zhifeng Chen, Adam Roberts, Maarten Bosma, Yanqi Zhou, Chung-Ching Chang, Igor Krivokon, Will Rusch, Marc Pickett, Kathleen S. Meier-Hellstern, Meredith Ringel Morris, Tulsee Doshi, Renelito Delos Santos, Toju Duke, Johnny Soraker, Ben Zevenbergen, Vinodkumar Prabhakaran, Mark Diaz, Ben Hutchinson, Kristen Olson, Alejandra Molina, Erin Hoffman-John, Josh Lee, Lora Aroyo, Ravi Rajakumar, Alena Butryna, Matthew Lamm, Viktoriya Kuzmina, Joe Fenton, Aaron Cohen, Rachel Bernstein, Ray Kurzweil, Blaise Aguera-Arcas, Claire Cui, Marian Croak, Ed Chi, and Quoc Le. 2022. Lamda: Language models for dialog applications. *CoRR*, abs/2201.08239.

Laura Weidinger, John Mellor, Maribeth Rauh, Conor Griffin, Jonathan Uesato, Po-Sen Huang, Myra Cheng, Mia Glaese, Borja Balle, Atoosa Kasirzadeh, Zac Kenton, Sasha Brown, Will Hawkins, Tom Stepleton, Courtney Biles, Abeba Birhane, Julia Haas, Laura Rimell, Lisa Anne Hendricks, William S. Isaac, Sean Legassick, Geoffrey Irving, and Iason Gabriel. 2021. Ethical and social risks of harm from language models. *CoRR*, abs/2112.04359.

Joseph Weizenbaum. 1983. Eliza — a computer program for the study of natural language communication between man and machine. *Commun. ACM*, 26(1):23–28.

Amir Hossein Yazdavar, Hussein S Al-Olimat, Monireh Ebrahimi, Goonmeet Bajaj, Tanvi Banerjee, Krishnaprasad Thirunarayan, Jyotishman Pathak, and Amit Sheth. 2017. Semi-supervised approach to monitoring clinical depressive symptoms in social media. In *Proceedings of the 2017 IEEE/ACM International Conference on Advances in Social Networks Analysis and Mining 2017*, pages 1191–1198.

Masking Morphosyntactic Categories to Evaluate Salience for Schizophrenia Diagnosis

Yaara Shriki
School of Computer Science
College of Management
Academic Studies
yaara.shriki@cs.colman.ac.il

Ido Ziv
Meuhedet Health Services
ido.z@meuhedet.co.il

Nachum Dershowitz
School of Computer Science
Tel Aviv University
nachum@tau.ac.il

Eiran Vadim Harel
Beer Yaakov Mental Heath Center
Beer Yaakov, Israel
eiran.harel@moh.gov.il

Kfir Bar
School of Computer Science
College of Management Academic Studies
kfirb@colman.ac.il

Abstract

Natural language processing tools have been shown to be effective for detecting symptoms of schizophrenia in transcribed speech. We analyze and assess the contribution of the various syntactic and morphological categories towards successful machine classification of texts produced by subjects with schizophrenia and by others. Specifically, we fine-tune a language model for the classification task, and mask all words that are attributed with each category of interest. The speech samples were generated in a controlled way by interviewing inpatients who were officially diagnosed with schizophrenia, and a corresponding group of healthy controls. All participants are native Hebrew speakers. Our results show that nouns are the most significant category for classification performance.

1 Introduction

Psychotic disorders such as schizophrenia are characterized by several symptoms, such as delusions, hallucinations, and thought disorders. Thought disorders are described as disturbances in the normal way of thinking, typically presented as various language impairments, such as disorganized speech, which is related to abnormal semantic associations between words (Aloia et al., 1998), and poverty of speech, a thought disorder that is associated with impairments in lexico-semantic retrieval (Nagels et al., 2016). Disorganized speech is divided into several markers, such as derailment, characterized by the usage of unrelated concepts in a conversation; tangentiality, which happens when providing oblique or irrelevant answers to a question; and incoherence, also known as "word salad", refers to speech that is incomprehensible at times due to multiple grammatical and semantic inaccuracies (Bar et al., 2019).

The diagnosis of schizophrenia is mostly based on a professional psychiatric review. However, some studies show that a computational linguistic analysis may help with diagnosis. Fraser et al. (1986), for example, demonstrated that by using a discriminant function analysis of linguistic variables it is possible to predict diagnoses with an accuracy rate of 79%.

There have been many attempts to study speech impairments that are related to thought disorders using a computational method. Some of those studies analyze the frequency of using different part-of-speech categories, such as nouns and verbs. For example, Obrębska and Obrębski (2007) reported a significantly lower frequency of adjectives in schizophrenic speech than in healthy control speech. On the other hand, they reported a higher frequency of verbs used by patients. Tang et al. (2020) measured a low frequency of adverbs in speech produced by patients with schizophrenia. Ziv et al. (2022) analyzed speech produced by Hebrew speaking patients with schizophrenia and reported low frequencies of words inflected in the third person or in the past tense. Aligned with previous work, they also reported lower frequencies of adverbs. It has been shown (Kircher et al., 2005) that patients with schizophrenia are produc-

Proceedings of the Eighth Workshop on Computational Linguistics and Clinical Psychology, pages 148 - 157
July 15, 2022 ©2022 Association for Computational Linguistics

ing grammatically simpler speech than healthy people. The results are not always consistent; Tang et al. (2021), for example, reported high frequencies of adverbs and adjectives in schizophrenic speech, in contrast to the reports made by other works. Until very recently, the large majority of those studies were conducted with English speaking patients.

One of the most popular technologies in natural language processing (NLP) is language modelling. A language model is essentially a function that assigns a probability to a given sequence of words occurring in a sentence. There are different ways to fit a language model to a certain distribution, typically using massive collections of texts. An autoregressive model conditions the probability of a word on the text that has already been seen in direction of reading. On the other hand, masked language models (MLM) are given the full sentence, while learning to assign probability to a randomly chosen hidden (masked) word. Such models are typically used as the basis for an algorithm that aims at solving a specific downstream task, such as sentiment analysis or document classification. In the first phase, the models are pre-trained for the word-probability assignment using a large unlabeled collection of texts, and later are fine-tuned on a labeled dataset for a specific downstream task. While the autoregressive models are more suitable for generation tasks, MLMs are typically the best option for fine-tuning on classification tasks.

This development of pre-trained language models provides us with the opportunity to examine the importance of certain morphosyntactic categories in speech of patients with schizophrenia, and compare it to that of a healthy control group. Specifically, we fine-tune an MLM to classify transcribed speech segments into patient or control categories, and examine its performance under extreme situations of hiding (masking) words that belong to a specific syntactic or morphological category.

While most existing techniques use some sort of counting method, in this study, we explore an alternative innovative way for assessing the salience of a specific category for detecting schizophrenic speech. We utilize the original masking technique of an MLM, by naturally masking out specific morphosyntactic categories and measure the performance of the model on a downstream classification task.

The experimental results show a decrease in prediction accuracy once nouns are masked, suggesting that nouns are more informative than other categories we tested for differentiating between patients and controls. Our participants are all native Hebrew speakers.

2 Related Work

Computational modeling has been studied in relation to cognitive disorders in order to fill the gap between theoretical models and biological evidence. Lanillos et al. (2020) reviews popular neural network models for autism spectrum disorder and schizophrenia, using different types of input. Both disorders are characterized by an altered perception of the world. According to this review, models of schizophrenia mainly concentrate on positive symptoms, such as hallucinations and delusional behavior (e.g., Hoffman and McGlashan (1997); Horn and Ruppin (1995)). However, there are also models that target other symptoms such as disturbances of attention (Cohen and Servan-Schreiber, 1992) and movement disorders (Yamashita and Tani, 2012).

The use of computational linguistic models has been applied to studying language abnormalities related to mental illness, specifically schizophrenia. Disorganized speech, including derailment, incoherence, and tangentiality, is among the common symptoms of schizophrenia being studied by researchers using computational methods (e.g., Bedi et al. (2015); Pauselli et al. (2018); Iter et al. (2018); Bar et al. (2019); Just et al. (2020)). Hitczenko et al. (2021) reviews computational methods that perform linguistic analysis of psychosis, focusing on three language abnormalities: disorganized speech, poverty of speech, and flat affect. Many studies have employed latent semantic analysis (LSA) and word embedding models (e.g., word2vec and GloVe) to measure disorganized speech. Typically, the embeddings are used to measure semantic similarity between words in the sentence, or between entire sentences or paragraphs, to assess semantic cohesion as a predictor for disorganized speech. In several studies (e.g., Elvevåg et al. (2007); Iter et al. (2018); Just et al. (2019)), psychosis patients scored significantly higher on disorganization than controls. However, Hitczenko et al. (2021) argues that the measures are not consistent across other studies.

As mentioned in the previous section, most of those works analyze transcribed speech spoken in

English, which is characterized by a relatively simple morphological system. Some recent studies have been exploring similar techniques applied to other languages, such as German (Just et al., 2020) and Hebrew (Bar et al., 2019). The latter have studied derailment, a symptom of thought disorder characterized by switching between topics and jumping from one disconnected thought to another. They measure derailment in speech through semantic similarity of adjacent words using their embeddings. It was found that patients with schizophrenia are more likely to derail than healthy controls, consistent with previous studies (Bedi et al., 2015; Iter et al., 2018). Further, they examine incoherence in schizophrenic patients, to see how they use adjectives and adverbs to describe specific nouns and verbs. Their analysis makes use of a dependency parser for Hebrew, which yields a word-dependency list for each sentence. Using dependencies, they discovered that the adjectives and adverbs used by the controls are more similar to those commonly used to describe the same nouns and verbs.

There are not many works that leverage language models to analyse text for detecting mental health symptoms, such as we do. In a recent work (Tang et al., 2021), BERT (Devlin et al., 2019), a large English language model, has been used to encode full sentences and compare the resulting embeddings of adjacent sentences for measuring tangentiality. Their results reflect increased tangentiality among patients with schizophrenia.

In our work, we use a language model as a tool for assessing the contribution of six morphosyntactic categories to the classification of transcribed speech into patients or controls.

3 Participants and Data Collection

We interviewed 49 males, aged 18–60, divided into control and patient groups, all speaking Hebrew as their first language. The patient group includes 23 inpatients from the Be'er Ya'akov–Ness Ziona Mental Health Center in Israel who were admitted following a diagnosis of schizophrenia. Diagnoses were made by a hospital psychiatrist according to the DSM-5 criteria (American Psychiatric Association DSM-5 Task Force, 2013) and a full psychiatric interview. Each participant was rewarded with approximately \$8. The control group includes 26 men, mainly recruited via an advertisement that we placed on social media. Exclusion criteria for all participants were as follows:

(1) participants whose mother tongue is not Hebrew;

(2) having a history of dependence on drugs or alcohol over the past year;

(3) having a past or present neurological illness; and

(4) using fewer than 500 words in total in their transcribed interview.

Additionally, the control group had to score below the threshold for subclinical diagnosis of depression and post-traumatic stress disorder (PTSD). Most of the control participants scored below the threshold for anxiety. Most of the patients scored above the threshold for borderline or mild psychosis symptoms on a standard measure.[1] See Section 3.1 for more information about the measures we use in this study.

The demographic characteristics of the two groups are presented in Table 1.

Patients were interviewed in a quiet room at the department where they are hospitalized by one of our professional team members, and the control participants were interviewed in a similar room outside the hospital. Each interview lasted approximately one hour. The interviews were recorded and later manually transcribed by a native Hebrew speaking student from our lab. All participants were assured of anonymity, and told that they are free to end the interview at any time.

After signing a written consent, each participant was asked to describe 14 images picked from the Thematic Appreciation Test (TAT) collection; the images were presented one by one. We used the TAT images identified with the following serial numbers: 1, 2, 3BM, 4, 5, 6BM, 7GF, 8BM, 9BM, 12M, 13MF, 13B, 14, and 3GF. All images are black and white, including a mixture of men and women, children and adults. Each picture stands by itself, presented alone and has no relation to the other pictures. The participants were asked to tell a brief story about each image based on four open questions: What led up to the event shown in the picture? What is happening in the picture at this moment? What are the characters thinking and feeling? What is the outcome of the story? The

[1]Our patient group is composed of inpatients who are undergoing treatment with medications; therefore, higher scores were not expected.

	Control	Patients	Statistics
N	26	23	
Age mean (*SD*)	25.46 (6.28)	33.15 (9.72)	$t = 3.38^{**}$
Education years mean (*SD*)	11.96 (0.15)	11.30 (1.15)	$t = 2.98^{**}$
Place of residence (frequencies)			χ^2 (3,55) = 8.84, $p = .03$
Southern Israel	1	7	
Central Israel	22	16	
Northern Israel	2	0	
Jerusalem	1	0	
Marital status (frequencies)			χ^2 (1,49) = 0.055, $p = .81$
Single	4	3	
Married	22	20	
Income (frequencies)			χ^2 (3,49) = 3.06, $p = .38$
Low	5	4	
Lower than average	6	4	
Average	9	13	
Higher than average	6	2	
PANSS positive subscale		8.91 ± 3.91	
PANSS negative subscale		7.82 ± 3.74	
PANSS total subscale		16.73 ± 6.23	

Table 1: Demographic characteristics by group. $^{**}p < .005$.

interviewer remained silent during the respondent's narration and offered no prompts or questions.

After describing the images, the participant was asked to answer four open questions, one by one. The four questions are listed in Table 2. As before, the interviewer remained silent during the respondent's narration and offered no prompts or questions.

Once all 18 components (14 image descriptions and 4 open questions) were answered, each participant was requested to fill in a demographic questionnaire as well as some additional questionnaires for assessing mental-health symptoms, which we describe next.

3.1 Symptom Assessment Measures

3.1.1 Control group

The control participants were assessed for symptoms of depression, PTSD, and anxiety.

Depression. Symptoms of depression were assessed using Beck's Depression Inventory–II (BDI–II) (Beck et al., 1996). The BDI–II is a 21-item inventory rated on a 4-point Likert-type scale (0 = "not at all" to 3 = "extremely"), with summary scores ranging between 0 and 63. Beck et al. (1996) suggested a preliminary cutoff value of 14 as an indicator for mild depression, as well as a threshold of 19 as an indicator for moderate depression. BDI–II has been found to demonstrate high reliability (Gallagher et al., 1982). We use a Hebrew version of BDI–II (Hasenson-Atzmon et al., 2016).

PTSD. Symptoms of PTSD were assessed using the PTSD checklist of the DSM–5 (PCL–5) (Weathers et al., 2013). The questionnaire contains twenty items that can be divided into four subscales, corresponding to the clusters B–E in DSM–5: intrusion (five items), avoidance (two items), negative alterations in cognition and mood (seven items), and alterations in arousal and reactivity (six items). The items are rated on a 5-point Likert-type scale (0 = "not at all" to 4 = "extremely"). The total score ranges between 0 and 80, provided along with a preliminary cutoff score of 38 as an indicator for PTSD. PCL–5 has been found to demonstrate high reliability (Blevins et al., 2015). We use a Hebrew translation of PCL-5 (Bensimon et al., 2013).

Anxiety. Symptoms of anxiety were assessed through the State Trait Anxiety Inventory (STAI) (Spielberger et al., 1970). The STAI questionnaire consists of two sets of twenty self-reporting measures. The STAI measure of state anxiety (S-anxiety) assesses how respondents feel "right now, at this moment" (e.g., "I feel at ease"; "I feel upset"), and the STAI measure of trait anxiety (T-anxiety) targets how respondents "generally feel" (e.g., "I am a steady person"; "I lack self-

ID	Question
1	Tell me as much as you can about your bar mitzvah.*
2	What do you like to do, mostly?
3	What are the things that annoy you the most?
4	What would you like to do in the future?

Table 2: Four open questions asked during the interview. *Bar mitzvah is a Jewish confirmation ceremony for boys who have reached the age of 13.

confidence"). For each item, respondents are asked to rate themselves on a 4-point Likert scale, ranging from 1 = "not at all" to 4 = "very much so" for S-anxiety, and from 1 = "almost never" to 4 = "almost always" for T-anxiety. Total scores range from 20 to 80, with a preliminary cutoff score of 40 recommended as indicating clinically significant symptoms for the T-Anxiety scale (Knight et al., 1983). STAI has been found to demonstrate high reliability (Barnes et al., 2002). We use a Hebrew translation of STAI (Saka and Gati, 2007).

3.1.2 Patients

Psychosis symptoms were assessed by the 6-item Positive And Negative Syndrome Scale (PANSS–6) (Østergaard et al., 2016). The original 30-item PANSS (PANSS–30) is the most widely used rating scale for schizophrenia, but it is relatively long for use in clinical settings. The items in PANSS–6 are rated on a 7-point scale (0 = "not at all" to 6 = "extremely"). The total score ranges from 0 to 36, with a score of 14 representing the threshold for mild schizophrenia, and a score between 10 and 14 defined as borderline disease or as remission. PANSS–30 has been found to demonstrate high reliability (Lin et al., 2018), while Østergaard et al. (2016) reported a high correlation between PANSS–6 and PANSS–30 (Spearman correlation coefficient = 0.86). We used the Hebrew version of PANSS–6 (Lin et al., 2018). The range of positive and negative symptoms are presented in Table 1.

4 Analysis

4.1 Preprocessing

We treat every response to any one of the 18 questions as a training/evaluation instance for our classifier. Overall we have 414 responses generated by patients, as well as 468 responses that were generated by controls. The responses are written in Hebrew, a morphologically rich Semitic language; Hebrew words are inflected for person, number, and gender, resulting in a relatively complicated word-production process. We preprocess each response using the Ben-Gurion University (BGU) morphological tagger (Adler and Elhadad, 2006), a context-sensitive morphological analyzer for Hebrew. The tagger displays morphosyntactic information for each word in the text, including part-of-speech tags, as well as information about person and number.

4.2 Classification Methodology

We use a Hebrew MLM to classify a response into the two groups, patients or controls. As mentioned before, MLMs are trained in two phases. During the first, also known as pre-training, the model is trained with a large set of text in which 15% of the input tokens are masked using a special mask token for which the model is trained to predict. In the second phase, also known as fine-tuning, the model is adapted for a downstream task using a relatively small set of annotated examples. For classification tasks, such as ours, the common practice is to add another neural dense layer connected to the output vector of the initial token. Therefore, we fine-tune a pre-trained language model using a portion of the dataset, and evaluate its performance on the remaining instances. To assess the contribution of different syntactic and morphological categories for the classification performance, we fine-tune the model several times individually, each time we mask all words of a selected category. We focus on four parts of speech including nouns, verbs, adverbs, and adjectives. Those are all considered as content words, rather than functional ones. In addition, we examine first-person and third-person words. Overall, we examine six morphosyntactic categories.

In all our experiments, we use AlephBERT (Seker et al., 2021), a pre-trained language model for Hebrew, to perform sequence classification using the Transformers library (Wolf et al., 2019). Specifically we use

`AutoModelForSequenceClassification` with the `alephbert-base` model code. The AlephBERT model was trained on data collected from three different Hebrew text sources: the OSCAR corpus (Ortiz Suárez et al., 2020), Hebrew tweets, and the Hebrew Wikipedia.

Given a category M, we begin each experiment by dividing the collection of responses into 80:20 train and test sets, respectively, by making sure the label distribution remains similar to the original dataset. We tokenize each response using the AlephBERT tokenizer, which was designed to truncate responses longer than the model's 512-token limitation. We proceed with the following three steps:

1. We iterate through all train and test responses and mask[2] all tokens that were attributed with M by the BGU Tagger. By design, the AlephBERT tokenizer may break words in the middle; therefore, to be more precise we mask all tokens that were broken from a word that was attributed with M by the BGU tagger. We then fine-tune the model on the masked train set and evaluate on the corresponding masked test set. We use accuracy as an evaluation metric.

2. As a control experiment, we mask tokens randomly by considering every token for masking using a Bernoulli trial with probability equals to the probability of occurrence of M. Same as before, we fine-tune the model on the modified train set and evaluate it on the modified test set.

3. We repeat this experiment 30 times, each time with a different random state, which affects the splitting to train and test sets, as well as on the random masking procedure, and calculate the average accuracy scores for both, M-based masking and random masking. After confirming the scores are normally distributed, we conduct a t-test in order to measure the impact of M-based masking by comparing its accuracy with the one achieved by random masking.

It should be noted that the random states that we use in the experiments are identical across different categories, to make sure that we use the same train/test splits in the 30 executions of each category.

5 Results

Figure 1 displays the probability of each morphosyntactic category to appear in the responses of patients and controls. All participants use more nouns and third-person words than verbs, adverbs, adjectives, and first-person words. The high frequency of third-person words is reasonable, since in most of the interview, the participants were asked to describe the situation as they interpret from a picture that was presented to them. Neither group uses a significant proportion of first or third person tokens. However, we can see that the inpatients use nouns and verbs slightly more often, whereas the controls use more adjectives and adverbs. The difference in adverbs has been confirmed to be statistically significant according to a Welch's unequal variances t-test (at $p < 0.0005$).

The classification results, under different masking conditions are summarized in Table 3. The table displays the difference between the mean classification accuracy of masking each morphosyntactic category (the Morph. Masking column), compared to a random masking of tokens with the same probability of occurrence (the Random Masking column). We run t-tests and provide the outcome statistics in the last two columns. The accuracy at the baseline level (i.e., no masking) is 84.4%. Standard deviations range between 5.5 and 6.5 percent for all accuracy measures. Unsurprisingly, most of the accuracy results listed in the table are below the baseline score. We expected that masking words at high rates may be detrimental for the classification performance. We do see some accuracy scores above the baseline score; however, the differences are minor and has no statistical significance.

We can clearly see the impact of masking nouns and adverbs on the classification performance. Especially when nouns are being masked, the accuracy decreases significantly compared to random masking at the same token-masking rate. The other categories do not show a significant decrease in accuracy compared to random masking at the same rate.

To confirm our results, we design another experiment in which all words in the text *except* nouns and adverbs, are masked. Like before, we compare the classification accuracy with a control model in which we use random masking at the same rate,

[2]With the special token `[MASK]`.

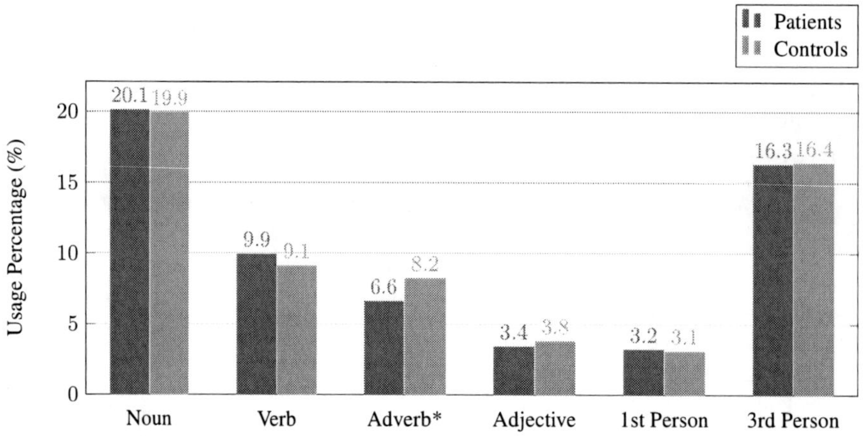

Figure 1: Usage percentage of selected syntactical and morphological categories. *$p < 0.0005$ (per Welch's unequal variance t-test).

as described before. In spite of the fact that we have masked more than 72% of the words in the text, the model has been able to achieve an accuracy of 82.8%, compared to 75% achieved by the random-masking model. This difference has been confirmed to be statistically significant by conventional standards, according to a t-test (at $p < 0.0005$). These results provide a consistent evidence that nouns and adverbs are more important than other categories for the classification task.

6 Discussion

We notice that nouns and adverbs make the biggest impact on the performance of the classifier, suggesting that those syntactic categories are the most informative to the model. Comparing with random masking of the same number of words, the accuracy drops significantly ($p < 0.0005$) when nouns are being masked. With adverbs, the difference in accuracy is less significant ($p = 0.058$). Based on the numbers assembled in Figure 1, we cannot attribute our findings to the frequency of usage of those categories. Whereas nouns are used more frequently than the other categories, adverbs are much less frequent. For adverbs, at least, we see a significant difference in the frequency of usage between the two groups; controls use them more. Adverbs are typically used in tandem with a verb; however, it turns out that the patients use slightly more verbs than the controls, although to an insignificant degree. Therefore, we believe that the significant difference in usage frequency of adverbs may be the reason for the impact that they make on classification performance.

As for nouns, we see no evidence for a usage frequency difference between the two groups. We believe that the reason for the impact made by masking nouns on the classification performance might be related to the importance of nouns in the syntactic tool set of patients with schizophrenia. Our results may suggest that the patients convey their messages more through nouns than through other linguistic categories. Nouns are considered the backbone of a language; it has been shown that English-speaking children acquire knowledge of nouns before verbs (Gentner, 1982). Nouns are considered easier to learn than verbs, probably due to their imageability (McDonough et al., 2011). Therefore, we presume that focusing more on nouns when conveying a message may be an indicator of poverty of speech. The way patients use nouns is slightly different from how controls do. This difference makes it easier for the model to predict schizophrenic symptoms. The source of the difference may be related to the type of nouns that they choose to use in a sentence, the similarity among the nouns in a sentence, or their syntactic relations with other words in the sentence. Since Hebrew is a highly inflected language, it could also be that patients inflect nouns differently than controls. We plan to further investigate the source of the difference in follow up work.

7 Ethical Considerations

This research was approved by the Helsinki Ethical Review Board (IRB) of the Be'er Ya'akov–Ness

Morph. Category	Morph. Masking	Random Masking	t	p
No masking (baseline)	84.4%	-	-	-
Noun	**82.2%**	**84.6%**	4.7809	$p < .0005$*
Verb	83.1%	84.0%	1.2646	$p = .2161$
Adverb	82.3%	83.5%	1.9739	$p = .0580$
Adjective	84.1%	84.9%	1.7963	$p = .0829$
First person	84.5%	84.9%	1.9598	$p = .0597$
Third person	83.2%	82.3%	-1.9527	$p = .0606$

Table 3: Accuracy scores under different masking conditions. *$p < 0.0005$.

Ziona Mental Health Center. Participants were guaranteed anonymity. The data was stored on a secured server, with limited access provided only to the authors of this paper.

Like with every other machine-learning model, there is a risk that the training data is unbalanced. Specifically, we do not intentionally balance the dataset for ethnicity or political affiliation. Moreover, this work is based on interviews with men only. Additionally, the language model that we use, AlephBERT, was trained on large and less controlled datasets. That may introduce some additional aspects of bias. Therefore, our study may harbor the danger of over-reliance on possibly biased machine tools.

We do not mean to suggest that an algorithm can or should be used to diagnose schizophrenia automatically. This study should not be considered as a building block for an apparatus that takes automatic decisions about topics related to mental health. Our intention is, rather, to use computational tools to identify and study the importance of various linguistic characteristics for diagnosing schizophrenia. Like other machine-learning applications, explainability is currently a problematic issue (what is it about the usage of nouns that contributes significantly to the model's success in classification?), and undue reliance on machine classification should be eschewed.

8 Conclusions

We studied the relative importance of several morphosyntactic categories for transcribed speech towards the classification task of distinguishing schizophrenia sufferers and controls. This was based on interviews of 23 male inpatients at a mental health center in Israel, officially diagnosed with schizophrenia, as well as 26 control participants; all are native Hebrew speakers. The interviews were manually transcribed and divided into individual responses that the participants provided for 18 discussion topics. Four topics were open-ended questions, and the rest were TAT images that were shown to the participants who were asked to describe the situation they see in the image.

We trained a natural-language-processing classifier by fine-tuning AlephBERT, a relatively large Hebrew language model, to distinguish between responses generated by patients and controls. To evaluate the contribution of different syntactic and morphological categories to the classification performance, we fine-tune the model each time by masking words of one specific category, and compare the classification performance with the same model trained on texts that were instead masked randomly for the same number of words. When the category-masked model performed more poorly than the randomly-masked model, we attribute it to an increased importance of the corresponding category. This new, masking method of evaluating the significance of linguistic features promises to be of use in many additional feature evaluation tasks.

Overall we examined six categories, and found (unsurprisingly) that nouns are the most important for distinguishing between patients and controls. We believe that it has to do with the idea of nouns being easier to capture in the mind due to their imageability. Given that nouns are used in comparable frequency by patients and controls, our findings reveal that the patients use nouns in a different way than do controls. We plan to investigate this further by looking more closely at the potential sources for this difference, in order to check how they may be related to poverty of speech.

Acknowledgements

This research was supported in part by grant #2168 from the Israeli Ministry of Science.

References

Meni Adler and Michael Elhadad. 2006. An unsupervised morpheme-based HMM for Hebrew morphological disambiguation. In *Proceedings of the 21st International Conference on Computational Linguistics and the 44th Annual Meeting of the Association for Computational Linguistics*, ACL-44, pages 665–672, USA. Association for Computational Linguistics.

Mark S. Aloia, Monica L. Gourovitch, David Missar, David Pickar, Daniel R. Weinberger, and Terry E. Goldberg. 1998. Cognitive substrates of thought disorder, II: Specifying a candidate cognitive mechanism. *American Journal of Psychiatry*, 155(12):1677–1684.

American Psychiatric Association DSM-5 Task Force. 2013. *Diagnostic and Statistical Manual of Mental Disorders: DSM-5*, volume 5. American Psychiatric Publishing, Washington, DC.

Kfir Bar, Vered Zilberstein, Ido Ziv, Heli Baram, Nachum Dershowitz, Samuel Itzikowitz, and Eiran Vadim Harel. 2019. Semantic characteristics of schizophrenic speech. In *Proceedings of the Sixth Workshop on Computational Linguistics and Clinical Psychology*, pages 84–93, Minneapolis, MN. Association for Computational Linguistics.

Laura L. B. Barnes, Diane Harp, and Woo Sik Jung. 2002. Reliability generalization of scores on the Spielberger state-trait anxiety inventory. *Educational and Psychological Measurement*, 62(4):603–618.

Aaron T. Beck, Robert A. Steer, Roberta Ball, and William F. Ranieri. 1996. Comparison of Beck Depression Inventories -IA and -II in psychiatric outpatients. *Journal of Personality Assessment*, 67(3):588–597.

Gillinder Bedi, Facundo Carrillo, Guillermo Cecchi, Diego Fernández Slezak, Mariano Sigman, Natália Mota, Sidarta Ribeiro, Daniel Javitt, Mauro Copelli, and Cheryl Corcoran. 2015. Automated analysis of free speech predicts psychosis onset in high-risk youths. *npj Schizophrenia*, 1:15030.

Moshe Bensimon, Stephen Zvi Levine, Gadi Zerach, Einat Stein, Vlad Svetlicky, and Zahava Solomon. 2013. Elaboration on posttraumatic stress disorder diagnostic criteria: A factor analytic study of PTSD exposure to war or terror. *Israel Journal of Psychiatry*, 50(2):84–90.

Christy A. Blevins, Frank W. Weathers, Margaret T. Davis, Tracy K. Witte, and Jessica L. Domino. 2015. The posttraumatic stress disorder checklist for *DSM-5* (PCL-5): Development and initial psychometric evaluation. *Journal of Traumatic Stress*, 28(6):489–498.

Jonathan D. Cohen and David Servan-Schreiber. 1992. Context, cortex, and dopamine: A connectionist approach to behavior and biology in schizophrenia. *Psychological Review*, 99(1):45.

Jacob Devlin, Ming-Wei Chang, Kenton Lee, and Kristina Toutanova. 2019. BERT: Pre-training of deep bidirectional transformers for language understanding. In *Proceedings of the 2019 Conference of the North American Chapter of the Association for Computational Linguistics: Human Language Technologies, Volume 1 (Long and Short Papers)*, pages 4171–4186, Minneapolis, MN. Association for Computational Linguistics.

Brita Elvevåg, Peter W. Foltz, Daniel R. Weinberger, and Terry E. Goldberg. 2007. Quantifying incoherence in speech: An automated methodology and novel application to schizophrenia. *Schizophrenia Research*, 93(1–3):304–316.

William I. Fraser, Kathleen M. King, Philip Thomas, and Robert E. Kendell. 1986. The diagnosis of schizophrenia by language analysis. *British Journal of Psychiatry*, 148(3):275–278.

Dolores Gallagher, Gloria Nies, and Larry W. Thompson. 1982. Reliability of the Beck Depression Inventory with older adults. *Journal of Consulting and Clinical Psychology*, 50(1):152–153.

Dedre Gentner. 1982. Why nouns are learned before verbs: Linguistic relativity versus natural partitioning. Technical Report 257, Center for the Study of Reading, University of Illinois at Urbana-Champaign.

Kelly Hasenson-Atzmon, Sofi Marom, Tamar Sofer, Lilac Lev-Ari, Rafael Youngmann, Haggai Hermesh, Jonathan Kushnir, and Haggai Hermesh. 2016. Cultural impact on SAD: Social anxiety disorder among Ethiopian and former Soviet Union immigrants to Israel, in comparison to native-born Israelis. *Israel Journal of Psychiatry*, 53(3):48–54.

Kasia Hitczenko, Vijay A. Mittal, and Matthew Goldrick. 2021. Understanding language abnormalities and associated clinical markers in psychosis: The promise of computational methods. *Schizophrenia Bulletin*, 47(2):344–362.

Ralph E. Hoffman and Thomas H. McGlashan. 1997. Synaptic elimination, neurodevelopment, and the mechanism of hallucinated "voices" in schizophrenia. *American Journal of Psychiatry*, 154(12):1683–1689.

David Horn and Eytan Ruppin. 1995. Compensatory mechanisms in an attractor neural network model of schizophrenia. *Neural Computation*, 7(1):182–205.

Dan Iter, Jong Yoon, and Dan Jurafsky. 2018. Automatic detection of incoherent speech for diagnosing schizophrenia. In *Proceedings of the Fifth Workshop on Computational Linguistics and Clinical Psychology: From Keyboard to Clinic*, pages 136–146.

Sandra Just, Erik Haegert, Nora Kořánová, Anna-Lena Bröcker, Ivan Nenchev, Jakob Funcke, Christiane Montag, and Manfred Stede. 2019. Coherence models in schizophrenia. In *Proceedings of the Sixth Workshop on Computational Linguistics and Clinical Psychology*, pages 126–136. Association for Computational Linguistics.

Sandra A. Just, Erik Haegert, Nora Kořánová, Anna-Lena Bröcker, Ivan Nenchev, Jakob Funcke, Andreas Heinz, Felix Bermpohl, Manfred Stede, and Christiane Montag. 2020. Modeling incoherent discourse in non-affective psychosis. *Frontiers in Psychiatry*, page 846.

Tilo T. J. Kircher, Tomasina M. Oh, Michael J. Brammer, and Philip K. McGuire. 2005. Neural correlates of syntax production in schizophrenia. *The British Journal of Psychiatry*, 186(3):209–214.

Robert G. Knight, Hendrika J. Waal-Manning, and George F. Spears. 1983. Some norms and reliability data for the state-trait anxiety inventory and the Zung self-rating depression scale. *British Journal of Clinical Psychology*, 22(4):245–249.

Pablo Lanillos, Daniel Oliva, Anja Philippsen, Yuichi Yamashita, Yukie Nagai, and Gordon Cheng. 2020. A review on neural network models of schizophrenia and autism spectrum disorder. *Neural Networks*, 122:338–363.

Ching-Hua Lin, Huey-Shyan Lin, Shih-Chi Lin, Chao-Chan Kuo, Fu-Chiang Wang, and Yu-Hui Huang. 2018. Early improvement in PANSS-30, PANSS-8, and PANSS-6 scores predicts ultimate response and remission during acute treatment of schizophrenia. *Acta Psychiatrica Scandinavica*, 137(2):98–108.

Colleen McDonough, Lulu Song, Kathy Hirsh-Pasek, Roberta Michnick Golinkoff, and Robert Lannon. 2011. An image is worth a thousand words: Why nouns tend to dominate verbs in early word learning. *Developmental science*, 14(2):181–189.

Arne Nagels, Paul Fährmann, Mirjam Stratmann, Sayed Ghazi, Christian Schales, Michael Frauenheim, Lena Turner, Tobias Hornig, Michael Katzev, Rüdiger Müller-Isberner, Michael Grosvald, Axel Krug, and Tilo Kircher. 2016. Distinct neuropsychological correlates in positive and negative formal thought disorder syndromes: The thought and language disorder scale in endogenous psychoses. *Neuropsychobiology*, 73(3):139–147.

Monika Obrębska and Tomasz Obrębski. 2007. Lexical and grammatical analysis of schizophrenic patients' language: A preliminary report. *Psychology of Language and Communication*, 11(1):63–72.

Pedro Javier Ortiz Suárez, Laurent Romary, and Benoît Sagot. 2020. A monolingual approach to contextualized word embeddings for mid-resource languages. In *Proceedings of the 58th Annual Meeting of the Association for Computational Linguistics*, pages 1703–1714, Online. Association for Computational Linguistics.

Soren Dinesen Østergaard, Ole Michael Lemming, Ole Mors, Christoph U. Correll, and Per Bech. 2016. PANSS-6: A brief rating scale for the measurement of severity in schizophrenia. *Acta Psychiatrica Scandinavica*, 133(6):436–444.

Luca Pauselli, Brooke Halpern, Sean D. Cleary, Benson S. Ku, Michael A. Covington, and Michael T. Compton. 2018. Computational linguistic analysis applied to a semantic fluency task to measure derailment and tangentiality in schizophrenia. *Psychiatry Research*, 263:74–79.

Noa Saka and Itamar Gati. 2007. Emotional and personality-related aspects of persistent career decision-making difficulties. *Journal of Vocational Behavior*, 71(3):340–358.

Amit Seker, Elron Bandel, Dan Bareket, Idan Brusilovsky, Refael Shaked Greenfeld, and Reut Tsarfaty. 2021. *AlephBERT*: A Hebrew large pre-trained language model to start-off your Hebrew NLP application with. *arXiv preprint arXiv:2104.04052*.

Charles Donald Spielberger, Richard L. Gorsuch, and Robert E. Lushene. 1970. *STAI Manual for the State-Trait Anxiety Inventory ("self-evaluation questionnaire")*. Consulting Psychologist Press, Palo Alto.

Sunny Tang, Reno Kriz, Sunghye Cho, João Sedoc, Suh Jung Park, Jenna Harowitz, Mahendra Bhati, Raquel Gur, Daniel Wolf, and Mark Liberman. 2020. Decreased speech coherence captured by novel natural language processing methods in two cohorts of individuals with schizophrenia. *Biological Psychiatry*, 87(9):S379–S380.

Sunny X. Tang, Reno Kriz, Sunghye Cho, Suh Jung Park, Jenna Harowitz, Raquel E. Gur, Mahendra T. Bhati, Daniel H. Wolf, João Sedoc, and Mark Y. Liberman. 2021. Natural language processing methods are sensitive to sub-clinical linguistic differences in schizophrenia spectrum disorders. *npj Schizophrenia*, 7(1):1–8.

Frank W. Weathers, Brett T. Litz, Terence M. Keane, Patrick A. Palmieri, Brian P. Marx, and Paula P. Schnurr. 2013. The PTSD checklist for DSM-5 (PCL-5). Scale available from the National Center for PTSD at `www.ptsd.va.gov`.

Thomas Wolf, Lysandre Debut, Victor Sanh, Julien Chaumond, Clement Delangue, Anthony Moi, Pierric Cistac, Tim Rault, Rémi Louf, Morgan Funtowicz, Joe Davison, Sam Shleifer, Patrick von Platen, Clara Ma, Yacine Jernite, Julien Plu, Canwen Xu, Teven Le Scao, Sylvain Gugger, Mariama Drame, Quentin Lhoest, and Alexander M. Rush. 2019. Huggingface's transformers: State-of-the-art natural language processing. *arXiv preprint arXiv:1910.03771*.

Yuichi Yamashita and Jun Tani. 2012. Spontaneous prediction error generation in schizophrenia. *PLoS One*, 7(5):e37843.

Ido Ziv, Heli Baram, Kfir Bar, Vered Zilberstein, Samuel Itzikowitz, Eran V. Harel, and Nachum Dershowitz. 2022. Morphological characteristics of spoken language in schizophrenia patients—an exploratory study. *Scandinavian Journal of Psychology*, 63(2):91–99.

Measuring Linguistic Synchrony in Psychotherapy

Natalie Shapira, Dana Atzil-Slonim, Rivka Tuval Mashiach, Ori Shapira

nd1234@gmail.com
{dana.atzil, tuvalmr}@biu.ac.il
obspp18@gmail.com

Bar-Ilan University, Israel

Abstract

We study the phenomenon of linguistic synchrony between clients and therapists in a psychotherapy process. Linguistic Synchrony (LS) can be viewed as any observed interdependence or association between more than one person's linguistic behavior. Accordingly, we establish LS as a methodological task. We suggest a LS function that applies a linguistic similarity measure based on the Jensen-Shannon distance across the observed part-of-speech tag distributions (*JSDuPos*) of the speakers in different time frames. We perform a study over a unique corpus of 872 transcribed sessions, covering 68 clients and 59 therapists. After establishing the presence of client-therapist LS, we verify its association with therapeutic alliance and treatment outcome (measured using WAI and ORS), and additionally analyse the behavior of *JSDuPos* throughout treatment.

Results indicate that (1) higher linguistic similarity at the session level associates with higher therapeutic alliance as reported by the client and therapist at the end of the session, (2) higher linguistic similarity at the session level associates with higher level of treatment outcome as reported by the client at the beginnings of the next sessions, (3) there is a significant linear increase in linguistic similarity throughout treatment, (4) surprisingly, higher LS associates with lower treatment outcome. Finally, we demonstrate how the LS function can be used to interpret and explore the mechanism for synchrony.[1]

1 Introduction

When people interact, they tend to naturally coordinate their behavior over time. Interpersonal synchrony is defined as the degree to which the behaviors in an interaction are nonrandom and patterned in both timing and form (Bernieri and Rosenthal, 1991). When this pattern occurs, it is often associated with greater rapport between the conversational partners (Butler and Randall, 2013). Research has demonstrated the beneficial effect of synchrony across various interpersonal relationships, such as between spouses or friends, as well as between parents and their children (Feldman, 2012).

The growing acknowledgment of the importance of synchrony in interpersonal relationships has recently led psychotherapy researchers to address the impact of synchrony in the psychotherapeutic process as a way to predict better therapeutic outcomes (Koole and Tschacher, 2016; Paulick et al., 2018).

Recent studies have demonstrated synchrony between clients and therapists through different modalities (Wiltshire et al., 2020). For example, higher levels of body-movement synchrony have been tied to more positive therapeutic relationships and treatment outcomes (Ramseyer and Tschacher, 2011, 2014; Tschacher and Meier, 2020), vocal synchrony was associated with higher empathy ratings (Imel et al., 2014), and physiological arousal coordination has been tied to client-perceived therapist empathy (Marci et al., 2007). However, *linguistic* synchrony (LS) between client and therapist has received relatively little attention.

The words and language clients and therapists use in psychotherapy sessions reflect their internal thoughts and emotions and reveal important information about their interaction. Thus, many of the active ingredients of psychotherapy are found in the words and how they are uttered within psychotherapy sessions. Client and therapist LS may reflect their ability to work together in concert and their adjustment to each other's language over time, which may in turn lead to better therapeutic outcome.

With the increased amount of conversational texts accessible, applying natural language processing is an appealing step for mental health research (e.g. Sharma and De Choudhury, 2018; Zhang

[1] For code availability please contact authors.

Proceedings of the Eighth Workshop on Computational Linguistics and Clinical Psychology, pages 158 - 176
July 15, 2022 ©2022 Association for Computational Linguistics

and Danescu-Niculescu-Mizil, 2020). Indeed, transcripts of psychotherapy sessions have recently become more readily available thanks to advanced ASR transcription technology. These transcripts allow the analysis of LS in psychotherapy (see Section 2).

The few studies that have considered client-therapist LS have tended to focus on one session, and assessed it's association with therapy processes (e.g., Lord et al., 2015; Pérez-Rosas et al., 2017). The extent to which LS develops from session to session and its association with treatment outcome were yet to be explored in a statistically sound manner. Furthermore, a major criticism on studies on interpersonal synchrony concerns the lack of control for coincidental random synchrony Ramseyer and Tschacher (2010). Based on studies that distinguish genuine synchrony from pseudosynchrony in physiological data, the current study proposes a method to assess LS, that is adapted for sequences of texts (Section 4). Section 5 presents a LS function, inspired by previous work addressing LS.[2] We examine client-therapist LS throughout treatment (N = 74, average number of sessions = 12.56, a total of 872 transcripts), session by session, and the association between LS and treatment process and outcome.

In Section 6 we demonstrate the implications of the ability to measure LS. Synchrony is viewed as an important mechanism of change between the client and the therapist, which leads in turn to a better bond and to a better outcome (for review see Koole and Tschacher, 2016; Paulick et al., 2018). When applying the proposed LS function on our dataset, the method displays an association to quality of client-therapist relationship and treatment outcome (Section 6.1), as well as a significant linear change across treatment (Section 6.2). Additionally, we show how the LS function can be used to interpret and explore the mechanism for synchrony (Section 6.3). Finally, we discuss limitations and potential future work in Section 7.

2 Related Work

We focus on previous work researching LS in psychotherapy.

Lord et al. (2015) dealt with motivational interview training treatment (N=122), where each treatment has a single 20-min transcribed session. They measured synchrony between client and therapist with function word coordination on the ordered utterances in a session (Danescu-Niculescu-Mizil et al., 2012). They show that high empathy sessions display greater coordination of function words compared to low empathy sessions. Overall, average coordination of function words is notably higher in high empathy vs. low empathy sessions.

Pérez-Rosas et al. (2017) explored counseling interaction dynamics (N=276; each session with 5 annotation points) and their relation to counselor empathy during motivational interviewing.

The two latter studies were based on synchrony within a single session. Thus, they could not examine patterns of change across treatment. In addition, while these studies demonstrated the presence of LS in sessions characterized by high empathy between clients and therapists, they do not explore the association between LS and other treatment processes and outcome.

Althoff et al. (2016) measured how various linguistic aspects of written conversations (15,555), as opposed to spoken, correlate with outcomes. This dataset is much larger than in our study, however they analyze the counselor's point of view (N=408) (as opposed to dyads) and overlooked the synchrony across long-term treatment.

Borelli et al. (2019) examine how language style matching (LSM; Niederhoffer and Pennebaker, 2002), clients' relational histories, and symptoms were associated within treatment. On a pilot test using a small sample (N=7, sessions=4) they found that LSM values decrease over the course of treatment, and that greater client interpersonal problems prospectively predict lower early LSM in client-therapist dyads, which in turn predicts greater post treatment psychiatric distress.

Aafjes-van Doorn et al. (2020) demonstrate the clinical usefulness of the LSM and rLSM (Müller-Frommeyer et al., 2019) approach in psychotherapy outcome measures with a small sample (N=7, sessions=20). They also described a case study comparing LSM values to observer-rated measure of working alliance, and conclude that a larger-scale study is required for examining the relationship between synchrony and alliance and outcome.

3 Linguistic Synchrony Definition

Inspired by behavioral and physiological synchrony (Bernieri and Rosenthal, 1991; Palumbo et al.,

[2]As opposed to previous work addressing LS, our LS function does not rely on LIWC (Tausczik and Pennebaker, 2010) since it does not support Hebrew language. See Appendix A.4 for a comparison between the use of LIWC and our method.

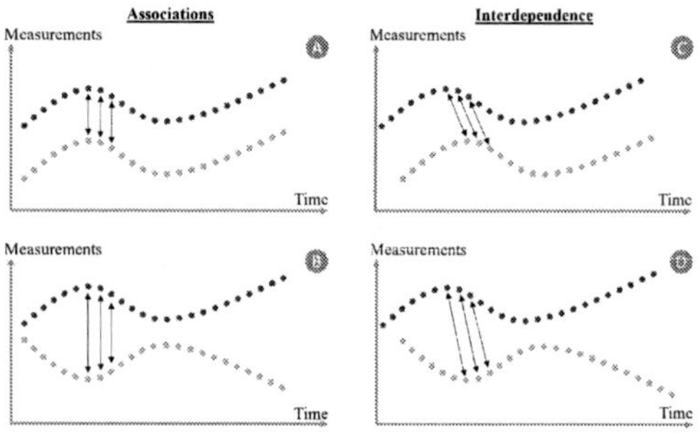

Figure 1: Illustrations of synchrony between repeated measures of two participants (blue and orange) as association (A,B) and interdependence (C,D), with a similarity (A,C) and complementary (B,D) behavior.

2017), **Linguistic Synchrony (LS)** can be viewed as *any observed association or interdependence between people's language dynamics, as indexed by their continuous spoken words, that are nonrandom or patterned in both timing and form.*

Association is a relationship between variables that makes them statistically dependent (e.g., as measured by correlation coefficient see Figure 1.A,B).

Interdependence is the state in which two or more variables rely on or react with one another such that one cannot change without affecting the other (VandenBos, 2007) (e.g., see Figure 1.C,D).

Language dynamics of a conversation are the changes in language use, for each participant individually, that can be captured over time by assessing utterances at a number of time points.

Non-random or patterned associations or interdependences are quantified by adopting an approach by Ramseyer and Tschacher (2010), "surrogate test", that distinguishes genuine synchrony from pseudosynchrony, which may arise due to random coincidence. This definition outlines the statistical tests required to show that a function can indeed measure LS, by pairing texts with non-original replacements and showing a significant difference in the synchrony measure.

4 Formalizing a Task

With respect to the LS definition, we formalize a task, for finding a function that measures LS, as follows:

Given a sample $[[(c^i_j, t^i_j)]^{m_i}_{j=1}]^n_{i=1}$ of n pairs (e.g., clients and their therapists) with m_i repeated measures (e.g., sessions in treatment) of lingual texts (i.e., c^i_j, t^i_j are each a written or transcribed text sequence) from a population P, function $f : L \to \mathbb{R}$ is said to be *Measuring Linguistic Synchrony (MLS)* within P, where L is a list of text pairs, if for a set of random texts r^i_j, the sample of values $[f([(c^i_j, t^i_j)]^{m_i}_{j=1})]^n_{i=1}$ statistically significantly differs from the generated sample of values $[f([(c^i_j, r^i_j)]^{m_i}_{j=1})]^n_{i=1}$ and $[f([(r^i_j, t^i_j)]^{m_i}_{j=1})]^n_{i=1}$.[3]

Intuitively, we would like to find a function that is able to recognize that a given list of text pairs has a non-random dependence.

To capture more than just "*any* observed interdependence or association", defined are two additional *pseudosynchrony* tests that use *surrogates* in place of the random texts.

Within challenge: a text c^i_j is paired with a different text contained within t^i's list of repeated measures. Formally, $[f([(c^i_j, t^i_j)]^{m_i}_{j=1})]^n_{i=1}$ statistically significantly differs from $[f([(c^i_j, t^i_{l_j})]^{m_i}_{j=1})]^n_{i=1}$ where $l_j \in \{1, \ldots, m_i\}$ and $l_j \neq j$.

Between challenge: a text c^i_j is paired with any text not contained within t^i's list of repeated measures. Formally, $[f([(c^i_j, t^i_j)]^{m_i}_{j=1})]^n_{i=1}$ statistically significantly differs from $[f([(c^i_j, t^{k_j}_{l_j})]^{m_i}_{j=1})]^n_{i=1}$ s.t. $k_j \in \{1, \ldots, n\}$, $k_j \neq i$ and $l_j \in \{1, \ldots, m_{k_j}\}$.

Linguistic synchrony. Populations A and B have different levels of *synchrony* with respect to $f \in MLS$ if f values on population A are statistically significantly different from f values on population B.

Synchrony direction. In order to determine the direction of the synchrony, i.e., whether low or high values of f will be considered as synchrony, we compare the f values of the original sample (i.e., $[f([(c^i_j, t^i_j)]^{m_i}_{j=1})]^n_{i=1}$) to the f values of the *surrogates* sample. If f values are lower for the original sample than for the surrogate sample, then lower f values imply higher synchrony. Correspondingly, if the f values are higher in the original sample, then higher f values imply higher synchrony.

Task objective. The objective is to find an MLS function that maximizes the *magnitude* – the strength of synchrony – typically represented by the effect size of the statistical test. In addition, the MLS function should strive to reveal an aspect with which synchrony can be expressed.[4]

[3] In social sciences, as opposed to exact sciences, a measurement is not required to obey a well-defined unit of measure.

[4] An important goal in synchrony research is to provide an interpretation for the observed synchrony. E.g., for synchrony in autism, there are diagnostic tools that assess social skills

We emphasize that synchrony is a change that occurs over time, as opposed to similarity that is measured at a single point. Additionally, synchrony may be expressed through, e.g., complementary behavior (Ackerman and Bargh, 2010; Chartrand and Lakin, 2013) or coordination that can be observed in a non-aligned manner, e.g., shifting content or aggregating several samples together (Figure 1).

Limitation. There exist outlier MLS functions that meet all requirements of the task definition, but do not actually measure synchrony. For example, a function that internally stores the full sample ($[[(c_j^i, t_j^i)]_{j=1}^{m_i}]_{i=1}^{n}$) and simply returns 1 if a given pair ($(c_j^i, t_j^i)]_{j=1}^{m_i}$) appears in the sample and 0 otherwise. A function with a reasonable description length (e.g., memory use) would not allow such functionality. Moreover, proposing such a function does not serve the purpose of synchrony research. Another example is a function that randomly chooses a value that happens to correctly distinguish between an actual pair and a surrogate pair. Such behaviour is not statistically expected.

We next present an LS function that exposes linguistic similarity over time, and in Appendix A.3, a different function that exhibits complementary behavior.

5 Exemplifying Solution

Adhering to the formalized conception of MLS, we next lay out a use case brought from psychotherapy research. First, the data we use is described, then a candidate MLS function is presented, and finally the function is tested for MLS.

5.1 Dataset Description

We employ a dataset of a total of 872 psychotherapy session transcripts, in Hebrew, from 74 different dyads (client-therapist pairs), constructed by 68 clients and 59 therapists. A treatment of a dyad is composed of several sessions (Mean=12.56; SD=4.93). For the purposes of this study, we referred only to verbal text and punctuation, marked by how they were heard (comma as a short pause in speech) and not by how proper sentences should be written.[5] Prior to each session, clients self-reported[6] their functioning, measured using the ORS questionnaire (Miller et al., 2003), which is considered to be an indicator for progress in treatment (see Appendix A.2.1). After each session, therapists and clients reported their perspective for the quality of the relationship during the session, measured by the WAI questionnaire (Horvath and Greenberg, 1989) (see Appendix A.2.2). We note that this dataset is an order of magnitude larger than those used in the few previous works dealing with psychotherapy text analysis (see Section 2).

5.2 Candidate Synchrony Function

Algorithm 1: Lingual distance of client's (c) and therapist's (t) texts list (size=m)

1 *candidateMLS*(c,t,m);
2 **for** $j \leftarrow 1$ *to* m **do**
3 $\quad cPos_j, tPos_j \leftarrow pos(c_j), pos(t_j)$;
4 $\quad cuPos_j \leftarrow prDis(cPos_j)$;
5 $\quad tuPos_j \leftarrow prDis(tPos_j)$;
6 $\quad JSDuPos_j \leftarrow jsd(cuPos_j, tuPos_j)$;
7 **end**
8 return: $average(JSDuPos)$

We present Algorithm 1 as a candidate MLS function.[7] *candidateMLS*, receives as input lists C^d and T^d (d represents specific dyad name) both of size m_d, of a client's and matching therapist's transcribed sessions. The client's and therapist's texts are paired by sessions. I.e., each list element contains the client's or therapist's utterances from a single session, $c_j^d \in C^d$ ($t_j^d \in T^d$) is a concatenation of all client's (therapist's) sentences within session number j, and c_j^d and t_j^d are from the same session, for each session j.

Inspired by previous work addressing LS, the *candidateMLS* function converts each element in the two lists to a probability distribution of unigram part-of-speech (POS) tags (see Appendix A.4 for the relation between LSM categories used in previous works and POS tags). In line 3 of Algorithm 1, the *pos* function[8] extracts the POS tags from the client's (therapist's) text c_j (t_j) in session j and

such as eye contact or speech turn coordination in conversation. Accordingly, this allows planning respective interventions that address these social skills (e.g., Hopkins et al., 2011). For LS, we provide a possible interpretation in Section 6.3.

[5]For further details about participants, treatment, transcriptions procedure and ethical concerns, see Appendix A.1.

[6]Note that there are biases related to subjective self-reports (Kazdin, 2008). Nevertheless, it is common to build upon such self-reports for psychotherapy research.

[7]An additional candidate synchrony function is presented in Appendix A.3, which measures complementary behavior and applies correlation for computing the magnitude of synchrony.

[8]We used YAP (More and Tsarfaty, 2016) for Hebrew POS tagging.

stores the resulting sequences in $cPos_j$ ($tPos_j$). In lines 4 and 5, the $prDis$ function converts the $cPos_j$ and $tPos_j$ POS sequences to their distributions, and stores them in $cuPos_j$ and $tuPos_j$ respectively. In line 6, the jsd function calculates the Jensen-Shannon Distance[9] (JSD) (Fuglede and Topsoe, 2004) between distributions $cuPos_j$ and $tuPos_j$ (method denoted *JSDuPos*). Finally, *candidateMLS* outputs the average of *JSDuPos*$_j$ values ($j \in [1, m_d]$), providing a synchrony score for dyad d, where a lower score means higher synchrony.

Note the difference between *JSDuPos* and *candidateMLS*. *JSDuPos* is a measure of linguistic **similarity** between the client and the therapist that is calculated for each **session** separately. A lower *JSDuPos* value indicates a closer distance between texts and therefore a higher similarity. *candidateMLS* is a measure of linguistic **synchrony** between the client and the therapist, that is calculated for a **treatment**. A lower *candidateMLS* value indicates lower synchrony (see Section 3 for an explanation on synchrony direction and Section 5.3 on how we determined the direction for our function).

As *JSDuPos* is an interpretable measure of linguistic similarity, it is useful for psychologists to better understand mechanisms of change throughout treatment, i.e., by viewing changes in use of part of speech, as demonstrated in Section 6.3. Furthermore, this function does not require training data, as opposed to data-hungry similarity methods (e.g. Bevendorff et al., 2020; Boenninghoff et al., 2020), which is pertinent in domains where data is rather scarce. Other measures, such as those used for authorship attribution (Koppel et al., 2009; Stamatatos, 2009; Juola, 2008; El and Kassou, 2014), are appealing MLS candidate functions, and we advocate future research to inspect such options.

5.3 Synchrony Function Evaluation

To assess whether the candidate function meets the MLS criteria, we test the *Within* and *Between* challenges, using the corpus of client-therapist conversations from Section 5.1.

The paired sequences of the conversations are as follows: each dyad i ($i \in [1, 74]$) has m_i sessions $S^i_{1:m_i}$. For each session $s^i_j \in S^i_{1:m_i}$ we separated the utterances of the client c^i_j and the utterances of the therapist t^i_j, producing sequences of texts $C^i_{1:m_i}$ and $T^i_{1:m_i}$. The whole corpus can be described as $[[(c^i_j, t^i_j)]^{m_i}_{j=1}]^{74}_{i=1}$.

Within-experiment: (1) For each dyad i: (1.1) Calculate *candidateMLS* on the client's $C^i_{1:m_i}$ and the corresponding therapist's $T^i_{1:m_i}$ to get synchrony magnitude value v^i. (1.2) Choose random permutation $perm(T^i_{1:m_1})$, and calculate *candidateMLS* on the client's $C^i_{1:m_1}$ and $perm(T^i_{1:m_1})$, to get result w^i. Due to non-normally distributed data, (2) Compute Wilcoxon signed-rank one-tail test[10] and Cohen's d (Cohen, 1988) on vectors $V = [v^i]$ and $W = [w^i]$, expecting values of V to be significantly lower than values of W.

This experiment is repeated 100 times on different permutations. All experiments yielded a significant superiority (Dror et al., 2020) of genuine synchrony versus pseudosynchrony ($p < 0.05$) with a small effect size (average Cohen's d = 0.12). V (M = 0.174; SD = 0.034) exists in a lower level compared to W (surrogate session) (M = 0.179; SD=0.035).

Between-experiment: (1) For each dyad i: (1.1) Compute v^i as described above. (1.2) For each $c^i_j \in C^i_{1:m_i}$, randomly choose, with replacements, a different therapist session t^k_l ($i \neq k$) from the entire set of therapists sessions $[[t^i_j]^{m_i}_{j=1}]^{74}_{i=1}$ and calculate *candidateMLS* on $C^i_{1:m_i}$ with the randomly generated therapist sequence, to get result b^i. (2) Compute Wilcoxon signed-rank one-tail test and Cohen's d on vectors $V = [v^i]$ and $B = [b^i]$.

On 100 different experiments (different replacements), all trials yielded a significant superiority of genuine synchrony versus pseudosynchrony ($p < 0.05$) with a very large effect size (Sawilowsky, 2009) (average Cohen's d = 1.459). V exists in a lower level compared to B (surrogate therapist) (M = 0.218; SD = 0.028).

Both *Within-* and *Between*-challenge tests pass, indicating that the candidate function meets the MLS criteria. Results are depicted in Figure 2.

[9] Jensen-Shannon Divergence is based on Kullback-Leibler Divergence with a simple manipulation that makes it symmetric (instead of measuring the relative entropy between the two distributions, measure the average of the entropies between each of the distributions and their average distribution) and thus maintains the triangular inequality. JS-Distance is the root of JS-Divergence. We chose *distance* over *divergence* since distance is the more common preference in the literature (1,850,000 search results in Semantic Scholar vs. 239,000).

[10] Based on previous studies, we hypothesize that the synchrony direction is inversely proportional to the similarity in our function. Thus, we expect lower *candidateMLS* values in the original text-pair sample compared to the surrogate sample.

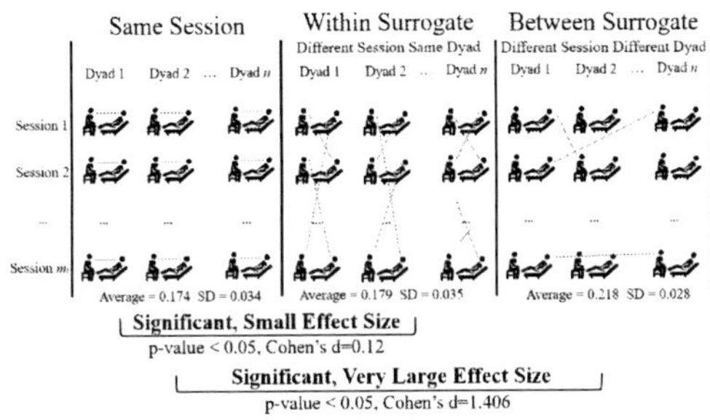

Figure 2: The degree of synchrony in conversations between therapist and client compared to pseudosynchrony in conversations that did not take place.

	Session Level			Dyad Level		
Variable	Obs.	M(SD)	Range	Obs.	M(SD)	Range
JSDuPos	871	0.17 (0.05)	0.08-0.9	74	0.17 (0.03)	0.11-0.3
ORS	860	24.4 (7.96)	0.3-40	74	24.5 (6.41)	10.15-38.24
C_WAI	823	50.89 (23.82)	4-84	74	49.48 (23.02)	9.5-84
T_WAI	831	41.69 (18.61)	0-74	74	40.33 (17.88)	8.75-67.61

ORS = Outcome Rating Scale; WAI = Working Alliance Inventory;
C = Client; T = Therapist; Obs. = Observations

Table 1: Descriptive statistics of treatment measurements (processes and outcome) and of our *JSDuPos* function. *JSDuPos* = JS-Distance between probability distributions of unigram POS-tags.

6 Implications of the Candidate Function

Psychology research puts forth much effort in trying to understand the synchrony phenomenon and mechanism (Section 1). In addition, studies show a link between client-therapist synchrony and treatment processes and outcomes (Sections 2). Thus, we examine the relationship between LS and treatment measures through the candidate function (Section 6.1), analyze the change of *JSDuPos* over the course of treatments (Section 6.2), and demonstrate what can be extracted from the function to further understand LS (Section 6.3).

6.1 Associations with Treatment Process and Outcome

Hypothesis 1: We expect that *JSDuPos* and *candidateMLS*, both associate with treatment process and outcome.

(Hypothesis 1a) A lower *JSDuPos* value in a session, i.e., higher linguistic similarity, associates with: (1) a higher level of alliance between therapist and client as reported by both therapist and client at the end of the session, and (2) a higher level of treatment outcome as reported by the client at the beginnings of the current and next sessions. I.e., $JSDuPos(c_s^d, t_s^d)$ correlates with Client_WAI$_s^d$, Therapist_WAI$_s^d$, ORS$_s^d$ and ORS$_{s+1}^d$.

(Hypothesis 1b) A lower *candidateMLS* value of a treatment, i.e., higher LS, associates with: (1) a higher level of alliance between the client and therapist as reported both by client and therapist at the end of each session in the treatment, and (2) a higher level of treatment outcome as reported by the client at the beginning of each session. I.e., $candidateMLS(C^d, T^d)$ correlates with average values of Client_WAI^d, average of Therapist_WAI^d and average of ORS^d.

Results: The descriptive statistics – means, standard deviations and ranges for all the variables – are presented in Table 1.

To examine *(Hypothesis 1a)* we conducted a multilevel model (MLM) test[11] (Bolger and Laurenceau, 2013) that predicts a session's treatment process/outcome value with the corresponding *JSDuPos* (dyad mean-centered) value. Multilevel models allow estimation of two levels (a within-dyad level and a between-dyad level) and accommodate non-balanced data (see Bolger and Laurenceau) as in our case (i.e., sessions nested within dyads and dyads have different numbers of sessions). We used two-level MLM and not three-level MLM (sessions nested within dyads nested within therapists) because of the limited number of clients per therapist.

To examine *(Hypothesis 1b)*, the same multilevel model test factors in the *candidateMLS* value (as a grand mean center of *JSDuPos* dyad values, denoted $meanJSDuPos$).

The mixed-level equation is as follows:

$$\begin{aligned} &Treatment\_Measure_s^d = \\ &(\gamma_0^0 + u_0^d) \\ &+ (\gamma_1^0 + u_1^d) JSDuPos_s^d \\ &+ (\gamma_2^0) meanJSDuPos^d + e_s^d \end{aligned} \quad (1)$$

s.t. *Treatment_Measure* $\in$ {ORS, Client_WAI, Therapist_WAI}. $Treatment\_Measure_s^d$ for a dyad d in session s is predicted by the sample's intercept (γ_0^0), by dyad d's deviation from this intercept (u_0^d), by the average (i.e., fixed) effects (γ_1^0, γ_2^0) of the predictors, by this client's deviation from the fixed effects (i.e., the random effects: (u_0^d, u_1^d)), and by a level-1 residual term quantifying the session's deviation from these effects (i.e., the random effect at level 1, e_s^c).

We note that to examine the prospective association between the MLS and treatment outcome as

[11] Using the R *lme4* library (Bates, 2010), *lmer* function.

reported by the client at the beginning of the *next* session (ORS^d_{s+1}), Equation 1 was computed with the next session index (index $s+1$ instead of s), as follows:

$$\begin{aligned} ORS^d_{s+1} &= \\ (\gamma^0_0 &+ u^d_0) \\ &+ (\gamma^0_1 + u^d_1)\mathit{JSDuPos}^d_s \\ &+ (\gamma^0_2)\mathit{meanJSDuPos}^d + e^d_s \end{aligned} \tag{2}$$

As can be seen in Table 2, consistent with *Hypothesis 1a*, a lower *JSDuPos* value (higher linguistic similarity) in a session associates with a higher level of alliance between the client and therapist as reported both by client and therapist at the end of each session (supporting *(Hypothesis 1a)* (1)), and a higher level of treatment outcome as reported by the client at the beginning of the next session (partially supporting *(Hypothesis 1a)* (2)). However, not consistent with *Hypothesis 1b*, a lower *candidateMLS* value (higher linguistic synchrony) in a treatment associates with a lower level of treatment outcome as reported by the client at the beginning of both the current session and the next session. Although the results of the model predicting ORS was statistically significant, the direction was opposite to the hypothesis. In addition, *candidateMLS* did not show associations with alliance of both client and therapist (i.e., *(Hypothesis 1b)* failed to reject the null hypothesis).

6.2 Similarity Increase throughout Treatment

In order to better understand the synchrony mechanism, we examine the change in similarity between client and therapist over the course of a treatment. Since all previous studies that examine LS were based on a single session or a small scale dataset (i.e., could not examine change over time), the following hypothesis will be tested in an exploratory manner.

Hypothesis 2 (exploratory): We expect an increase in linguistic similarity throughout treatment.

Results: To examine the extent in which similarity changes throughout a treatment, a linear growth-curve analysis is conducted over the *JSDuPos* values of treatments.[12] Growth curve models typically refer to statistical methods that allow the estimation of patterns of change over time (the most basic feature of an intensive longitudinal outcome) (Bolger et al., 2003).

Results show a significant linear change across treatment. Specifically, the time trend was negative

[12]Using the R *nlme* library, *lme* function.

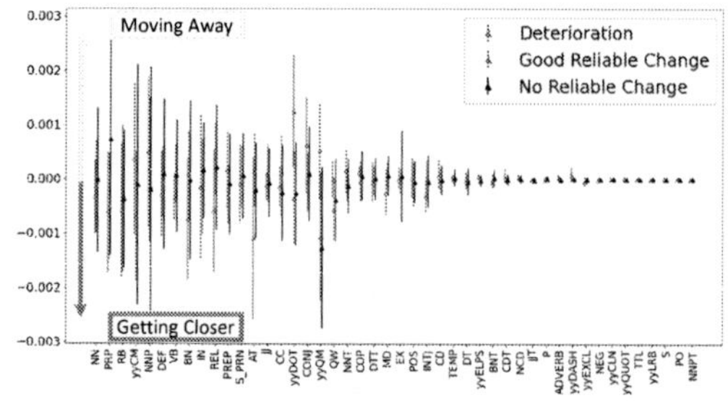

Figure 3: Average and standard deviation of changes in part-of-speech tag frequencies from session to session by all clients and therapists, viewed separately for three groups of dyads divided according to treatment outcome. On average over all treatments with good reliable change, question-mark (yyQM) is the tag for which therapists and corresponding clients move closer the most over a treatment.

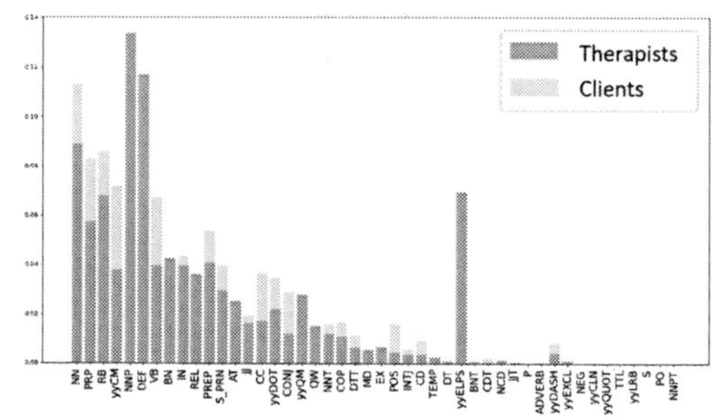

Figure 4: The most asynchronous treatment with frequencies of POS tags in a client's (orange) and therapist's (purple) transcriptions. The three major sources of asynchrony, with the highest frequency gap, are the parts-of-speech NNP (proper noun singular), DEF (morphological determiner) and yyELPS (ellipsis).

($b = -0.001$, $SE = 0.0002$,$t = -4.854$, $p < 0.001$), indicating that on average client-therapist linguistic similarity was higher (*JSDuPos* was lower) in the later stages of therapy compared to the initial stages. See Figure 5 in the appendices for a visualization of the constant decrease in *JSDuPos* over time.

6.3 Utility of LS in Treatment

In this section we will demonstrate how the LS mechanism can be further explored.[13] As shown in Section 6.2, *JSDuPos* values decrease over treatment. We explore in what sense the client and therapist become closer in terms of changes in POS

[13]There are no clinical recommendations here but rather a demonstration of the benefits of an interpretable synchrony function. In the current study it is not possible to examine the causality relation between synchrony and outcome.

Predictors	Previous Week ORS		Next Week ORS		Client_WAI		Therapist_WAI	
	Estimates (Std. Err)	95% CI (t value)	Estimates (Std. Err)	95% CI (t value)	Estimates (Std. Err)	95% CI (t value)	Estimates (Std. Err)	95% CI (t value)
(Intercept)	24.50*** (0.711)	[23.11, 25.90] (34.439)	24.71*** (0.744)	[23.26, 26.17] 33.22	49.50*** (2.674)	[44.26, 54.74] (18.51)	40.36*** (2.077)	[36.29, 44.43] (19.433)
Session *JSDuPos*	-8.41 (70171)	[-22.46, 5.65] (-1.172)	-17.16*** (4.868)	[-26.70, -7.62] (-3.525)	-30.90*** (8.811)	[-48.17, -13.63] (-3.507)	-25.76*** (7.446)	[-40.35, -11.17] (-3.46)
Dyad *meanJSDuPos*	64.54*** (21.987)	[21.44, 107.63] (2.935)	47.74* (21.350)	[5.89, 89.58] (2.236)	-72.70 (83.612)	[-236.58, 91.17] (-0.87)	-79.80 (65.122)	[-207.44, 47.84] (-1.225)
Observations	859		849		822		830	
Conditional R (ICC)	0.646 (0.62)		0.680 (0.67)		0.931 (0.93)		0.920 (0.92)	

Note. ***p<0.001; **p<0.01; *p<0.05; ORS = Outcome Rating Scale; WAI = Working Alliance Inventory
JSDuPos = Jensen–Shannon-Distance between Probability Distribution over Unigram POS-tag;
meanJSDuPos = Result of the synchrony function (*candidateMLS*) which is the average *JSDuPos* for each dyad.

Table 2: Associations between similarity (*JSDuPos*) or synchrony (*meanJSDuPos*), and treatment measurements – outcome (ORS) or process (WAI).

tag distributions over a treatment, using two approaches.

In the first approach, we analyze the changes in use of POS tags in treatment in three different groups of dyads (of the 74 available): those with a good reliable change in treatment, those with a reliable deterioration, and those with no reliable change.[14] Then, for each POS tag p and for each dyad d in its group, for a sequence of sessions $s_1^d, s_2^d, ..., s_{n_d}^d$ we compute the distances $\delta_1^{d,p}, ..., \delta_{n_d}^{d,p}$ where $\delta_i^{d,p}$ is computed as the absolute difference between the client's frequency of p and the therapist's frequency of p in session i. We then compute the difference in distances between consecutive sessions of the treatment $\Delta_i^{d,p} = \delta_i^{d,p} - \delta_{i-1}^{d,p}$. The score for this treatment and POS tag is then $score^{d,p} = \sum_2^{n_d} \frac{\Delta_i^{d,p}}{n_d - 1}$, i.e., the average of the differences in the sequence of sessions. Finally, for each POS tag separately, we calculate the average and standard deviation of scores of all dyads within their group. A lower value for a POS tag means the clients' and corresponding therapists' tag frequency becomes more similar overall.

As seen in Figure 3, in the dyads with a good reliable change, the POS tag frequencies of clients and therapists moved towards each other in the question-mark (yyQM) and question (QW) tags. When zooming in from part-of-speech- to the lexical-level, i.e., analysing frequencies of question *words*, we found the biggest change in the "what" token. Throughout the treatment, the frequency of "what" increases for clients ($+0.1\%$) while decreasing for therapists (-0.1%). See also Figure 6 in the Appendices for separate client and therapist points of view of a similar analysis.

Another approach for exploring the LS mechanism is by analyzing the contributors that influence the magnitude of synchrony within a specific treatment. We demonstrate this through a case study from our data in the treatment with the lowest synchrony value as calculated with *candidateMLS* (highest average *JSDuPos* scores). This treatment was also considered unsuccessful as measured by ORS. Figure 4 shows the POS tag distribution of the whole treatment for a client-therapist dyad. The differences in the tag distributions may hint at reasons for the unsuccessful treatment. Here we see that the therapist uses some POS tags far more often than the client. For example, there is a frequent use of ellipses (yyELPS), indicating many silent moments. Accordingly, these tags can expose behavior that may have gone unnoticed.

7 Discussion and Future Work

In Section 5.2 we propose a function that is able to measure LS, based on a similarity approach. Future work may assess LS functions that apply different similarity methods. Additionally, new LS functions should examine other forms of synchrony such as coordination and accommodation.

The field of *Authorship Attribution* (Koppel et al., 2009; Stamatatos, 2009; Juola, 2008; El and Kassou, 2014), for example, may inspire development of new LS functions. This field relies on features of complexity measures (e.g., average word length, average number of words in a sentence),

[14]The ORS has a Reliable Change Index (RCI = 5 points) that identifies when change is clinically significant and attributable to therapy. (Low et al., 2012)

syntax, taxonomies, morphological analysis, orthographic/syntactic errors, idiosyncrasies and others. These may be adapted for measuring LS as well.

We note that in this work we describe synchrony as it is commonly referred to in psychology. This definition does not discriminate between *intrinsic* synchrony and *extrinsic* synchrony. Two bodies synchronize intrinsically when they directly influence each other. For example, the moon's motion synchronizes with sea levels due to the gravitational force exerted by the moon on the sea. In other cases, an *external* constituent impacts the two bodies in such a way that they synchronize independently. For example, two clocks are in synchrony with each other as a result of the time specified by an independent source. In the case of LS, the use of linguistic features by two "synchronized" speakers may be due to an outside cause, like a seasonal use of words. When discovering synchrony with a measuring function, the underlying root cause remains unknown. More research should be conducted in order to reveal the confounding variables of synchrony.

8 Conclusion

Researching synchrony enhances our understanding of the mechanisms of change in psychotherapy treatment. Language, in particular, reveals important information about the interaction between a client and a therapist. Following previous work on synchrony research, we formally define a task for measuring *linguistic* synchrony, and describe two tests for quantifying the quality of a function that measures LS. We suggest a function, consisting of a similarity component inspired by methods used in Psychology research, that satisfies the definition and tests. The function and its component are shown to correlate with measures of psychotherapy process and outcome and show a significant linear increase across treatment. Furthermore, we demonstrate how this function can be interpreted for understanding the interaction between the client and therapist throughout treatment. While this non-standard task of Linguistic Synchrony can strongly contribute to analysis in Psychology, we also generally see it as an intriguing challenge to undertake in comparative textual analysis.

9 Ethical Considerations

This study was approved by an Institutional Review Board and was conducted ethically in accordance with the World Medical Association Declaration of Helsinki. The procedures were part of the routine assessment and monitoring process in the clinic. Informed written consent was obtained from all participants at the outset of this study. Participants are asked to provide written consent that their data will be used for research. They are informed that at any time they may request to terminate their participation in the research and / or request that the content of the recordings be deleted without jeopardizing treatment. All data collected was anonymized and only then exposed to a very small number of researchers, as agreed upon by the participants. More information is avaialbale in Appendix A.1.

Acknowledgements

We would like to thank Yoav Goldberg, Daniel Juravski, Tslil Ofir, Ayelet Meiri, Adar Paz, Dana Stolowicz-Melman, Yarden Menashri Sinai, Almog Simchon, Rotem Dror, Victoria Basmov, Shoval Sadde, Reut Tsarfaty, Nir Milstein, Oded Mayo, Alon Tomashin, Maya Sabag, Shahar Siegman, the synchrony PhD forum members in the Psychology department, and BIU-NLP lab members for helpful discussions and contributions, each in their own way. We thank the anonymous reviewers for their insightful comments and suggestions. This project was partially funded by the Israel Science Foundation (grants 1348/15 and 1278/16); the European Research Council (ERC) under the European Union's Horizon 2020 research and innovation program, grant agreement No. 802774 (iEXTRACT); and the Computer Science department of Bar-Ilan University.

References

Katie Aafjes-van Doorn, John Porcerelli, and Lena Christine Müller-Frommeyer. 2020. Language style matching in psychotherapy: An implicit aspect of alliance. *Journal of Counseling Psychology*, 67(4):509.

Joshua M Ackerman and John A Bargh. 2010. Two to tango: Automatic social coordination and the role of felt effort.

Aviad Albert, Brian MacWhinney, Bracha Nir, and Shuly Wintner. 2013. The hebrew childes corpus: transcription and morphological analysis. *Language resources and evaluation*, 47(4):973–1005.

Tim Althoff, Kevin Clark, and Jure Leskovec. 2016. Large-scale analysis of counseling conversations: An application of natural language processing to mental health. *Transactions of the Association for Computational Linguistics*, 4:463–476.

Douglas M Bates. 2010. lme4: Mixed-effects modeling with r.

Frank J Bernieri and Robert Rosenthal. 1991. Interpersonal coordination: Behavior matching and interactional synchrony.

Janek Bevendorff, Bilal Ghanem, Anastasia Giachanou, Mike Kestemont, Enrique Manjavacas, Ilia Markov, Maximilian Mayerl, Martin Potthast, Francisco Rangel, Paolo Rosso, et al. 2020. Overview of pan 2020: Authorship verification, celebrity profiling, profiling fake news spreaders on twitter, and style change detection. In *International Conference of the Cross-Language Evaluation Forum for European Languages*, pages 372–383. Springer.

Matthew D Blagys and Mark J Hilsenroth. 2000. Distinctive features of short-term psychodynamic-interpersonal psychotherapy: A review of the comparative psychotherapy process literature. *Clinical psychology: Science and practice*, 7(2):167–188.

Benedikt Boenninghoff, Julian Rupp, Robert M Nickel, and Dorothea Kolossa. 2020. Deep bayes factor scoring for authorship verification. *arXiv preprint arXiv:2008.10105*.

Niall Bolger, Angelina Davis, and Eshkol Rafaeli. 2003. Diary methods: Capturing life as it is lived. *Annual review of psychology*, 54(1):579–616.

Niall Bolger and Jean-Philippe Laurenceau. 2013. *Intensive longitudinal methods: An introduction to diary and experience sampling research.* Guilford Press.

Jessica L Borelli, Lucas Sohn, BingHuang A Wang, Kajung Hong, Cindy DeCoste, and Nancy E Suchman. 2019. Therapist–client language matching: Initial promise as a measure of therapist–client relationship quality. *Psychoanalytic Psychology*, 36(1):9.

Emily A Butler and Ashley K Randall. 2013. Emotional coregulation in close relationships. *Emotion Review*, 5(2):202–210.

Tanya L Chartrand and Jessica L Lakin. 2013. The antecedents and consequences of human behavioral mimicry. *Annual review of psychology*, 64:285–308.

J Cohen. 1988. Statistical power analysis for the behavioral sciences, 2nd edn. á/l.

Cristian Danescu-Niculescu-Mizil, Lillian Lee, Bo Pang, and Jon Kleinberg. 2012. Echoes of power: Language effects and power differences in social interaction. In *Proceedings of the 21st international conference on World Wide Web*, pages 699–708.

Rotem Dror, Lotem Peled-Cohen, Segev Shlomov, and Roi Reichart. 2020. Statistical significance testing for natural language processing. *Synthesis Lectures on Human Language Technologies*, 13(2):1–116.

Sara El Manar El and Ismail Kassou. 2014. Authorship analysis studies: A survey. *International Journal of Computer Applications*, 86(12).

Fredrik Falkenström, Robert L Hatcher, Tommy Skjulsvik, Mattias Holmqvist Larsson, and Rolf Holmqvist. 2015. Development and validation of a 6-item working alliance questionnaire for repeated administrations during psychotherapy. *Psychological Assessment*, 27(1):169.

Ruth Feldman. 2012. Bio-behavioral Synchrony: A Model for Integrating Biological and Microsocial Behavioral Processes in the Study of Parenting. *Parenting*, 12(2-3):154–164.

Bent Fuglede and Flemming Topsoe. 2004. Jensen-shannon divergence and hilbert space embedding. In *International Symposium onInformation Theory, 2004. ISIT 2004. Proceedings.*, page 31. IEEE.

Ingrid Maria Hopkins, Michael W Gower, Trista A Perez, Dana S Smith, Franklin R Amthor, F Casey Wimsatt, and Fred J Biasini. 2011. Avatar assistant: improving social skills in students with an asd through a computer-based intervention. *Journal of autism and developmental disorders*, 41(11):1543–1555.

Adam O Horvath and Leslie S Greenberg. 1989. Development and validation of the working alliance inventory. *Journal of counseling psychology*, 36(2):223.

Zac E Imel, Jacqueline S Barco, Halley J Brown, Brian R Baucom, John S Baer, John C Kircher, and David C Atkins. 2014. The association of therapist empathy and synchrony in vocally encoded arousal. *Journal of counseling psychology*, 61(1):146.

Molly E Ireland and James W Pennebaker. 2010. Language style matching in writing: Synchrony in essays, correspondence, and poetry. *Journal of personality and social psychology*, 99(3):549.

Patrick Juola. 2008. *Authorship attribution*, volume 3. Now Publishers Inc.

Alan E Kazdin. 2008. Evidence-based treatment and practice: new opportunities to bridge clinical research and practice, enhance the knowledge base, and improve patient care. *American psychologist*, 63(3):146.

Sander L Koole and Wolfgang Tschacher. 2016. Synchrony in psychotherapy: A review and an integrative framework for the therapeutic alliance. *Frontiers in psychology*, 7:862.

Moshe Koppel, Jonathan Schler, and Shlomo Argamon. 2009. Computational methods in authorship attribution. *Journal of the American Society for information Science and Technology*, 60(1):9–26.

Sarah Peregrine Lord, Elisa Sheng, Zac E Imel, John Baer, and David C Atkins. 2015. More than reflections: empathy in motivational interviewing includes language style synchrony between therapist and client. *Behavior therapy*, 46(3):296–303.

DC Low, SD Miller, and B Squire. 2012. The outcome rating scales (ors) & session rating scales (srs): Feedback informed treatment in child and adolescent mental health services (camhs). *Norwich: Norfolk & Suffolk NHS Foundation Trust.*

Carl D Marci, Jacob Ham, Erin Moran, and Scott P Orr. 2007. Physiologic correlates of perceived therapist empathy and social-emotional process during psychotherapy. *The Journal of nervous and mental disease*, 195(2):103–111.

Mary Ann Marcinkiewicz. 1994. Building a large annotated corpus of english: The penn treebank. *Using Large Corpora*, page 273.

Erhard Mergenthaler and Charles Stinson. 1992. Psychotherapy transcription standards. *Psychotherapy research*, 2(2):125–142.

Scott D Miller, BL Duncan, J Brown, JA Sparks, and DA Claud. 2003. The outcome rating scale: A preliminary study of the reliability, validity, and feasibility of a brief visual analog measure. *Journal of brief Therapy*, 2(2):91–100.

Amir More and Reut Tsarfaty. 2016. Data-driven morphological analysis and disambiguation for morphologically rich languages and universal dependencies. In *Proceedings of COLING 2016*.

Lena C Müller-Frommeyer, Niels AM Frommeyer, and Simone Kauffeld. 2019. Introducing rlsm: An integrated metric assessing temporal reciprocity in language style matching. *Behavior Research Methods*, 51(3):1343–1359.

Kate G Niederhoffer and James W Pennebaker. 2002. Linguistic style matching in social interaction. *Journal of Language and Social Psychology*, 21(4):337–360.

Richard V Palumbo, Marisa E Marraccini, Lisa L Weyandt, Oliver Wilder-Smith, Heather A McGee, Siwei Liu, and Matthew S Goodwin. 2017. Interpersonal autonomic physiology: A systematic review of the literature. *Personality and Social Psychology Review*, 21(2):99–141.

Jane Paulick, Anne-Katharina Deisenhofer, Fabian Ramseyer, Wolfgang Tschacher, Kaitlyn Boyle, Julian Rubel, and Wolfgang Lutz. 2018. Nonverbal synchrony: A new approach to better understand psychotherapeutic processes and drop-out. *Journal of Psychotherapy Integration*, 28(3):367.

Verónica Pérez-Rosas, Rada Mihalcea, Kenneth Resnicow, Satinder Singh, and Lawrence An. 2017. Understanding and predicting empathic behavior in counseling therapy. In *Proceedings of the 55th Annual Meeting of the Association for Computational Linguistics (Volume 1: Long Papers)*, pages 1426–1435.

Slav Petrov, Dipanjan Das, and Ryan McDonald. 2011. A universal part-of-speech tagset. *arXiv preprint arXiv:1104.2086*.

Fabian Ramseyer and Wolfgang Tschacher. 2010. Nonverbal synchrony or random coincidence? how to tell the difference. In *Development of multimodal interfaces: Active listening and synchrony*, pages 182–196. Springer.

Fabian Ramseyer and Wolfgang Tschacher. 2011. Nonverbal synchrony in psychotherapy: coordinated body movement reflects relationship quality and outcome. *Journal of consulting and clinical psychology*, 79(3):284.

Fabian Ramseyer and Wolfgang Tschacher. 2014. Nonverbal synchrony of head-and body-movement in psychotherapy: different signals have different associations with outcome. *Frontiers in psychology*, 5:979.

Shlomo S Sawilowsky. 2009. New effect size rules of thumb. *Journal of Modern Applied Statistical Methods*, 8(2):26.

Natalie Shapira, Dana Atzil-Slonim, Daniel Juravski, Moran Baruch, Dana Stolowicz-Melman, Adar Paz, Tal Alfi-Yogev, Roy Azoulay, Adi Singer, Maayan Revivo, et al. 2021. Hebrew psychological lexicons. In *Proceedings of the Seventh Workshop on Computational Linguistics and Clinical Psychology: Improving Access*, pages 55–69.

Eva Sharma and Munmun De Choudhury. 2018. Mental health support and its relationship to linguistic accommodation in online communities. In *Proceedings of the 2018 CHI conference on human factors in computing systems*, pages 1–13.

Jonathan Shedler. 2010. The efficacy of psychodynamic psychotherapy. *American psychologist*, 65(2):98.

David V Sheehan, Yves Lecrubier, K Harnett Sheehan, Patricia Amorim, Juris Janavs, Emmanuelle Weiller, Thierry Hergueta, Roxy Baker, and Geoffrey C Dunbar. 1998. The mini-international neuropsychiatric interview (mini): the development and validation of a structured diagnostic psychiatric interview for dsm-iv and icd-10. *The Journal of clinical psychiatry*.

Efstathios Stamatatos. 2009. A survey of modern authorship attribution methods. *Journal of the American Society for information Science and Technology*, 60(3):538–556.

Richard F Summers and Jacques P Barber. 2009. *Psychodynamic therapy: A guide to evidence-based practice*. Guilford Press.

Yla R Tausczik and James W Pennebaker. 2010. The psychological meaning of words: Liwc and computerized text analysis methods. *Journal of language and social psychology*, 29(1):24–54.

Wolfgang Tschacher and Deborah Meier. 2020. Physiological synchrony in psychotherapy sessions. *Psychotherapy Research*, 30(5):558–573.

Gary R VandenBos. 2007. *APA dictionary of psychology*. American Psychological Association.

Travis J Wiltshire, Johanne Stege Philipsen, Sarah Bro Trasmundi, Thomas Wiben Jensen, and Sune Vork Steffensen. 2020. Interpersonal coordination dynamics in psychotherapy: A systematic review.

Justine Zhang and Cristian Danescu-Niculescu-Mizil. 2020. Balancing objectives in counseling conversations: Advancing forwards or looking backwards. *arXiv preprint arXiv:2005.04245*.

A Appendices

A.1 Dataset Description

A.1.1 Clients

The dataset was drawn as a sample from a broader pool of clients who received individual psychotherapy at a university training outpatient clinic, located in a central city in Israel. Data were collected naturalistically between August 2014 and August 2016 as part of the clinic's regular practice of monitoring clients' progress. From an initial sample of 180 clients who provided their consent to participate in the study, 34 (18.88%) dropped out (deciding onesidedly to end treatment before the planned termination date). Clients were selected from the larger sample to match two criteria: (1) treatment duration of at least 15 sessions, and (2) full data including audio recordings to be used for the transcriptions and session-by-session questionnaires available for each client. These criteria corresponded to our analytic strategy of detecting within-client associations between linguistic features and session processes and outcomes. Clients were also excluded, based on the M.I.N.I. 6.0 (Sheehan et al., 1998) if they were diagnosed as severely disturbed, either due to a current crisis, had severe trauma and accompanying post- traumatic stress disorder, a past or present psychotic or manic diagnosis, and/or current substance abuse. Based on these criteria we excluded 77 (42.7%) clients. Thus, of the total sample, the data for 68 (38.33%) clients who met the abovementioned inclusion criteria were transcribed, for a total of 872 transcribed sessions.

The clients were all above the age of 18 (M_{age}=39.06, SD=13.67, range=20–77), majority of whom were women (58.9%). Of the clients, 53.5% had at least a bachelor's degree, 53.5% reported being single, 8.9% were in a committed relationship, 23.2% were married and 14.2% were divorced or widowed. Clients' diagnoses were established based on the Mini International Neuropsychiatric Diagnostic Interview for Axis I DSM-IV diagnoses (MINI 5.0; Sheehan et al., 1998). Of the entire sample, 22.9% of the clients had a single diagnosis, 20.0% had two diagnoses, and 25.7% had three or more diagnoses. The most common diagnoses were comorbid anxiety and affective disorders[15] (25.7%), followed by other comorbid disorders (17.1%), anxiety disorders (14.3%), and affective disorders (5.7%). A sizable group of clients (31.4%) reported experiencing relationship concerns, academic/occupational stress, or other problems but did not meet criteria for any Axis I diagnosis.

A.1.2 Therapists and Therapy

Clients were treated by 59 therapists in various stages of their clinical training. Clients were assigned to therapists in an ecologically valid manner based on real-world issues, such as therapist availability and caseload. Most therapists treated one client each (47 therapists), but some (10) treated two clients and (2) more. Each therapist received one hour of individual supervision every two weeks and four hours of group supervision on a weekly basis. All therapy sessions were audiotaped for supervision. Supervisors were senior clinicians. Individual and group supervision focused heavily on reviewing audiotaped case material and technical interventions designed to facilitate the appropriate use of therapist interventions. Individual psychotherapy consisted of once- or twice-weekly sessions. The language of therapy was Modern Hebrew (MH). The dominant approach in the clinic includes a short-term psychodynamic psychotherapy treatment model (e.g.,Blagys and Hilsenroth,2000; Shedler, 2010; Summers and Barber, 2009). The key features of the model include: (a) a focus on affect and the experience and expression of emotions, (b) exploration of attempts to avoid distressing thoughts and feelings, (c) identification of recurring themes and patterns, (d) an emphasis on past experiences, (e) a focus on interpersonal experiences, (f) an emphasis on the therapeutic relationship, and (g) exploration of wishes, dreams, or fantasies (Shedler, 2010). On average, treatment length was 37 sessions (SD = 23.99, range = 18–157). Treatment was open- ended in length, but given that psychotherapy was provided by clinical trainees at a university-based outpatient community clinic, the treatment duration was often restricted to be 9 months.

A.1.3 Transcriptions

To capture the treatment processes from session to session, and since the transcription process is highly expensive, transcriptions were conducted alternately (i.e., sessions 2, 4, 6, 8 and so on until

[15]The following DSM-IV diagnoses were assessed in the affective disorders cluster: major depressive disorder, dysthymia and bipolar disorder. The following DSM-IV diagnoses were assumed in the anxiety disorders cluster: panic disorder, agoraphobia, generalized anxiety disorder and social anxiety disorder.

one session before the last session). In cases where material was incomplete (such as the quality of the recordings, or the questionnaires for a specific session), the next session was transcribed instead. The transcriber team was composed of seven transcribers, all of whom were graduate students in the University's psychology department. The transcribers went through a one day training workshop and monthly meetings were held throughout the transcription process to supervise the quality of their work. The training included specific guidelines on how to handle confidential and sensitive information and the transcribers were instructed to replace names by pseudonyms and to substitute any other identifying information. The transcription protocol followed general guidelines, as described in (Mergenthaler and Stinson, 1992), and in (Albert et al., 2013). The word forms, the form of commentaries, and the use of punctuation were kept as close as possible to the speech presentation. Everything was transcribed, including word fragments as well as syllables or fillers (such as "ums", "ahs", "uh huhs" and "you know"). The audiotape was transcribed in its entirety and provided a verbatim account of the session. The transcripts included elisions, mispronunciations, slang, grammatical errors, non- verbal sounds (e.g., laughs, cry, sighs), and background noises. The transcription rules were limited in number and simple (for example, each client and therapist utterances should be on a separate line; each line begins with the specification of the speaker) and the format used several symbols to indicate comments (such as [...] to indicate the correct form when the actual utterance was mispronounced, or <number of minutes of silence >). The transcripts were proofread by the research coordinator. The final transcripts could be processed by human experts or automatically by computer.

There were 872 transcripts in total (the mean transcribed sessions per client was 12.56; SD=4.93) Each transcript incorporated metadata such as the client's code, which allowed the client data to be linked across sessions and for hierarchical analysis. The transcriptions totaled about four million words over 150,000 talk turns (i.e., switching between speakers). On average, there were 5800 words in a session, of which 4538 (78%; SD=1409.62; range 416-8176) were client utterances and 1266 (22%; SD=674.99; range 160-6048) were therapist utterances with a mean of 180.07 (SD=95.37; range 30-845) talk turns per session.

A.1.4 Procedure and Ethical Considerations

The procedures were part of the routine assessment and monitoring process in the clinic. All research materials were collected after securing the approval of the authors' university ethics committee. Only clients that gave their consent to participate were included in the study. Clients were told that they could choose to terminate their participation in the study at any time without jeopardizing treatment. The clients completed the ORS before each therapy session and the WAI after each session. The therapist completed the WAI after each therapy session. The sessions were audiotaped and transcribed according to a protocol described above. All data collected was anonymized and only then exposed to a very small number of researchers, as agreed upon by the participants.

A.1.5 Missing Data

In the concurrent session-level models, from the transcribed sessions (872), 860 had functioning (ORS), 831 had therapist's therapeutic alliance (T_WAI) and 823 had client's therapeutic alliance (C_WAI). One transcription was detected with errors. Sessions with missing or faulty data were excluded from the analysis.

A.2 Outcome & Process Measurements

A.2.1 Outcome Rating Scale (ORS; (Miller et al., 2003))

The ORS is a 4-item visual analog scale developed as a brief alternative to the OQ-45. The scale is designed to assess change in three areas of client functioning that are widely considered to be valid indicators of progress in treatment: functioning, interpersonal relationships, and social role performance. Respondents complete the ORS by rating four statements on a visual analog scale anchored at one end by the word "Low" and at the other end by the word "High". This scale yields four separate scores between 0 and 10 that sum to one score ranging from 0 to 40, with higher scores indicating better functioning. The ORS has strong reliability estimates (α=0.87-0.96) and moderate correlations between the ORS items and the OQ-45 subscale and total scores (ORS total - OQ-45 total: r = 0.59).

A.2.2 Working Alliance Inventory (WAI; (Horvath and Greenberg, 1989)

The WAI is a self report questionnaire (both for therapist and client). It is one of the most widely investigated common factors that was found positively correlated to treatment outcome in psychotherapy. It includes items ranging from 0 ("not at all") to 5 ("completely") to evaluate three components (1) agreement on treatment goals, (2) agreement on therapeutic tasks and (3) a positive emotional bond between client and therapist (Falkenström et al., 2015)

A.3 Complementing Behavior as Synchrony

Synchrony may be observed through complementing behavior, where the actions of one party influences the second party in a complementing manner, e.g., if a rise of an occurrence of a feature in the first party directly causes a proportional decline for the second party, and vice-versa, yielding a negative correlation.

We show here that the number of words spoken by the participants in the sessions renders such behavior. As one participant talks more within a session, the other naturally talks less. Since all psychotherapy sessions have a fixed length of one hour, we can comparably measure this effect across all sessions.

Algorithm 2: Client's (c) and therapist's (t) word count in sessions (size=m) correlation

```
candidateMLS-2(c,t,m);
for j ← 1 to m do
    cWC_j ← wordCount(c_j);
    tWC_j ← wordCount(t_j);
end
return: pearsonr(cWC, tWC)
```

We propose MLS function *CandidateMLS-2* (Algorithm 2) which receives as input lists C^d and T^d of size m_d of a client's and the matching therapist's transcribed speech within each of their sessions (m_d is the number of sessions within a specific dyad, d). Each list element contains the clients'/therapists' utterances from a single session, and $c_j^d \in C^d$ and $t_j^d \in T^d$ are from the same session, for each session j. The algorithm converts each element in the lists to the word-count-number. Finally, the algorithm outputs the Pearson coefficient correlation between the new lists.

A surrogate test (as describe in Section 5.3) produces significant separation both at the between-surrogate ($p < 0.05$ with large effect size, Cohen's d = 0.953) and within-surrogate ($p < 0.05$ with large effect size, Cohen's d = 1.038). These results shows that *CandidateMLS-2* is indeed MLS, notably featuring *complementing* synchrony.

A.4 LSM vs. POS

The LSM method (Ireland and Pennebaker, 2010) takes advantage of word categories defined in LIWC, see Table 3. LIWC was not translated to a Hebrew version. Languages behave differently and it is therefore impossible to produce a perfect translation. For example, in Hebrew there is no use of articles (for the challenges in the Hebrew translation process see Shapira et al., 2021).

Since a Hebrew LIWC version is not available, an alternative approach is to apply *part-of-speech* categories that can be loosely mapped to LIWC categories used in the LSM method. Part-of-speech (POS Marcinkiewicz, 1994) is a linguistic category of words that have similar grammatical properties, i.e., words assigned with the same part-of-speech tag play a similar role within the grammatical structure of sentences (for the multilingual efforts to create a universal POS tagset see Petrov et al., 2011).[16] The POS categories can express the way things are said rather than the content itself ("how" versus "what"). Extraction of POS tags is a common procedure in natural language processing, and relevant tools exist in Hebrew (e.g., YAP; More and Tsarfaty, 2016, see Table 4).

There is a loose relationship between LIWC categories used by LSM and the POS categories.

- The **Auxiliary** category in LIWC contains the words that fall under the COP POS category, but COP represents any copula (אוגד) which is not always a verb in Hebrew. In addition there is an intersection with the MD POS category (e.g., could).
- The **Conjunction** LIWC category can be mapped to the POS categories CONJ, CC, TEMP and REL. CONJ is for the coordinating conjunction ו (and); TEMP is for the subordinating conjunctions that precede time clauses e.g., כש (when); REL is for the relative clauses ה, ש (that); CC is for the rest of conjunctions, both coordinating and subordinating.

[16] For the universal POS tags see https://universaldependencies.org/u/pos/

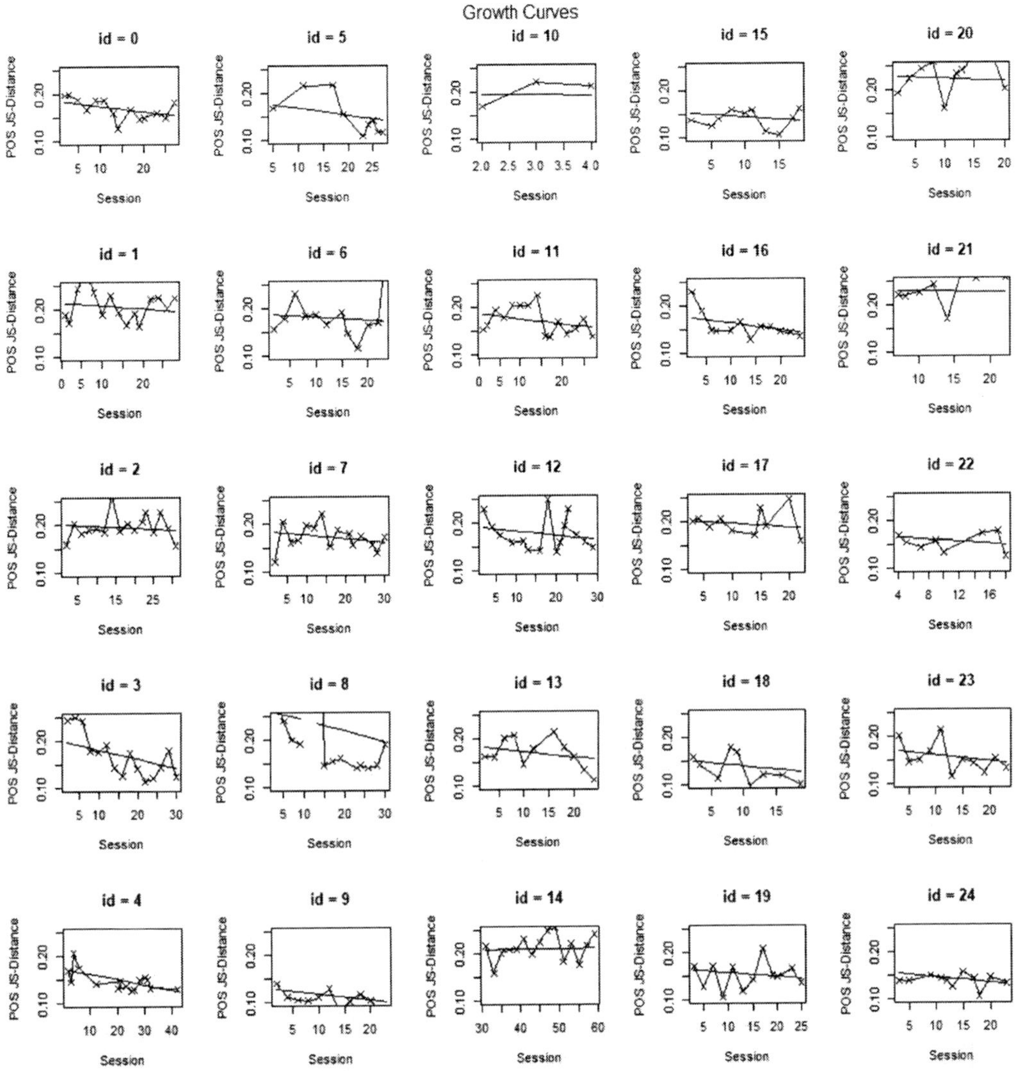

Figure 5: Growth Curves of 25 sampled dyads of the 74 available. There is a decrease of 0.001 units (i.e., slope) of JS-Distance between Probability Distribution over Unigram POS-tag in each session throughout treatment, indicating an increase in linguistic similarity. Results are statistically significant with $p<0.0001$.

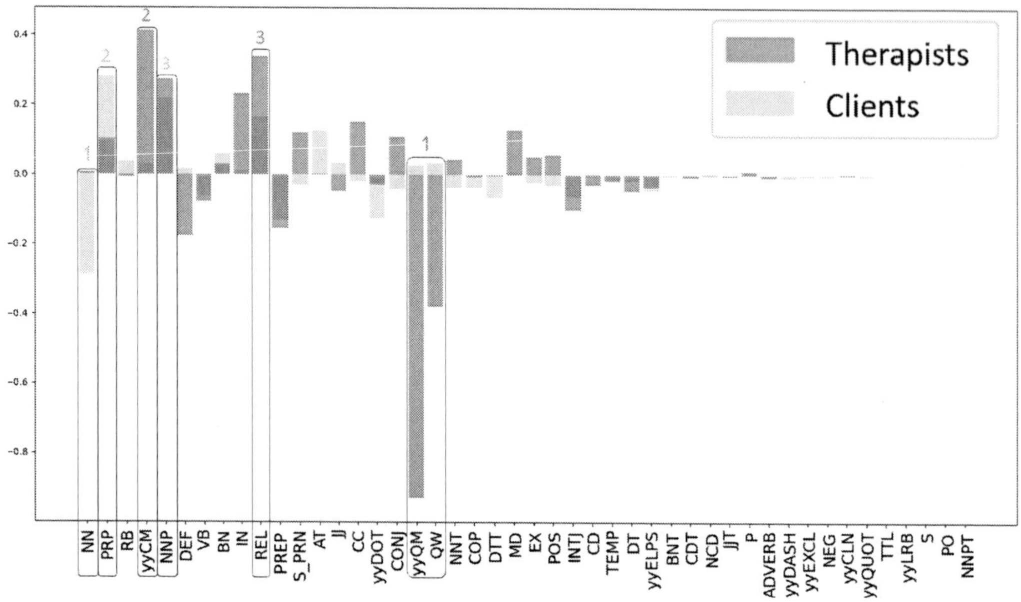

Figure 6: The sum of POS tag frequency changes between consecutive sessions for all clients (orange) and therapists (purple). A positive (negative) value means an overall increase (decrease) in the frequency of a POS tag throughout treatments. The three major changes in treatments for therapists are (1) decrease in questions (yyQM, QM) while for clients this increases, (2) increase in commas i.e., short break (yyCM), similarly to clients, (3) increase in "that" (REL), also similar to clients' behavior. The three for clients are: (1) decrease in nouns (NN) while for therapists this increases, (2) increase in personal pronouns (PRP), as for therapists, and (3) increase in names (NNP) like for therapists. Overall, the therapists change throughout treatment more than the clients do.

- There is no POS category for the LIWC category **High-Frequency Adverbs**, but there is a POS category, RB, for general adverbs.
- The POS category PRP intersects with the LIWC categories **Personal** and **Impersonal Pronouns**. The POS category S_PRN is fully contained in the LIWC category of **Personal Pronouns** but only for single first person.
- The LIWC category **Negations** is partially represented by the POS category NEG.
- **Prepositions** with the LIWC categories can be mapped to the POS categories PREPOSITION and IN.
- **Quantifiers** with the LIWC categories can be mapped to the POS categories DT and DTT.
- In Hebrew there is no use of **Articles**.

In our study we used all possible POS categories.

LIWC LSM Categories

Category	Examples of Words in Lexicon
Articles	a, an, the
Auxiliary Verbs	ain't, am, are, ...
Conjunctions	also, and, as, but, ...
High-Frequency Adverbs	about, absolutely, actually, again, ...
Impersonal Pronouns	another, anybody, if, itself, ...
Personal Pronouns	he, him, ...
Prepositions	about, above, along, ...
Quantifiers	add, alot, all, few, ...
Negations	not, no, never, ...

Table 3: LSM categories by LIWC. In some versions there are slight differences regarding the included markers (e.g., in linguistic style coordination Danescu-Niculescu-Mizil et al., 2012, the negation marker is not included).

YAP POS-tags

Tag	Examples of Hebrew Words in Tag (Translation)
ADVERB	כ (about)
AT	את (term used to indicate a direct object)
BN	מתרוצצת (scampering), רוצה (wanting), ...
BNT	לובשי (wearing), ...
CC	כאילו (like), אבל (but), אם (if), ...
CD	אחת (one), 44, ...
CDT	שני (two), ...
CONJ	ו (and)
COP	הייתי (was), היא (is), ...
DEF	ה (the)
DT	איזשהו (some), איזשהי (some)
DTT	שום (any), כל (all), ...
EX	יש (exist), אין (not exist)
IN	בשביל (for), אצל (at), ...
INTJ	נא (please)
JJ	קשה (hard), בטוח (safe), ...
JJT	עומסי (load), ...
MD	נוכל (could), תוכלי (could), צריכה (need), ...
NCD	40, 30%, ...
NEG	לאו (not)
NN	ארץ (country), קניון (mall), משהו (somthing), ...
NNP	חולון (Holon), צרפת (France), ...
NNPT	פלמח (Palmach)
NNT	קרית (a first part in names of cities and neighborhoods), ...
POS	של (of)
PREPOSITION	ל (to), ב (at), ...
PRP	הוא (he), זה (it), אני (I), ...
QW	למה (why), מי (who), איפה (where), ...
RB	רק (only), מאוד (really), מהר (quickly), ...
REL	ש (that)
S_PRN	את (you), היא (she), אני (I), ...
TEMP	כש (when)
TTL	אדון (Mr.), ...
VB	להתלבש (to dress), נפלו (fall), ...
yyCLN	:
yyCM	,
yyDASH	-
yyDOT	.
yyELPS	...
yyEXCL	!
yyLRB	(
yyQM	?
yyQUOT	"
yyRRB	)

Table 4: POS-tags by Hebrew parser YAP.
For the full list and meanings see `https://nlp.biu.ac.il/~rtsarfaty/onlp/hebrew/postags`

Nonsuicidal Self-Injury and Substance Use Disorders: A Shared Language of Addiction

Salvatore Giorgi[1,2], McKenzie Himelein-Wachowiak[2], Daniel Habib[2], Lyle H. Ungar[1], Brenda Curtis[2]
[1]University of Pennsylvania, [2]National Institute on Drug Abuse
sgiorgi@sas.upenn.edu, brenda.curtis@nih.gov

Abstract

Nonsuicidal self-injury (NSSI), or the deliberate injuring of one's body without intending to die, has been shown to exhibit many similarities to substance use disorders (SUDs), including population-level characteristics, impulsivity traits, and comorbidity with other mental disorders. Research has further shown that people who self-injure adopt language common in SUD recovery communities (e.g., "clean", "relapse", "addiction," and celebratory language about sobriety milestones). In this study, we investigate the shared language of NSSI and SUD by comparing discussions on public Reddit forums related to self-injury and drug addiction. To this end, we build a set of LDA topics across both NSSI and SUD Reddit users and show that shared language across the two domains includes SUD recovery language in addition to other themes common to support forums (e.g., requests for help and gratitude). Next, we examine Reddit-wide posting activity and note that users posting in *r/selfharm* also post in many mental health-related subreddits, while users of drug addiction related subreddits do not, despite high comorbidity between NSSI and SUDs. These results show that while people who self-injure may contextualize their disorder as an addiction, their posting habits demonstrate comorbidities with other mental disorders more so than their counterparts in recovery from SUDs. These observations have clinical implications for people who self-injure and seek support by sharing their experiences online.

1 Introduction

Nonsuicidal self-injury (NSSI), or the intentional injuring of one's body without aiming to die for reasons outside of social norms, causes significant morbidity (Nock, 2010). Lifetime prevalence of NSSI is estimated to range from 5-6% among adults to 17-18% among adolescents (Swannell et al., 2014) while the prevalence of NSSI among adolescents with psychiatric disorders is thought to be much higher (Nock and Prinstein, 2004; Glenn and Klonsky, 2013).

Qualitative, quantitative, mixed-methods, and psychometric studies have pointed towards the addictive features of NSSI and shared characteristics between NSSI and substance use disorders (SUDs; Brown and Kimball, 2013; Davis and Lewis, 2019). Populations who self-injure may resemble populations with SUDs in personality (MacLaren and Best, 2010) or impulsivity traits (Dir et al., 2013). NSSI and SUD are often comorbid with anxiety, depressive, and psychotic disorders (Guvendeger Doksat et al., 2017) and with each other, with one study reporting that approximately 60% of people who self-injure met criteria for a SUD (Nock et al., 2006). Both substance use and NSSI are used to avoid and/or cope with feelings of psychological distress (Chawla and Ostafin, 2007), especially among adolescents (Peterson et al., 2008). Although there is debate as to whether or not NSSI is an addiction in a clinical sense or experienced as intensely as SUD (Victor et al., 2012), addiction models of NSSI have been proposed (Faye, 1995; Buser and Buser, 2013; Blasco-Fontecilla et al., 2016). There is also evidence that those who endorse more addictive features of self-injury harm themselves more frequently and more severely (Martin et al., 2013), have higher levels of internalized anger (Nixon et al., 2002), and are at increased risk for accidentally harming to a greater extend than intended (Buser et al., 2017) and attempting suicide (Csorba et al., 2009). Some have urged clinicians to consider addictive features of NSSI when treating people who self-injure (Blasco-Fontecilla et al., 2016), making the addictive aspects of NSSI a valuable research target. Previous work has also identified the adoption of language used in SUD recovery circles Alcoholics Anonymous (AA) and Narcotics Anonymous (NA) in NSSI communities. This "addiction

Proceedings of the Eighth Workshop on Computational Linguistics and Clinical Psychology, pages 177 - 183
July 15, 2022

language" (i.e., phrases such as "relapse", "recovery" and celebration of time without self-injury) has been found on NSSI message boards (Whitlock et al., 2006), Facebook groups (Niwa and Mandrusiak, 2012), and LiveJournal (Davis and Lewis, 2019).

This study builds on the work of Himelein-Wachowiak et al. (2022), who investigated the use of "addiction language" and experiences of addiction in the *r/selfharm* subreddit. This was done through a text-based annotation process where experts in addiction and recovery adapted the Diagnostic and Statistical Manual of Mental Disorders Fifth Edition (DSM-5) criteria for SUD to NSSI as a method for measuring the symptoms and severity of addiction to NSSI for Reddit users. Results showed that over three-quarters of the sample met the criteria for addiction and 86% used "addiction language".This work also builds on a large body of research using Reddit as a tool for mental health applications (De Choudhury and De, 2014), which includes depression (Pirina and Çöltekin, 2018), anxiety (Shen and Rudzicz, 2017), suicide (Zirikly et al., 2019), substance use (Lu et al., 2019), schizophrenia (Zomick et al., 2019), and DSM-5 evaluations (Gaur et al., 2018).

The purpose of this study is to further evaluate experiences of addiction by examining shared "addiction language" between NSSI and SUD communities using automated methods. We begin by directly comparing NSSI and SUD subreddits through a set of LDA topics estimated over a corpus of Reddit comments and examine how themes of addiction and recovery are used across both communities. While previous studies have identified "addiction language" in NSSI communities, to our knowledge, none have directly compared NSSI and SUD communities. We also identify where users of these subreddits are posting across Reddit in order to identify common communities. We end by discussing the clinical implications of our findings.

2 Data

Self-injury We looked at posts from the *r/selfharm* subreddit from the Pushshift Reddit Data set (Baumgartner et al., 2020). We focused on *r/selfharm* based on the fact that it had the highest number of posts and users and the most diverse discussion around NSSI. See Supplemental Materials for descriptions of other self-injury related subreddits as well as temporal trends in post histories.

Substance Use In order to compare the *r/selfharm* users to those posting in SUD recovery communities, we gather the posting activity from all users who have posted in the following subreddits: *r/addiction*, *r/alcoholism*, *r/opiatesrecovery*, *r/leaves*, *r/stopdrinking*, and *r/redditorsinrecovery*. These were manually selected due to high post volume and focus on recovery (vs. drug use itself).

Our data set consisted of both comments and submissions (i.e., the first post in a Reddit thread) across 2019 in order to match the temporal span of Himelein-Wachowiak et al. (2022). Across both data sets, we removed any accounts with the word "bot" in the user handle, after manually inspecting the account to confirm that the account is indeed a bot, as well as deleted posts, deleted accounts, and moderators. We also removed any redditor who posted in both *r/selfharm* and one or more SUD subreddits, in order to remove the possibility that common users will drive shared language. To identify redditors who are active in their respective communities, we remove any redditor with less than 10 comments, resulting in 2,470 *r/selfharm* who together posted 77,414 comments. We then identified a matched sample of 2,470 SUD redditors (posting 77,424 comments), approximately matched on both comment and submission counts.

3 Methods

Task 1: Shared Language To examine shared language across NSSI and SUD subreddits, we estimate a set of Content Specific LDA topics (Zamani et al., 2020). Content Specific LDA (CSLDA) is a method for estimating LDA topics across a thematically concise corpus and has been previously used to model conversations around excessive drinking, diabetes, and Black Lives Matter tweets (Giorgi et al., 2020; Griffis et al., 2020; Giorgi et al., 2022). CSLDA contains a preprocessing pipeline that identifies words related to the theme in question (i.e., NSSI and SUD) by comparing this data to a background corpus of general language (i.e., data from *r/AskReddit*). This removes language that is specific to Reddit as opposed to being NSSI or SUD related. We create CSLDA topics across a combined corpus of comments from NSSI and SUD subreddits. See Supplemental Materials for full details of the CSLDA pipeline.

We use the Mallet Java software wrapper within the DLATK Python package (McCallum, 2002; Schwartz et al., 2017). All default settings are used,

	25 Topics		50 Topics		75 Topics		100 Topics	
alpha	TU	Coh.	TU	Coh.	TU	Coh.	TU	Coh.
1	0.98	0.46	0.85	0.38	0.72	0.33	0.61	0.28
3	0.99	0.31	0.90	0.37	0.72	0.32	0.65	0.34
5	0.99	0.41	0.90	0.36	0.78	0.37	0.66	0.36

Table 1: Topic quality as measured through Topic Uniqueness (TU) and Coherence (Coh.).

and we evaluate a range of $\alpha \in \{1, 3, 5\}$, a prior on the number of topics per document, and topic set sizes $K \in \{25, 50, 75, 100\}$. All topics are quantitatively and qualitatively evaluated. Quantitative evaluation consists of two metrics: coherence and topic uniqueness. Coherence measures semantic similarity between the words in the topic using Normalized Pointwise Mutual Information (NPMI; Syed and Spruit, 2017). This is calculated for each topic and then averaged across all topics. Topic uniqueness (TU), a measure of topic diversity, is inversely proportional to the number of times a set of L keywords is repeated across a set of K topics (Nan et al., 2019). Thus, a topic set with high TU means that the representative keywords are rarely repeated across topics. While past research has used a value of $L = 10$ (Nan et al., 2019), we set $L = 30$ to be more conservative (with a large L, the probability of a given word appearing in more than one topic will increase, thus decreasing TU). TU ranges between 1 and $1/K$, and therefore we normalize TU to be between 0 and 1, since we are evaluating topic sets of sizes K.

Qualitative evaluation consisted of manually inspecting topics for three criteria: (1) breadth of themes, (2) minimal thematic overlap, and (3) a single topic contains a single theme. Note that (2) and (3) are similar to TU and coherence.

Task 2: Posting Activity Here, we look at all posts (i.e., submissions and comments) across the whole of Reddit in 2019 from our disjoint samples of NSSI and SUD redditors. For the NSSI redditors, we gather 1,019,796 of their posts in subreddits other than *r/selfharm*. For the SUD redditors, we gather 927,733 posts to subreddits other than the 6 addiction subreddits used to collect the sample. We then reported the most frequently visited subreddits for both the NSSI and SUD samples and calculated the percentage of users posting in each.

4 Results

Task 1: Shared Language In Table 1, we evaluate the CSLDA topics. Here we see Topic Uniqueness (TU) decrease as the number of topics grows. This is to be expected since as the number of topics grows one can expect words to be shared across a larger number of topics. TU also increases with α within a fixed topic set (i.e., column-wise). Coherence shows no clear pattern across α or topic set size. Through the qualitative evaluation, $K = 50$ topics with $\alpha = 5$ were chosen as the most interpretable. Since TU is high across the $K = 50$ topic sets and coherence is reasonable with $\alpha = 5$ (i.e., neither the highest nor the lowest value across all topic sets), we proceed with this topic set.

Figure 1 shows the average user-level topic usage across all 50 topics, ordered by the difference between the NSSI (green) and SUD (blue) groups. In total, we see 9 out of 50 topics include addiction or recovery-related keywords within the top 10 highest weighted words in the topic. We do not include topics that contain both "clean" and "cut" since "clean" most likely does not refer to "staying clean" in the recovery sense. Additionally, we see that the most similar topic is addiction-related ("addiction", "addicted"), as well as the 6th ("recovery", "relapsed", "clean"). Notably, the remaining addiction topics are the least similar (i.e., towards the

NSSI users		SUD users	
Subreddit	% Users	Subreddit	% Users
AskReddit	52.8	AskReddit	47.7
depression$^\diamond$	37.8	pics	21.9
MadeOfStyrofoam†	34.4	funny	21.8
SuicideWatch$^\diamond$	33.9	aww	21.3
SelfHarmScars†	29.0	todayilearned	17.0
teenagers	26.2	AmItheAsshole	16.3
memes	21.1	Showerthoughts	16.2
Showerthoughts	18.7	mildlyinteresting	15.1
aww	16.8	relationship_advice	14.9
AmItheAsshole	15.5	news	14.9
unpopularopinion	15.3	worldnews	14.5
funny	14.4	politics	14.4
wholesomememes	14.0	gifs	13.6
dankmemes	13.6	trashy	12.8
mentalhealth$^\diamond$	13.5	unpopularopinion	12.8
2meirl4meirl	13.0	interestingasfuck	12.6
Anxiety$^\diamond$	12.6	gaming	12.6
mildlyinteresting	12.6	PublicFreakout	12.4
offmychest$^\diamond$	12.0	videos	12.1
relationship_advice	11.8	tifu	11.2

Table 2: Most popular subreddits, defined as the percentage of users within each group that post in a given subreddit (% Users). † and ◇ are self-injury and mental health related subreddits, respectively.

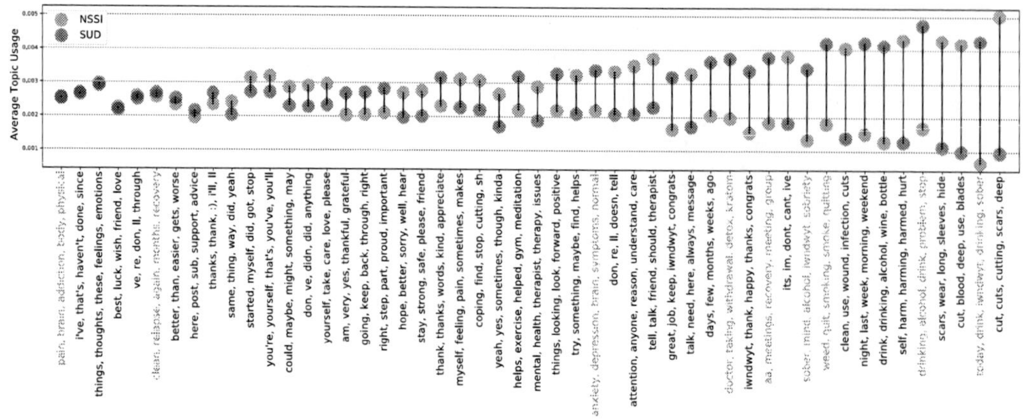

Figure 1: Differences in average user-level topic frequency. The top 5 weighted words from each topic are visualized with red topics containing an addiction keyword (e.g., "sober" and "relapse") in the top 10 weighted words.

right side of the figure). These later topics contain either substance use keywords (and are, thus, used more often by SUD users) or contain words like "withdrawal" which was one of the least prevalent DSM-5 symptoms (6% of users) found in Himelein-Wachowiak et al. (2022).

There are also a number of topics that do not contain recovery keywords but seem to be related to addiction and recovery such as mentions of starting/stopping ("started", "stop") and getting easier ("gets", "easier"). While not addiction or recovery related, we do see shared language of gratitude ("thanks" and "thank"), support ("better", "easier", "gets" as well as "support", "advice"), emotions ("thoughts", "feelings", "emotions"), and coping.

Task 3: Posting Activity Table 2 shows the 20 subreddits in which the highest percentage of NSSI and SUD redditors are also posting. Not surprisingly, we see that NSSI redditors are posting in other NSSI related subreddits: *r/MadeOfStyrofoam* and *r/SelfHarmScars*. On the other hand, SUD redditors are not posting in other substance-related subreddits outside of the six used to collect data. We also note that there are a number of mental health related subreddits in which NSSI redditors are posting, which is in line with common NSSI comorbidities: *r/depression*, *r/SuicideWatch*, and *r/Anxiety*. We do not see similar posting activity in mental health subreddits among SUD redditors despite high comorbidity between SUD and both depression and anxiety (Conway et al., 2006).

5 Conclusion

In this work, we directly compared language across large online communities dedicated to discussing NSSI and SUD recovery. We showed that there is indeed a shared language of addiction between these two communities, evidenced by equivalent usage of topics related to addiction ("addiction", "addicted") and recovery ("recovery", "relapsed", "clean"). To our knowledge, this is the first study using automated methods to quantify addiction language in NSSI communities, as well as the first to directly compare language between online NSSI and SUD recovery forums. We also examined Reddit-wide posting activity and showed that, while NSSI redditors posted in a number of mental health related subreddits, SUD redditors did not even though both NSSI and SUD are comorbid with many of the same mental health disorders.

One limitation of our study is the high comorbidity between NSSI and SUD: 60% of adolescents engaging in NSSI meeting criteria for SUDs (Nock et al., 2006). Thus, the shared addiction language may be a result of NSSI redditors also having and discussing SUDs. We attempted to control for this by excluding redditors who are posting in both NSSI and SUD subreddits. Additionally, Himelein-Wachowiak et al. (2022) noted that only 2% of NSSI redditors explicitly mentioned having a SUD.

Despite this limitation, our results suggest that the adoption of addiction and recovery language in NSSI communities may provide psychological benefit to the users and help them cope with self-injury. Himelein-Wachowiak et al. (2022) posits that alignment with SUD may buffer against self-stigma and encourage adoption of common SUD recovery strategies. In a similar study, Pritchard et al. (2021) suggest that people who self-injure use addiction messages to convey the difficulty in stopping, as well as to caution those considering NSSI as a coping strategy. Our results also suggest that NSSI redditors seek support in similar communities for other mental health concerns, perhaps with the goals of broadening their support network or seeking specific advice for a separate mental disorder. The lack of mental health cross posting among SUD redditors may imply a need for more discussion of cormorbid mental disorders among SUD recovery communities as well as greater engagement with people dealing with other mental disorders. Regardless, NSSI and SUD recovery communities share similar language of support ("yourself", "take", "care", "love") and encouragement ("better", "easier", "gets"), illustrating the broad psychological benefits of sharing intimate experiences with empathetic others online, regardless of the particular mental health concern.

6 Ethical Considerations

NSSI communities and their members tend to refer to NSSI as "self harm." In this paper, we use "self-injury" or the acronym NSSI as it is more specific to the behavior in question ("self harm" could also include suicide attempts) as well as the term most frequently found in recent literature. However, papers we cite may use the terms "deliberate self harm" (DSH) or "self-mutilation." For a review and discussion of the most appropriate language to use when referring to people who self-injure, see (Hasking et al., 2021).

There are a number of ethical considerations when using sensitive data. Since Reddit data is publicly available, this study was deemed non-human subjects research and exempt from approval of an Institutional Review Board. Despite this official classification, the data used throughout is indeed human generated and reflects the lived experiences, intimate feelings, and personal struggles of the authors. Related, there are issues regarding informed consent when using public data. Online communities such as *r/selfharm* are intimate and personal spaces, where consensual sharing happens between community members and not with researchers who collect the data. For a full discussion of related issues, we recommend the work of Chancellor et al. (2019), which identifies conflicting representations of humans in "human centered machine learning." There are also issues of privacy; while Reddit is anonymous, there are risks of revealing sensitive information or the identities of the accounts used in the study (Proferes et al., 2021). As such, we only report aggregate information throughout the manuscript, and we have chosen to not publicly release any of the data used in this study. Finally, there are some egregious use cases with this data. For example, given the cross posting between NSSI and mental health forums, one could imagine ads for anti-depressants being targeted to the redditors in this study.

One must also consider researchers' well-being when working with data of this type. Spending time with sensitive and potentially triggering data can be emotionally challenging for researchers. As such, researchers should also consent to working with this type of data and continue to consent throughout the life of the project. To help with these issues, our research group held one-on-one and group sessions to discuss triggering content and mental and emotional fatigue experienced while working on this and similar projects.

Acknowledgements

This research was supported by the Intramural Research Program of the NIH, NIDA. We would like to thank Elise Bragard, Amy Kwarteng, Destiny Schriefer, Chase Smitterberg, and Kenna Yadeta for their help in the annotation process, as well as Garrick Sherman, Amanda Devoto, and the CLPsych reviewers for their thoughtful feedback.

References

Jason Baumgartner, Savvas Zannettou, Brian Keegan, Megan Squire, and Jeremy Blackburn. 2020. The pushshift reddit dataset. In *Proceedings of the International AAAI Conference on Web and Social Media*, volume 14, pages 830–839.

Hilario Blasco-Fontecilla, Roberto Fernández-Fernández, Laura Colino, Lourdes Fajardo, Rosa Perteguer-Barrio, and Jose De Leon. 2016. The addictive model of self-harming (non-suicidal and suicidal) behavior. *Frontiers in Psychiatry*, 7:8.

Tiffany B Brown and Thomas Kimball. 2013. Cutting to live: A phenomenology of self-harm. *Journal of Marital and Family Therapy*, 39(2):195–208.

Trevor J Buser and Juleen K Buser. 2013. Conceptualizing nonsuicidal self-injury as a process addiction: Review of research and implications for counselor training and practice. *Journal of Addictions & Offender Counseling*, 34(1):16–29.

Trevor J Buser, Juleen K Buser, and Corrine C Rutt. 2017. Predictors of unintentionally severe harm during nonsuicidal self-injury. *Journal of Counseling & Development*, 95(1):14–23.

Stevie Chancellor, Eric PS Baumer, and Munmun De Choudhury. 2019. Who is the" human" in human-centered machine learning: The case of predicting mental health from social media. *Proceedings of the ACM on Human-Computer Interaction*, 3(CSCW):1–32.

Neharika Chawla and Brian Ostafin. 2007. Experiential avoidance as a functional dimensional approach to psychopathology: An empirical review. *Journal of clinical psychology*, 63(9):871–890.

Kevin P Conway, Wilson Compton, Frederick S Stinson, and Bridget F Grant. 2006. Lifetime comorbidity of dsm-iv mood and anxiety disorders and specific drug use disorders: results from the national epidemiologic survey on alcohol and related conditions. *The Journal of clinical psychiatry*, 67(2):10343.

Janos Csorba, Elek Dinya, Paul Plener, Edit Nagy, and Eszter Páli. 2009. Clinical diagnoses, characteristics of risk behaviour, differences between suicidal and non-suicidal subgroups of hungarian adolescent outpatients practising self-injury. *European Child & Adolescent Psychiatry*, 18(5):309–320.

Sarah Davis and Christopher Alan Lewis. 2019. Addiction to self-harm? the case of online postings on self-harm message boards. *International journal of mental health and addiction*, 17(4):1020–1035.

Munmun De Choudhury and Sushovan De. 2014. Mental health discourse on reddit: Self-disclosure, social support, and anonymity. In *Eighth international AAAI conference on weblogs and social media*.

Allyson L Dir, Kenny Karyadi, and Melissa A Cyders. 2013. The uniqueness of negative urgency as a common risk factor for self-harm behaviors, alcohol consumption, and eating problems. *Addictive behaviors*, 38(5):2158–2162.

Pamela Faye. 1995. Addictive characteristics of the behavior of self-mutilation. *Journal of psychosocial nursing and mental health services*, 33(6):36–39.

Manas Gaur, Ugur Kursuncu, Amanuel Alambo, Amit Sheth, Raminta Daniulaityte, Krishnaprasad Thirunarayan, and Jyotishman Pathak. 2018. " let me tell you about your mental health!" contextualized classification of reddit posts to dsm-5 for web-based intervention. In *Proceedings of the 27th ACM International Conference on Information and Knowledge Management*, pages 753–762.

Salvatore Giorgi, Sharath Chandra Guntuku, McKenzie Himelein-Wachowiak, Amy Kwarteng, Sy Hwang, Muhammad Rahman, , and Brenda Curtis. 2022. Twitter corpus of the #blacklivesmatter movement and counter protests: 2013 to 2021. *Proceedings of the International AAAI Conference on Web and Social Media.*

Salvatore Giorgi, David B Yaden, Johannes C Eichstaedt, Robert D Ashford, Anneke EK Buffone, H Andrew Schwartz, Lyle H Ungar, and Brenda Curtis. 2020. Cultural differences in tweeting about drinking across the us. *International Journal of Environmental Research and Public Health*, 17(4):1125.

Catherine R Glenn and E David Klonsky. 2013. Nonsuicidal self-injury disorder: an empirical investigation in adolescent psychiatric patients. *Journal of clinical child & adolescent Psychology*, 42(4):496–507.

Heather Griffis, David A Asch, H Andrew Schwartz, Lyle Ungar, Alison M Buttenheim, Frances K Barg, Nandita Mitra, and Raina M Merchant. 2020. Using social media to track geographic variability in language about diabetes: Infodemiology analysis. *JMIR diabetes*, 5(1):e14431.

Neslim Guvendeger Doksat, Oguzhan Zahmacioglu, Arzu Ciftci Demirci, Gizem Melissa Kocaman, and Ayten Erdogan. 2017. Association of suicide attempts and non-suicidal self-injury behaviors with substance use and family characteristics among children and adolescents seeking treatment for substance use disorder. *Substance use & misuse*, 52(5):604–613.

Penelope A Hasking, Mark E Boyes, and Stephen P Lewis. 2021. The language of self-injury: a data-informed commentary. *The Journal of Nervous and Mental Disease*, 209(4):233–236.

McKenzie Himelein-Wachowiak, Salvatore Giorgi, Amy Kwarteng, Destiny Schriefer, Chase Smitterberg, Kenna Yadeta, Elise Bragard, Amanda Devoto, Lyle Ungar, and Brenda Curtis. 2022. Getting "clean" from nonsuicidal self-injury: Experiences of addiction on the subreddit r/selfharm. *Journal of behavioral addictions*, 11(1):128–139.

John Lu, Sumati Sridhar, Ritika Pandey, Mohammad Al Hasan, and Georege Mohler. 2019. Investigate transitions into drug addiction through text mining of reddit data. In *Proceedings of the 25th ACM SIGKDD International Conference on Knowledge Discovery & Data Mining*, pages 2367–2375.

Vance V MacLaren and Lisa A Best. 2010. Nonsuicidal self-injury, potentially addictive behaviors, and the five factor model in undergraduates. *Personality and Individual Differences*, 49(5):521–525.

Jodi Martin, Paula F Cloutier, Christine Levesque, Jean-François Bureau, Marie-France Lafontaine, and Mary K Nixon. 2013. Psychometric properties of the functions and addictive features scales of the ottawa self-injury inventory: A preliminary investigation using a university sample. *Psychological assessment*, 25(3):1013.

Andrew Kachites McCallum. 2002. Mallet: A machine learning for language toolkit (2002).

Feng Nan, Ran Ding, Ramesh Nallapati, and Bing Xiang. 2019. Topic modeling with wasserstein autoencoders. In *Proceedings of the 57th Annual Meeting of the Association for Computational Linguistics*, pages 6345–6381.

Kendra D Niwa and Michael N Mandrusiak. 2012. Self-injury groups on facebook. *Canadian Journal of Counselling and Psychotherapy*, 46(1).

Mary K Nixon, Paula F Cloutier, and Sanjay Aggarwal. 2002. Affect regulation and addictive aspects of repetitive self-injury in hospitalized adolescents. *Journal of the American Academy of Child & Adolescent Psychiatry*, 41(11):1333–1341.

Matthew K Nock. 2010. Self-injury. *Annual review of clinical psychology*, 6:339–363.

Matthew K Nock, Thomas E Joiner Jr, Kathryn H Gordon, Elizabeth Lloyd-Richardson, and Mitchell J Prinstein. 2006. Non-suicidal self-injury among adolescents: Diagnostic correlates and relation to suicide attempts. *Psychiatry research*, 144(1):65–72.

Matthew K Nock and Mitchell J Prinstein. 2004. A functional approach to the assessment of self-mutilative behavior. *Journal of consulting and clinical psychology*, 72(5):885.

John Peterson, Stacey Freedenthal, Christopher Sheldon, and Randy Andersen. 2008. Nonsuicidal self injury in adolescents. *Psychiatry (Edgmont)*, 5(11):20.

Inna Pirina and Çağrı Çöltekin. 2018. Identifying depression on Reddit: The effect of training data. In *Proceedings of the 2018 EMNLP Workshop SMM4H: The 3rd Social Media Mining for Health Applications Workshop & Shared Task*, pages 9–12, Brussels, Belgium. Association for Computational Linguistics.

Tyler R Pritchard, Chelsey A Fedchenko, and Stephen P Lewis. 2021. Self-injury is my drug: the functions of describing nonsuicidal self-injury as an addiction. *The Journal of Nervous and Mental Disease*, 209(9):628–635.

Nicholas Proferes, Naiyan Jones, Sarah Gilbert, Casey Fiesler, and Michael Zimmer. 2021. Studying reddit: A systematic overview of disciplines, approaches, methods, and ethics. *Social Media+ Society*, 7(2):20563051211019004.

H Andrew Schwartz, Salvatore Giorgi, Maarten Sap, Patrick Crutchley, Lyle Ungar, and Johannes Eichstaedt. 2017. Dlatk: Differential language analysis toolkit. In *Proceedings of the 2017 Conference on Empirical Methods in Natural Language Processing: System Demonstrations*, pages 55–60.

Judy Hanwen Shen and Frank Rudzicz. 2017. Detecting anxiety through Reddit. In *Proceedings of the Fourth Workshop on Computational Linguistics and Clinical Psychology — From Linguistic Signal to Clinical Reality*, pages 58–65, Vancouver, BC. Association for Computational Linguistics.

Sarah V Swannell, Graham E Martin, Andrew Page, Penelope Hasking, and Nathan J St John. 2014. Prevalence of nonsuicidal self-injury in nonclinical samples: Systematic review, meta-analysis and meta-regression. *Suicide and Life-Threatening Behavior*, 44(3):273–303.

Shaheen Syed and Marco Spruit. 2017. Full-text or abstract? examining topic coherence scores using latent dirichlet allocation. In *2017 IEEE International conference on data science and advanced analytics (DSAA)*, pages 165–174. IEEE.

Sarah Elizabeth Victor, Catherine Rose Glenn, and Elisha David Klonsky. 2012. Is non-suicidal self-injury an "addiction"? a comparison of craving in substance use and non-suicidal self-injury. *Psychiatry research*, 197(1-2):73–77.

Janis L Whitlock, Jane L Powers, and John Eckenrode. 2006. The virtual cutting edge: the internet and adolescent self-injury. *Developmental psychology*, 42(3):407.

Mohammadzaman Zamani, H. Andrew Schwartz, Johannes Eichstaedt, Sharath Chandra Guntuku, Adithya Virinchipuram Ganesan, Sean Clouston, and Salvatore Giorgi. 2020. Understanding weekly COVID-19 concerns through dynamic content-specific LDA topic modeling. In *Proceedings of the Fourth Workshop on Natural Language Processing and Computational Social Science*, pages 193–198, Online. Association for Computational Linguistics.

Ayah Zirikly, Philip Resnik, Özlem Uzuner, and Kristy Hollingshead. 2019. CLPsych 2019 shared task: Predicting the degree of suicide risk in Reddit posts. In *Proceedings of the Sixth Workshop on Computational Linguistics and Clinical Psychology*, pages 24–33, Minneapolis, Minnesota. Association for Computational Linguistics.

Jonathan Zomick, Sarah Ita Levitan, and Mark Serper. 2019. Linguistic analysis of schizophrenia in Reddit posts. In *Proceedings of the Sixth Workshop on Computational Linguistics and Clinical Psychology*, pages 74–83, Minneapolis, Minnesota. Association for Computational Linguistics.

Overview of the CLPsych 2022 Shared Task: Capturing Moments of Change in Longitudinal User Posts

Adam Tsakalidis[1,2], **Jenny Chim**[1], **Iman Munire Bilal**[2,3], **Ayah Zirikly**[5], **Dana Atzil-Slonim**[6], **Federico Nanni**[2], **Philip Resnik**[8], **Manas Gaur**[7], **Kaushik Roy**[7], **Becky Inkster**[2,4], **Jeff Leintz**[9], **Maria Liakata**[1,2,3]
[1]Queen Mary University of London (UK), [2]The Alan Turing Institute (UK), [3]University of Warwick (UK), [4]University of Cambridge (UK), [5]Johns Hopkins University (US), [6]Bar Ilan University (Israel), [7]The Artificial Intelligence Institute (Columbia, US), [8]University of Maryland (US), [9]NORC at the University of Chicago (US)

Abstract

We provide an overview of the CLPsych 2022 Shared Task, which focusses on the automatic identification of *'Moments of Change'* in longitudinal posts by individuals on social media and its connection with information regarding mental health . This year's task introduced the notion of longitudinal modelling of the text generated by an individual online over time, along with appropriate temporally sensitive evaluation metrics. The Shared Task consisted of two subtasks: (a) the main task of capturing changes in an individual's mood (drastic changes-'Switches'- and gradual changes -'Escalations'- on the basis of textual content shared online; and subsequently (b) the subtask of identifying the suicide risk level of an individual – a continuation of the CLPsych 2019 Shared Task– where participants were encouraged to explore how the identification of changes in mood in task (a) can help with assessing suicidality risk in task (b).

1 Introduction

Increasingly the clinical community are looking for new and better diagnostic measures and tools for monitoring mental health conditions. Over the past decade, there has been a surge in methods at the intersection of NLP and mental health, showing that signals for the diagnosis of certain conditions can be found in language. However, most research tasks have been defined on the basis of classifying individuals (e.g., on the basis of suicide risk (Shing et al., 2018; Zirikly et al., 2019) or on the basis of having a mental health condition or not (Coppersmith et al., 2015)), thus lacking the longitudinal aspect of monitoring an individual's mood and well-being in real-time.

Through this shared task we follow Tsakalidis et al. (2022) to introduce the problem of assessing changes in a person's mood over time on the basis of their linguistic content. For the purpose of the task we focus on posting activity in online social media platforms. In particular, given an individual's posts over a certain period in time, we aim: (a) at capturing those sub-periods during which an individual's mood deviates from their baseline mood – a post-level sequential classification task; (b) leveraging this task to help us assess the suicide risk level of the individual – a user-level classification task (Shing et al., 2018) & a continuation of the 2019 Shared Task (Zirikly et al., 2019). Thus, this year's shared task consists of two subtasks: (A) the main task of identifying mood changes in an individual's online posts over time and (B) assessing the suicide risk level of the invididual, where ideally participants will have been able to establish a connection between tasks A and B. This paper makes the following contributions:

- We introduce tasks A and B and provide a detailed description
- We describe the datasets used for these tasks.
- We provide an overview of the secure data enclave environment used for the shared task.
- We provide an overview of participating team selection, evaluation strategy and discussion of results, paving the way for future approaches.
- We present the limitations of the current set up and provide suggestions for future organisers.

2 Task Definitions

Task A involves capturing 'Moments of Change' (MoC) in posts by individuals on social media over time. In particular, following Tsakalidis et al. (2022), given a sequence of chronologically ordered posts between two dates (*'timeline'*) made by an individual on an online social media platform, we aim to capture the post(s) – or the sequence(s) of posts – in the timeline indicating that the individual's mood has shifted in one of the following ways: *(a) Switch* – the individual's mood shifts

{a.tsakalidis,m.liakata}@qmul.ac.uk

Proceedings of the Eighth Workshop on Computational Linguistics and Clinical Psychology, pages 184 - 198
July 15, 2022 ©2022 Association for Computational Linguistics

suddenly from positive to negative (or vice versa); and *(b) Escalation* – the individual's mood gradually progresses from neutral/negative (positive) to very negative (positive). Both sudden and gradual changes in individuals' mood over time are important for monitoring mental health conditions (Lutz et al., 2013; Shalom and Aderka, 2020) and constitute one of the dimensions to measure in psychotherapy (Barkham et al., 2021). By definition, this task is temporally sensitive, since the goal is to classify each post in a given timeline as belonging to a Switch (IS), belonging to an Escalation (IE) or not being part of either mood shift (O) – with the majority of the posts expected to be (O).

Task B is a continuation of the work by Shing et al. (2018) and Zirikly et al. (2019). Given the posts of an individual, the aim is to classify their suicide risk into (a) no risk, (b) low, (c) moderate or (d) severe level. Due to the very low number of users of (a) and (b) in our data, we have merged the no/low classes leading to a 3-label user classification task. Participants were encouraged to use insights from Task A in solving Task B.

3 Dataset

Dataset creation for the two tasks (§3.4) involved data collection & data relabelling (§3.1), timeline extraction (§3.2) and annotation (§3.3).

3.1 Data Collection

As our ultimate goal is to find the connection between Moments of Change (MoC) in individuals' longitudinal online data (Task A) and other information regarding the individuals' level of risk (Task B), we wanted to repurpose as much as possible existing mental health datasets (Losada and Crestani, 2016; Losada et al., 2020; Shing et al., 2018; Zirikly et al., 2019) by annotating MoC within them. We also collected a new dataset from Reddit annotated for both MoC and suicidality risk. Our final dataset consists of:

Reddit-UMD. The UMD-Suicidality dataset (Shing et al., 2018; Zirikly et al., 2019) consists of 38K posts by 245 Reddit users who have posted in the *r/SuicideWatch* subreddit (and an equal number of control users who do not feature in our tasks). We have labelled the content generated by these individuals with MoC and relabelled the users' risk level for consistency across datasets.

Reddit-New. We collected a new dataset from Reddit, in two steps: we first collected all public Reddit posts in any mental health-related subreddit (MHS) between 2015-2021 (incl.) and then obtained the posting history for 83K users with at least 10 posts in MHS (for the list of MHS, refer to Appendix A).

	Reddit-UMD	Reddit-New	eRisk++	Total
Timelines	90	139	27	256
Users	77	83	26	186
Posts	2,399	3,089	717	6,205
Duration	~2 months	~2 months	(varies)	

Table 1: Dataset overview

eRisk++. We obtained the eRisk dataset (Losada and Crestani, 2016; Losada et al., 2020) upon signing a data use agreement.It contains Reddit posts and comments made by 41 users with and 299 users without self-harm conditions. Inspection of posts by the 299 users showed they were irrelevant for our tasks and so we focussed on the 6,927 posts and comments by the 41 users.[1]

3.2 Timeline Extraction

For each dataset, we extracted user timelines to allow annotation of MoC (Task A), while ensuring that these timelines also contain the information required for Task B (i.e., all associated users' posts in *r/SuicideWatch* are included in the timelines). Table 1 provides an overview of the datasets.

Reddit-UMD. We ordered each user's posts chronologically, identified their posts in *r/SuicideWatch* and defined a user timeline as t days around each such post. Upon experimentation t was set to 30. We extracted 156 timelines of [10,125] posts each, so that annotation was manageable, corresponding to 126 users. These timelines were manually inspected internally by two researchers asked to judge the suitability of the former for Task A. Timelines were thus independently labelled as 'good', 'medium', or 'bad' (Cohen's κ=.66).[2] We only kept 90 timelines that (a) were labelled as 'good' by both annotators and (b) contained all of the user's posts on *r/SuicideWatch* so that we could follow the same annotation for Task B as in Shing et al. (2018).

To inform subsequent data collection we analysed what constitutes a 'good' timeline in Reddit-UMD. For this we trained a Logistic Regression learning to separate between 'good' and 'bad' timelines. We used the timeline-level features

[1] As opposed to Reddit-New and Reddit-UMD, the eRisk dataset contains posts *and* comments made by the users on Reddit. For consistency, we will refer to all of them as 'posts'.

[2] Details of the annotation are provided in Appendix B.

[#posts, % of posts in MHS, and % of posts in r/SuicideWatch, r/depression and r/AskReddit[3]], further accompanied by the average difference (in terms of #posts) between two postings on the same subreddit. We found that the % of posts in MHS is the most predictive feature, with 95% of the 'bad' timelines containing less than 17% MHS posts, whereas 99% of the 'good' timelines have contain less than 82%. We use this information to select 'good' timelines for the Reddit-New dataset.

Reddit-New. Following our notion of 'good' timelines in Reddit-UMD we looked for two-month periods within which the user had at least 10 and no more than 125 posts, at least (most) 17% (82%) of which is posted on a MHS. 150 such timelines were selected at random (from an overall of 1,114) and annotated internally for quality (good/medium/bad), similarly to Reddit-UMD – this time by a single annotator, given the high agreement achieved in Reddit-UMD, resulting into 139 'good' timelines (83 users). Interestingly, one timeline in Reddit-New was identical to another one present in Reddit-UMD – signalling a consistency between the collection process of the two datasets – and hence removed from Reddit-New on our final processing.

eRisk++. Two annotators with experience in mental health research on social media independently reviewed 103 timelines to check suitability for task A. 91 timelines were labeled either as 'good' or 'medium' (Cohen's κ=.78). For consistency with the other datasets, we kept the 15 timelines (14 users) having at least (most) 10 (125) posts.

Upon inspecting the resulting datasets, we found that there was a disproportionate representation of 'low' and 'no' risk users based on the labelling provided in (Shing et al., 2018; Zirikly et al., 2019). To mitigate this, we enriched the eRisk++ dataset with 12 timelines by 12 users from UMD-Suicidality, who had been labelled as 'no'/'low' risk in Zirikly et al. (2019). Though we did not use their associated suicidality risk labels, this step ensured a fairer representation of users for capturing MoC (task A).

3.3 Annotation

Task A. We hired four annotators (2 native English, 2 fluent English language speakers), two of whom had previous experience with performing task A on a different dataset (TalkLife), and provided them with the guidelines from Tsakalidis et al. (2022). Briefly, the task involves reading one timeline at a time in an annotation interface and labelling (a) the first post that signals a 'Switch' (IS) in an individual's mood, along with the respective duration of the Switch (range of consecutive posts), as well as (b) the post signalling the 'peak' (most intense posts) of an 'Escalation' (IE) in an individual's mood, along with the respective range of consecutive posts that belong to the same Escalation. The training of the two non-experienced annotators involved annotating timelines from TalkLife that were previously annotated by the two experienced annotators, measuring their agreement and discussing cases of disagreement in iterative cycles, until reaching an agreement level similar to that in Tsakalidis et al. (2022). Subsequently, the four annotators were provided with 10 separate timelines extracted from UMD Suicidality for training purposes, and disagreements in their annotations were discussed in two meetings. Finally, they were provided with the 255 timelines that have been used in the current Shared Task.

Task B. We worked with four Clinical Psychology experts, all of whom are fluent English language speakers. The experts were provided with the guidelines by Shing et al. (2018), which focus on the task of classifying the suicide risk level (no/low/moderate/severe risk) of an individual, solely on the basis of their *r/SuicideWatch* posts. An annotation interface was developed, where the experts could view and assign a single label to an individual based on up to 5 *r/SuicideWatch* posts made by the individual within the Reddit-New and Reddit-UMD datasets. Our experts re-annotated the suicidality risk of users in Reddit-UMD to provide annotation consistency between the two datasets. [4] For users with more than 5 posts on *r/SuicideWatch*, the annotation was performed in several passes, with the most 'severe' label being finally assigned to the respective individual (Shing et al., 2018). We completed two training rounds with the experts, where they discussed disagreements in their labelling and clarified points especially concerning the distinction between 'moderate' and 'severe' cases.

[3]We selected these 3 subreddits on the basis of being present in at least 20% of the timelines.

[4]We did not use the users from the eRisk++ dataset for Task B: the information on the type of subreddit where a post was shared was not available in eRisk and the remaining 12 timelines from UMD-Suicidality (part of eRisk++) were incorporated at a latter stage.

	Task B: #users					Task A: #posts			
	N/A	Low	Mod.	Sev.	Total	IS	IE	O	Total
Train	22	11	55	61	149	327	773	4,043	5,143
Test	4	3	14	15	36	83	208	762	1,052

Table 2: Summary of the data for both tasks.

	Half	Majority	Perfect
Switch (IS)	.451	.264	.129
Escalation (IE)	.550	.309	.122
None (O)	.920	.832	.692

Table 3: IAA for Task A per agreement threshold.

3.4 Resulting Dataset

Task A. Following Tsakalidis et al. (2022), we assess the inter-annotator agreement (IAA) based on the Intersection over Union for each label independently. The majority agreement (see Table 3) is lower than the agreement in Tsakalidis et al. (2022) (.30/.50/.89 for IS/IE/O, respectively), primarily because in the latter there were 3 annotators employed (requiring 2/3 to agree) whereas here a majority requires agreement between 3/4). A post receives the label assigned to it by the majority. In the case of ties the least populous class receives the label (e.g. if 'IS' ('IE') is chosen over 'O'. In the rare (64 cases overall) of a tie between 'IS' and 'IE', we assigned the label 'IE' given its higher prior.
Task B. The agreement between the expert annotators was considerably lower than that reported in Shing et al. (2018) (Krippendorff's α .43 vs .81), primarily for two reasons: (a) in this dataset, there was only one user assigned 'no risk', which is the easiest category to identify even for non-experts; (b) the experts in Shing et al. (2018) had a background on suicidality whereas our clinical psychologists have broader expertise. Most cases of disagreement involved 'moderate' vs 'severe', or 'low' vs 'moderate' as opposed to 'low' vs 'severe'. We used the majority label for each user and in case of ties the highest level of risk assigned was chosen. We split the data into train and test sets (80/20) preserving the distribution of labels in the two sets. Subsequently, all 204/51 timelines from users in our train/test split, were assigned to the respective set (see Table 2).

4 Working in a Secure Environment

The CLPsych shared task 2021 (Macavaney et al., 2021) was the first to be conducted in a secure environment to provide a high level of safety for sensitive data. We have also opted for carrying out this year's shared task in the same secure environment and continue efforts in protecting highly sensitive data. NORC is an independent non-profit research institution at the University of Chicago who provide the NORC Data Enclave(r), chosen both this year and last for the shared task. Compared to other solutions (see for instance Arenas et al. (2019)) the NORC Data Enclave(r) (hereafter, 'DE') does not rely on dedicated laptops but solely on a browser interface over HTTPS channels and Citrix HDX technology, making the setup of a shared task more feasible. All teams (see §5) signed a data use agreement (DUA) and terms and conditions (T&C) with NORC before being provided with instructions to set up multi-factor authentication for login, procedures for requesting the ingress in the DE of written code, libraries, models or additional data and procedures for technical support. All ingress of information into the DE requires thorough system scans and human review to ensure the safety and integrity of the Enclave.

After login authorized users can access a secure virtual machine within the DE. Although all applications and data run on servers in the NORC data center, the user interface is a familiar full Windows 10 virtual desktop. The DE is a closed environment: it does not have access to the internet and all functionalities for moving data in and out of the virtual space are disabled. This Citrix-based technology is configured to prevent users from downloading output from the remote server to an external machine. Similarly, other security protection features prevent the user from using the "cut and paste" feature in Windows to move data from the Citrix session into an Excel spreadsheet residing on the local computer. In addition, the user is prevented from printing the data on a local computer. There is documentation regarding the virtual environment and how to securely connect to the dedicated DE Cluster on Amazon Web Services (AWS). To connect to the cluster (via ssh) users rely on PuTTY and on the dedicated machine they can find a dedicated Python 3.9.1 environment with all requested libraries available (see §5). Users can both run code and submit batch jobs using the Slurm cluster management while also monitoring the budget available for computational experiments. Following last year's suggestions, we ensured participants would be able to use Jupyter Notebooks to implement code on the cluster through ssh tunneling and by opening the notebook in the browser of the Windows machine. At the end of the Shared Task, each team was to inform NORC to egress the predictions for the test set.

Due to an unprecedented technical issue out of NORC's control, several teams faced issues with running their code a week prior to the system submissions deadlineTo avoid eliminating the teams despite their continuous efforts throughout the Shared Task, we decided to distribute the data outside the DE during the last few days on the basis of the signed DUA. To ensure fairness, we asked all teams (i.e., not only the ones affected) to let us know if they would like to receive the data outside the enclave to help them with the system submission. We made it clear that those submitting their results within the DE would feature separately in our evaluation (see Tables 4-5), since they had more limited resources at their disposal.

5 Call for Participation – Teams Selection

We invited teams to register their interest in the shared task by providing details such as team members, motivation, related background, experience and NLP skills. We also asked for their requirements in terms of programming languages, libraries and pre-trained language models to prepare the set up in the DE. Given our limited resources pertaining to the functional costs of using the DE, we were limited to accepting 15 teams (~50 members) for participating in the Shared Task. Therefore, we compiled a list of criteria that were given to two internal reviewers, along with the (anonymised) registrations of interest. The criteria were related to (a) the relevance of the team's background/current work to the shared task, (b) their motivation and likelihood of committing to the task and (c) details provided wrt technical requirements (see Appendix C for the complete guidelines). Based on the reviewers' assessments, we selected 13/37 teams to participate and asked another five applicant teams to be merged together into two groups, so as to accommodate as many requests as possible (one team was formed by three individual applicants, and another individual applicant was merged into a two-member team), leading to the acceptance of 18/37 requests (53 individuals).

6 Evaluation metrics

Task A. Following Tsakalidis et al. (2022), besides the common post-level evaluation metrics (Precision, Recall, F1) – per class and macro-averaged – we report two sets of timeline-level metrics based on work in change-point detection (van den Burg and Williams, 2020) and image segmentation (Arbelaez et al., 2010), emphasizing respectively performance at the level of a timeline and the prediction of regions of change.

Firstly, working on each timeline and label type independently, we calculate Recall $R_w^{(l)}$ (Precision $P_w^{(l)}$) by counting as "correct" a model prediction for label l if the prediction falls within a window of w posts around a post labelled as l in our ground truth – however, a post's predicted label can only be counted as 'correct' only once (at most). By increasing the value of w, we perform a less strict evaluation of a model. Results are macro-averaged for each label independently across all timelines.

Secondly, we assess model performance on the basis of its ability to capture *regions of change*. For each true region $R_{GS}^{(l)}$ within a timeline, we define its overlap $O(R_{GS}^{(l)}, R_M^{(l)})$ with each predicted region $R_M^{(l)}$ as the intersection over union between the two sets. Finally, we retrieve recall- and precision-based *coverage* metrics (again, macro-averaged across all timelines for each label independently:

$$C_r^{(l)}(M \rightarrow GS) = \frac{1}{\sum_{R_{GS}^{(l)}} |R_{GS}^{(l)}|} \sum_{R_{GS}^{(l)}} |R_{GS}^{(l)}| \cdot max_{R_M^{(l)}} \{O(R_{GS}^{(l)}, R_M^{(l)})\},$$

$$C_p^{(l)}(M \rightarrow GS) = \frac{1}{\sum_{R_M^{(l)}} |R_M^{(l)}|} \sum_{R_M^{(l)}} |R_M^{(l)}| \cdot max_{R_{GS}^{(l)}} \{O(R_{GS}^{(l)}, R_M^{(l)})\}.$$

Ideally we want to see a system performing well on both window based and coverage metrics.

Task B. We use standard classification metrics (Precision, Recall and F1) for each user-based class label and macro-averaged. Due to the low number of users in the 'Low' class on the test set, we also report micro-averaged metrics; however, these are added for completeness purposes in our analysis (i.e., the teams were guided to improve their performance on a per-class and macro-average basis).

7 Shared Task Results

This section outlines the submissions by each team. For Task A, we also provide the results of three baselines: the majority classifier, a logistic regression (LR) trained on tfidf features, and BERT trained using the focal loss on a related but separate dataset on the same task (Tsakalidis et al., 2022). For Task B, we include the majority classifier and a LR trained on tfidf features from users' posts.

7.1 Overview

Task A. Each team was allowed to submit up to three sets of test results. Nine teams submitted their

		DE	Post-level Evaluation macro-avg P	R	F1	IS P	R	F1	IE P	R	F1	O P	R	F1	Coverage-based Metrics macro-avg C_p	C_r	IS C_p	C_r	IE C_p	C_r	O C_p	C_r
Baseline	Majority		–	.333	.280	–	.000	.000	–	.000	.000	.724	1.000	.840	–	.142	–	.000	–	.000	.489	.426
	LR-tfidf		.545	.495	.492	.222	.024	.044	.569	.514	.540	.844	.948	.893	.378	.425	.111	.008	.284	.504	.738	.762
	BERT$_f$-TalkLife		.523	.386	.380	.091	.012	.022	.723	.163	.267	.754	.983	.853	.260	.204	.025	.007	.226	.094	.529	.513
System Submissions	BLUE		.505	.495	.499	.175	.171	.173	.484	.433	.457	.855	.882	.868	**.499**	.378	.500	.028	.299	.395	**.699**	.712
	IIITH		.520	**.600**	.519	.206	.524	.296	.402	.630	.491	.954	.647	.771	.347	.405	.254	**.356**	.249	.373	.536	.486
	LAMA		.552	.535	.524	.166	**.354**	.226	.609	.389	.475	.882	.861	.871	.376	.441	.253	.373	.193	.244	.680	.706
	NLP-UNED	✓	.493	.518	.501	.189	.293	.230	.414	.471	.440	.876	.791	.832	.306	.401	.244	.304	.134	.330	.541	.569
	UArizona	✓	.525	.507	.510	.142	.220	.172	.561	.423	.482	.872	.879	.876	.418	.416	.368	.248	.202	.285	.682	.716
	UoS		.689	.625	.649	.490	.305	.376	.697	.630	.662	.881	.940	.909	.506	.503	**.453**	.343	.369	**.450**	.695	**.717**
	uOttawa-AI		.505	.530	.512	.213	.244	.227	.402	.553	.466	**.899**	.793	.842	.348	.453	.272	.317	.176	.417	.595	.625
	WResearch	✓	.625	.579	.598	.362	.256	.300	.646	.553	.596	.868	.929	.897	.472	.503	.406	.318	.307	.467	.703	.725
	WWBP-SQT-lite		.508	.509	.508	.231	.220	.225	.440	.462	.451	.852	.845	.848	.336	.376	.270	.224	.186	.321	.551	.583

Table 4: Task A – System evaluation, with first, **second** and third highest scores (as well as the highest scores for submissions within the DE) being highlighted. Only the best submission for each team is shown, selected separately on the basis of macro-avg F1 (Post-level Evaluation) and F1=2·C_p·C_r/(C_p+C_r), macro-based (Coverage-based).

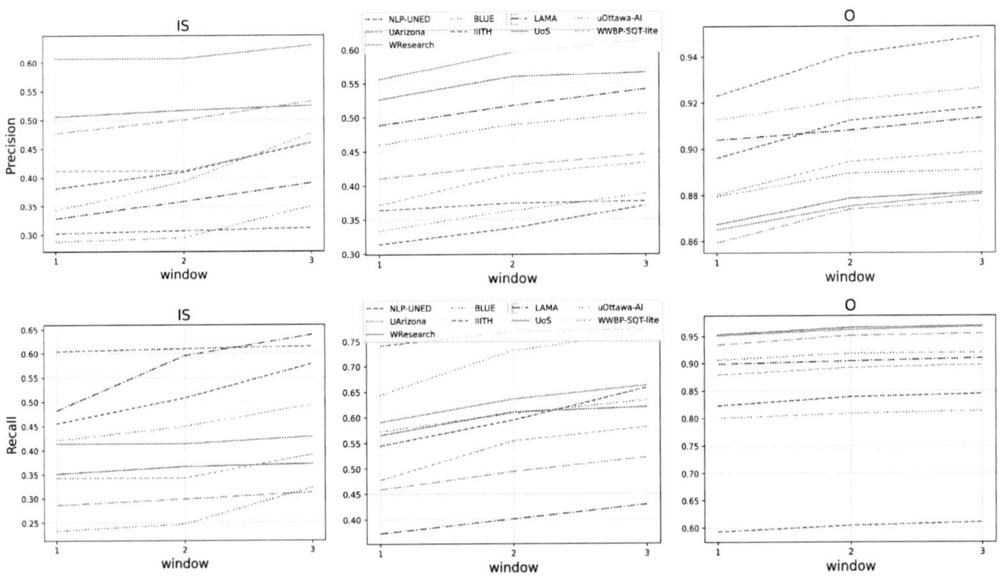

Figure 1: Timeline-level Precision P_w and Recall R_w of the submitted systems. Only the best performing submission by each team is shown (selected on the basis of F1=2·P_1·C_1/(P_1+R_1), macro-based).

predictions – an overview of the best results per team/metric is shown in Table 4 and Fig. 1. The two best-performing teams (one submitting within and one outside the DE) incorporated a longitudinal component in their models, either in a multi-task setting (UoS) or in an emotionally-informed seq2seq-based approach (WResearch), demonstrating the importance of temporally-sensitive modelling as opposed to classifying each post in isolation. The class imbalance problem was tackled by several teams either via balancing the instances (e.g., LAMA, uOttawa) or via weighted loss functions, notably by IIITH who achieved high recall for IS/IE. Time-related information was incorporated by UArizona, a proximity-based approach was followed by NLP-UNED, an ensemble on emotional and non-emotional features was chosen by BLUE, whereas WWBT-SQT-lite achieved high accuracy (albeit post-deadline) by using different combinations of consecutive post representations.

Task B. Each team was allowed to make a single submission; a second submission was allowed only for teams making use of their predictions from Task A. Seven teams submitted and two teams further took up the challenge of leveraging Task A (see Table 5). The teams that took up this challenge did not demonstrate (important) performance gains. However, the best-performing teams (in average, macro-terms) used some information from Task A, either by focusing mostly on posts labelled as MoC (WResearch) or by jointly learning the two tasks (UoS). The ranking of the teams differs when considering the micro-F1, due to the low number of 'low' risk users. Here IIITH and NLP-UNED, along with WResearch, were ranked amongst the top, being particularly effective in capturing 'severe' and 'moderate' cases, respectively.

7.2 Summary of System Submissions

BLUE (Bucur et al., 2022) explored a variety of feature representation approaches for Task A: (a) Emotion-aware embeddings and (b) non-emotion embeddings (e.g., tfidf, GloVe). They experimented with different combinations of algorithms and features sets, with the most notable performance achieved by a majority voting-based model over an ensemble of predictions obtained by LR, SVM, and Adaptive Boosting classifiers trained on (a), which ranked them second in macro-avg precision-oriented coverage (.499).

IIITH (Boinepelli et al., 2022) used transformers for representing the user's posts before feeding them to an LSTM for Task A. They tuned their model using the weighted cross-entropy loss function, yielding very high recall for the two minority classes (see post-level results for IS/IE in Table 4). For Task B, they fine-tuned RoBERTa on the training data, tackling the class imbalance with weighted random sampling and producing the outcome label through majority voting. The team came second (third) in this task on micro-F1 (macro-F1), achieving the best scores for the 'Severe' class (see Table 5, 'Severe').

LAMA (AlHamed et al., 2022) tackled the data imbalance problem by undersampling posts with high sentiment polarity corresponding to the majority class. They adopted a post-level BERT and LSTM models that take into account the sequence of the previous posts for a given target post for Task A. BERT performed particularly well wrt the recall-oriented metrics for IS, leading to the third-best performance in terms of macro-F1 overall. Their models for Task B were Random Forests enriched with sentiment-related features and word frequencies of manually collected high-risk keywords.

NLP-UNED (Fabregat et al., 2022) completed all 5/5 submissions via the DE. In Task A, they analysed the encoded user posts via an Approximate Nearest Neighbour approach – labelling individual posts based on their proximity to others – achieving high recall-oriented scores for IE/IS and the highest macro-average timeline-level recall (for $w = 3$). For Task B, they represented each post on the basis of its proximity to each of the labels in Task A and fed the resulting sequence into a BiLSTM. Amongst the two submissions that leveraged Task A for performing Task B, NLP-UNED was marginally the best-performing in terms of F1.

UArizona (Culnan et al., 2022) completed their 2/2 submissions for Task A via the DE. They tested several variants of RoBERTa-based models, including (a) timeline-agnostic models that incorporate the time lag between consecutive posts and (b) models combining consecutive post vectors, either through concatenation or by passing them through an LSTM to extract the resulting states. They showcased that the incorporation of time boosts the performance of the model on IS cases, whereas they were consistently among the top-3 performing systems in macro-averaged, timeline-level precision.

UoS (Azim et al., 2022) achieved the highest scores for Task A in most metrics and across classes, as well as the second-highest macro-F1 for Task B. They first represent a post in different ways (merged), including its emotion/sentiment-based scores. Their approach involved an attention-based, multi-task BiLSTM operating at the timeline-level, with each post corresponding to a single timestep in the input/output for Task A, and additional outputs for the user's risk label for Task B at the timeline level (selecting the most 'severe' label across all timelines for the user's classification).

uOttawa-AI (Buddhitha et al., 2022) employed convolutional neural networks with global max-pooling and linear layers for multi-task learning. Task A was casted as two post-level binary tasks (i.e., (a) IS vs O and (b) IE vs O) using soft and hard parameter sharing, by also tackling the class imbalance through down-sampling the majority class. They achieved high recall-oriented metrics for capturing IE and were among the highest scoring teams wrt recall-oriented coverage. In Task B, the team experimented with the additional task of predicting self-declared mental health diagnoses using a separate dataset (Cohan et al., 2018).

WResearch (Bayram and Benhiba, 2022) completed 4/5 submissions in the DE. In Task A, they derived emotionally-informed vectors from pre-trained models and constructed abnormality vectors (i.e., differences in expected vs predicted vectors via a seq2seq model) and differences in the vectors of consecutive posts, using them as inputs to post-level classifiers that take into account the class imbalance. Their best performing submission used

		DE	macro-avg			micro-avg			Low			Moderate			Severe		
			P	R	F1	P	R	F1	P	R	F1	P	R	F1	P	R	F1
(a)	**Majority**		.156	.333	.213	.220	.469	.299	–	.000	.000	–	.000	.000	.469	1.000	.638
	LR-tfidf		.303	.338	.295	.413	.469	.406	.000	.000	.000	.429	.214	.286	.480	.800	.600
(b)	**IIITH**		.397	.408	.380	**.538**	.563	**.520**	.000	.000	.000	.625	.357	.455	.565	.867	.684
	LAMA		.306	.424	.298	.359	.344	.316	.167	.667	**.267**	.250	.071	.111	.500	.533	.516
	NLP-UNED (1)	✓	.361	.394	.369	.492	.531	.500	.000	.000	.000	.500	.714	.588	.583	.467	.519
	UoS		.618	**.427**	**.451**	.482	.469	.438	1.000	**.333**	.500	.375	.214	.273	.478	**.733**	.579
	uOttawa-AI		.329	.365	.344	.449	.500	.470	.000	.000	.000	.462	.429	.444	.526	.667	.588
	WResearch (1)		**.467**	.479	.465	.565	**.531**	.543	**.200**	**.333**	.250	.533	.571	.552	.667	.533	.593
	WWBP-SQT-lite		.346	.370	.354	.471	.500	.480	.000	.000	.000	.500	**.643**	**.563**	.538	.467	.500
(c)	**NLP-UNED (2)**	✓	.367	.387	.365	.497	.531	.497	.000	.000	.000	.600	.429	.500	.500	.733	.595
	WResearch (2)	✓	.367	.365	.362	.499	.500	.494	.000	.000	.000	.545	.429	.480	.556	.667	.606

Table 5: Task B - System Evaluation: (a) baselines, (b) system submissions, (c) systems utilising Task A.

XGBoost (Chen and Guestrin, 2016) and was consistently among the highest-scoring systems across metrics – and the best-performing from systems within the DE. In Task B, they used LR on n-grams and emotion bandwidth-based vectors extracted from the IS/IE posts for each user, achieving the highest averaged F1. They further leveraged the posts predicted for Task A as IS/IE via a timeline-level BiLSTM, assigning the most 'severe' label for a user based on their timeline classifications, without improvement in performance, however.

WWBP-SQT-lite (Ganesan et al., 2022) experimented with theoretically-motivated features and representations based on Human-aware Recurrent Transformers (Soni et al., 2022) and PCA-reduced RoBERTa. After the deadline the team also tested a version of PCA-reduced RoBERTa vectors, yielding very high accuracy when concatenating them with the previous post's vector and their difference, as features (macro-F1: .61, not reported in Table 4). For Task B the team used LR on user-level features (ngrams, theoretically motivated features), achieving the second-best results on separating the 'Moderate' cases of risk level.

8 Conclusion

We presented the overview of the CLPsych 2022 Shared Task, focusing on (A) capturing changes in an individual's mood as self-disclosed online and (B) classifying the individual's suicide risk level – as well as studying the link between the two tasks. The best results for (A) showcase the importance of taking into account the sequence-aware modelling of an individual's online shared content, whereas the link between the two tasks has been highlighted on the basis of the best results achieved for (B).

Following last year's setting (Macavaney et al., 2021), we utilised NORC's Enclave. Faced with challenges out of our and NORC's control, we provide directions for shared tasks on sensitive domains (§9). Our aim for the future is to emphasize the need for research on longitudinal tracking and modelling of a user's mental health, under a common experimental setting in a secure environment.

9 Recommendations for the Future

Organising a NLP shared task on highly sensitive datasets is an incredibly challenging effort that relies on the coordination and collaboration of many different actors. In addition to the very useful feedback given by last year's organisers (Macavaney et al., 2021), we have compiled an anonymous feedback questionnaire shared with the 39 members that had access to the DE or were the contact members of a team. In this section, we summarise the key insights gained from the teams' feedback (§9.1) and provide suggestions for future versions of Shared Tasks in such sensitive domains (§9.2).

9.1 Feedback from Participants

The questionnaire consists of 4 multiple choice questions and 2 free-text answers on (Q5) what they liked about this year's shared task vs (Q6) what needs improvement in future editions.

Overview & Q1 – *'My team managed to produce results'*: 18 members completed the feedback form (34% of all 53 participants; 46% of the 39 participants that the questionnaire was shared with), 17 of whom were members of teams that managed to submit their results (within or outside the DE).

Q2 – *'The task description was clear'* (completely disagree to completely agree, [1-5]): All 18 responses were between [3-5], with an average of 4.4/5.0. Based on Q6 shown below, there were two respondents for whom the annotation guidelines and/or resulting labels for Task A were unclear. Providing more examples in such longitudinal tasks from the beginning of the Shared Task can offer an improvement in this regard.

Q3 – *'Communication via slack was easy and efficient.'* (completely disagree to completely agree, [1-5]): Responses were between [3-5], with an average of 4.7/5.0, suggesting that an active communication channel can help participants along the way and is recommended for future editions.

Q4 – *'How was your experience with working on the Data Enclave?'* (5 pre-defined choices): 50% of the respondents said that they faced many difficulties, but would have managed to produce results within the DE nevertheless if there wasn't the major incident during test time (see §4); 4/18 respondents said that there were only some difficulties resulting in minor/medium loss in their productivity. We provide concrete suggestions to this effect in §9.2.

Q5 – *What did you like about the shared task?*: The 17 responses on Q5 can be categorised into two main topics: 13 commented positively on the task itself and 7 on the organisational aspect (quick responses from the organisers – see also Q3 – and working in a secure manner through NORC's DE).

Q6 – *'What did you mostly not like about this year's Shared Task? What issues did you face? How can we improve for the next year?*: Most of the 17 responses concerned issues around working within the DE – from inability to copy/paste to downloading resources. We compile a list of suggestions in §9.2. 2/17 respondents commented on the delay of providing the code (e.g., evaluation, baselines/results); 2/17 commented on the clarity of the annotations (see also Q2); 2/17 also commented on the tightness of deadlines, which were packed towards the end of the Shared Task to allow more time for model training – a wider time frame for future Shared Tasks is recommended. Isolated concerning points (1/17) included the small size of the dataset to reach conclusive outcomes (often a concern in this domain) and inability to perform a direct comparison between systems trained within vs outside the DE (tackled by highlighting the best-performing system for submissions within the DE).

9.2 Suggestions for future organisers

Secure Environment. Given the sensitive nature of data for the Shared Task, it is essential to be able to rely on a secure environment. Following CLPsych 2021, we opted for NORC and their DE. It is important that future organisers plan this collaboration in advance to make sure NORC has sufficient time to identify and secure enough resources and specific expertise to the project. The technical issue faced this year also highlights the need for a wider test-time period, to allow enough time for resolving such cases. Ideally there should be an ongoing collaboration with the DE so that any issues and the necessary expertise to overcome them are built during a sufficiently long period of time.

Libraries and Resources. It is crucial to have a clear pre-defined list of libraries, resources and dependencies (e.g., pre-trained models) that would need to be reviewed before being available in the DE. This means reaching out in advance to the teams and also planning for a trial period of 2 weeks where the teams can access part of the data and check their needs, live. The teams for instance encountered many issues with NLP libraries that required additional downloads of resources when used.[5] It is also important to keep track of the approved/installed libraries each year.

Communication and Peer Support. Following last year's suggestions, we wanted to avoid sending many similar requests to NORC, and try to provide a common setting for people to help each other. We relied on Slack by setting up two dedicated channels, which received very positive feedback and also facilitated the communication between the organisers and NORC. Participants helped each other e.g. in setting up the ssh tunneling for Jupyter Notebook or in identifying the specific issue to report back to NORC (which we have tried to do through a more coordinated effort, where one of the organisers would be the point of contact).

Preparation. Notes from last year's edition already highlighted the complexity of organizing the shared task and recommended more advance planning. Even with that in mind, core challenges remain due to the antithesis between two very different agendas: the intensive experimental work in a very limited time frame (the shared task) and a centralised, step-by-step highly controlled process (the DE). We believe that only through long-term collaboration with DEs such as NORC is it feasible to define a middle-ground working solution which can guarantee high level of security while supporting researchers to develop their solutions. Such collaboration requires the recognition of the importance of DEs by funding bodies and the need to fund long-term collaborations between DEs and research organisations.

[5] e.g., the NLTK tokenizer requires 13MB of Punkt Tokenizer Models, which are not accessible in the DE.

Ethical statement

Secure access to the shared task dataset was provided with IRB approval under University of Maryland, College Park protocol 1642625 and approval by the Biomedical and Scientific Research Ethics Committee (BSREC) at the University of Warwick (ethical application reference BSREC 40/19-20). Annotators were given contracts and paid fairly in line with University payscales. They were alerted about potentially encountering disturbing content and were advised to take breaks. The annotations are used to train and evaluate natural language processing models for recognising moments of change and linking them to suicidality risk, where the latter is provided by clinical psychology experts. Working with data on online platforms where individuals disclose personal information involves ethical considerations (Mao et al., 2011; Keküllüoğlu et al., 2020). Such considerations include careful analysis and data sharing policies to protect sensitive personal information. Potential risks from the application of NLP models in being able to identify moments of change in individuals' timelines are akin to those in earlier work on personal event identification from social media and the detection of suicidal ideation. Potential mitigation strategies include restricting access to the code base and annotation labels used for evaluation. In this shared task we have asked participants to sign DUA agreements and we opted for a secure data enclave environment to work in.

Acknowledgments

This work was supported by a UKRI/EPSRC Turing AI Fellowship to Maria Liakata (grant ref EP/V030302/1), the Alan Turing Institute (grant ref EP/N510129/1) and especially UKRI funding to promote collaboration between UK and US researchers. Aspects of this work were also supported by an Amazon Research Award and by the National Science Foundation under grant 2124270, and the effort also received internal financial support at NORC. The shared task organizers would like to express their gratitude to the anonymous users of Reddit whose data feature in this year's shared task dataset; to the annotators of the data for Task A, to the clinical experts from Bar-Ilan University who annotated the data for TaskB, the American Association of Suicidology; to all participants for their efforts and patience; to the NORC partners and personnel (especially co-author Jeff Leintz, Dariush Wilkowski, Julia Crothers, Bill Olesiuk and the Data Enclave Manager team) for their tremendous contributions and their willingness to put in a great amount of resources in setting up and managing the Enclave and enabling this year's shared task, especially given the short time frame, and finally to NAACL for its support for CLPsych.

References

Falwah AlHamed, Julia Ive, and Lucia Specia. 2022. Predicting moments of mood changes overtime from imbalanced social media data. In *Proceedings of the Eighth Workshop on Computational Linguistics and Clinical Psychology: Mental Health in the Face of Change*.

Pablo Arbelaez, Michael Maire, Charless Fowlkes, and Jitendra Malik. 2010. Contour detection and hierarchical image segmentation. *IEEE transactions on pattern analysis and machine intelligence*, 33(5):898–916.

Diego Arenas, Jon Atkins, Claire Austin, David Beavan, Alvaro Cabrejas Egea, Steven Carlysle-Davies, Ian Carter, Rob Clarke, James Cunningham, Tom Doel, et al. 2019. Design choices for productive, secure, data-intensive research at scale in the cloud. *arXiv preprint arXiv:1908.08737*.

Tayyaba Azim, Loitongbam Gyanendro Singh, and Stuart E. Middleton. 2022. Detecting moments of change and suicidal risks in longitudinal user texts using multi-task learning. In *Proceedings of the Eighth Workshop on Computational Linguistics and Clinical Psychology: Mental Health in the Face of Change*.

Michael Barkham, Wolfgang Lutz, and Louis G Castonguay. 2021. *Bergin and Garfield's handbook of psychotherapy and behavior change*. John Wiley & Sons.

Ulya Bayram and Lamia Benhiba. 2022. Emotionally-informed models for detecting moments of change and suicide risk levels in longitudinal social media data. In *Proceedings of the Eighth Workshop on Computational Linguistics and Clinical Psychology: Mental Health in the Face of Change*.

Sravani Boinepelli, Shivansh Subramanian, Abhijeeth Singam, Tathagata Raha, and Vasudeva Varma. 2022. Towards capturing changes in mood and identifying suicidality risk. In *Proceedings of the Eighth Workshop on Computational Linguistics and Clinical Psychology: Mental Health in the Face of Change*.

Ana-Maria Bucur, Hyewon Jang, and Farhana Ferdousi Liza. 2022. Capturing changes in mood over time in longitudinal data using ensemble methodologies. In *Proceedings of the Eighth Workshop on Computational Linguistics and Clinical Psychology: Mental Health in the Face of Change*.

Prasadith Buddhitha, Ahmed Husseini Orabi, Mahmoud Husseini Orabi, and Diana Inkpen. 2022. Multi-task learning to capture changes in mood over time. In *Proceedings of the Eighth Workshop on Computational Linguistics and Clinical Psychology: Mental Health in the Face of Change*.

Tianqi Chen and Carlos Guestrin. 2016. XGBoost: A scalable tree boosting system. In *Proceedings of the 22nd ACM SIGKDD International Conference on Knowledge Discovery and Data Mining*, KDD '16, pages 785–794, New York, NY, USA. ACM.

Arman Cohan, Bart Desmet, Andrew Yates, Luca Soldaini, Sean MacAvaney, and Nazli Goharian. 2018. SMHD: a large-scale resource for exploring online language usage for multiple mental health conditions. In *Proceedings of the 27th International Conference on Computational Linguistics*, pages 1485–1497, Santa Fe, New Mexico, USA. Association for Computational Linguistics.

Glen Coppersmith, Mark Dredze, Craig Harman, Kristy Hollingshead, and Margaret Mitchell. 2015. CLPsych 2015 shared task: Depression and PTSD on Twitter. In *Proceedings of the 2nd Workshop on Computational Linguistics and Clinical Psychology: From Linguistic Signal to Clinical Reality*, pages 31–39, Denver, Colorado. Association for Computational Linguistics.

John Culnan, Damian Y. Romero Diaz, and Steven Bethard. 2022. Exploring transformers and time lag features for predicting changes in mood over time. In *Proceedings of the Eighth Workshop on Computational Linguistics and Clinical Psychology: Mental Health in the Face of Change*.

Gildo Fabregat, Ander Cejudo, Juan Martinez-Romo, Alicia Pérez, Lourdes Araujo, Nuria Lebeña, Maite Oronoz, and Arantza Casillas. 2022. Approximate nearest neighbour extraction techniques and neural networks for suicide risk prediction in the CLPsych 2022 shared task. In *Proceedings of the Eighth Workshop on Computational Linguistics and Clinical Psychology: Mental Health in the Face of Change*.

Adithya V Ganesan, Vasudha Varadarajan, Juhi Mittal, Shashanka Subrahamanya, Matthew Matero, Nikita Soni, Sharath Chandra Guntuku, Johannes C. Eichstaedt, and H. Andrew Schwartz. 2022. WWBP-SQT-lite: Difference embeddings and multi-level models for moments of change identification in mental health forums. In *Proceedings of the Eighth Workshop on Computational Linguistics and Clinical Psychology: Mental Health in the Face of Change*.

Dilara Kekülluoğlu, Walid Magdy, and Kami Vaniea. 2020. Analysing privacy leakage of life events on twitter. In *Proceedings of the 12th ACM Conference on Web Science*.

David E. Losada and Fabio Crestani. 2016. A test collection for research on depression and language use. In *Experimental IR Meets Multilinguality, Multimodality, and Interaction - 7th International Conference of the CLEF Association, CLEF 2016, Évora, Portugal, September 5-8, 2016, Proceedings*, volume 9822 of *Lecture Notes in Computer Science*, pages 28–39. Springer.

David E. Losada, Fabio Crestani, and Javier Parapar. 2020. Overview of erisk 2020: Early risk prediction on the internet. In *Experimental IR Meets Multilinguality, Multimodality, and Interaction - 11th International Conference of the CLEF Association, CLEF 2020, Thessaloniki, Greece, September 22-25, 2020, Proceedings*, volume 12260 of *Lecture Notes in Computer Science*, pages 272–287. Springer.

Wolfgang Lutz, Torsten Ehrlich, Julian A. Rubel, Nora Hallwachs, Marie-Anna Röttger, Christine Jorasz, Sarah Mocanu, Silja Vocks, Dietmar Schulte, and Armita Tschitsaz-Stucki. 2013. The ups and downs of psychotherapy: Sudden gains and sudden losses identified with session reports. *Psychotherapy Research*, 23:14 – 24.

Sean Macavaney, Anjali Mittu, Glen Coppersmith, Jeff Leintz, and Philip Resnik. 2021. Community-level research on suicidality prediction in a secure environment: Overview of the clpsych 2021 shared task. In *Proc. of CLPsych*, pages 70–80.

Huina Mao, Xin Shuai, and Apu Kapadia. 2011. Loose tweets: An analysis of privacy leaks on twitter. WPES '11, page 1–12, New York, NY, USA. Association for Computing Machinery.

Jonathan G. Shalom and Idan M. Aderka. 2020. A meta-analysis of sudden gains in psychotherapy: Outcome and moderators. *Clinical Psychology Review*, 76:101827.

Han-Chin Shing, Suraj Nair, Ayah Zirikly, Meir Friedenberg, Hal Daumé III, and Philip Resnik. 2018. Expert, crowdsourced, and machine assessment of suicide risk via online postings. In *Proceedings of the Fifth Workshop on Computational Linguistics and Clinical Psychology: From Keyboard to Clinic*, pages 25–36, New Orleans, LA. Association for Computational Linguistics.

Nikita Soni, Matthew Matero, Niranjan Balasubramanian, and H. Schwartz. 2022. Human language modeling. In *Findings of the Association for Computational Linguistics: ACL 2022*, pages 622–636, Dublin, Ireland. Association for Computational Linguistics.

Adam Tsakalidis, Federico Nanni, Anthony Hills, Jenny Chim, Jiayu Song, and Maria Liakata. 2022. Identifying moments of change from longitudinal user text. In *Proc. of ACL*.

Gerrit JJ van den Burg and Christopher KI Williams. 2020. An evaluation of change point detection algorithms. *arXiv preprint arXiv:2003.06222*.

Ayah Zirikly, Philip Resnik, Özlem Uzuner, and Kristy Hollingshead. 2019. CLPsych 2019 shared task: Predicting the degree of suicide risk in Reddit posts. In

Proceedings of the Sixth Workshop on Computational Linguistics and Clinical Psychology, pages 24–33, Minneapolis, Minnesota. Association for Computational Linguistics.

A Reddit New: Data Collection

We used the Pushshift API (`https://reddit-api.readthedocs.io/en/latest/`) to crawl the posts from the following subreddits for Reddit-New: Agoraphobia, HealthAnxiety, autism, hardshipmates, rant, Anxiety, Needafriend, bipolar, lonely, rapecounseling, Anxietyhelp, StopSelfHarm, bipolarreddit, mentalhealth, schizophrenia, BPD, SuicideWatch, bulimia, mentalillness, socialanxiety, COVID19_support, addiction, depression, offmychest, survivorsofabuse, EDAnonymous, adhd, depression_help, panicparty, traumatoolbox, EatingDisorderHope, alcoholism, eating_disorders, psychoticreddit, trueoffmychest, EatingDisorders, anxietysupporters, foreveralone, ptsd, unsentletters.

B Timeline Selection Criteria

When selecting informative timelines, the internal annotators independently classified them into the following categories.

- **Good:** Timelines comprise posts that clearly indicate user mood or at least 1 moment of change in mood.
- **Medium:** Timelines comprise posts from which user mood is challenging to infer. The individual may disclose information about their own life events, but such discussions are objective in tone.
- **Bad:** Timelines comprise posts that do not provide indicators of the user's own mood. If there are posts by the user on subreddits related to mental health, these posts do not clearly relate to the user's own mood (e.g., words of encouragement for other users, cross-posted content shared with intent to help other users rather than themselves).

C Team Selection Assessment Criteria

In this section, we outline the assessment criteria used for selecting the teams for participate in the Shared Task. The guidelines were given to two annotators internally, who achieved a high agreement (Pearson correlation ρ=.83).

CLPsych 2022 Shared Tak: Registration of Interest

Guidelines for Reviewing

Aim: We have received applications (Registration of Interest) from 37 teams to participate in the CLPsych Shared Task 2022 (https://clpsych.org/sharedtask2022/). The goal of this reviewing process is to review the submitted applications on the basis of the main questions outlined below.

Registration of Interest Data: Each of the 37 teams that registered their interest provided us with the following information:

1. Timestamp
2. Team name (brief, no spaces)
3. Team Members (provide all names, comma-separated)
4. Main Contact (name)
5. Main Contact (email)
6. Main Contact (Affiliation(s))
7. Tell us why you are interested in participating
8. Tell us about your background, experience and NLP skills
9. Which programming languages (and corresponding version) are you planning to use? (if other, please specify)
10. Which software libraries do you expect to use? (one per line)
11. Do you plan to use a pre-trained model (such as GloVe, BERT, T5, etc.)? If so, please specify the version and the software library that you plan to use it with. (one per line)
12. Confirmation

We anonymised the list presented above and provided you with the following:

1. Number of participants in the team
2. Tell us why you are interested in participating (question 7 form the list above)
3. Tell us about your background, experience and NLP skills (question 8)
4. Which programming languages [...]? (question 9)
5. Which software libraries [...] (question 10)
6. Do you plan to use a pre-trained model [...] (question 11)

The reviewing task will be done solely on the basis of the responses given by each time on questions 2-6. For each team, *please read carefully the responses given by the team to all of the 5 questions prior to assessing their application.* The reason is that even though a reviewing criterion (see below) might seem explicitly related to a particular question (e.g., Criterion 1 seems to be clearly linked to the third question), the responses to the other questions might provide additional information for the team (e.g., the response to the second question might provide you with additional information for Criterion 1).

Assessment Criteria

For each of the three reviewing criteria presented below, please provide your score (half scores, such as "2.5", are also allowed), your confidence and a justification of your rating.

Criterion 1: Team Background

- Does the background/current work of the team match the requirements of the task? Please rate between 1-5 (half scores allowed):
  - 5: The team has worked/works on similar longitudinal/sequential NLP tasks on mental health.
  - 4: The team has worked/works on similar NLP tasks with a longitudinal or sequential component.
  - 3: The team has worked/works with NLP methods on the mental health domain, though without a sequential/longitudinal component.
  - 2: The team has worked/works with NLP methods, though outside of the mental health domain and without a sequential/longitudinal component .
  - 1: The team has some/no experience with NLP tasks and methods.
- Please justify/comment on your score:
- How confident are you on your assessment?
  - Very
  - Moderately
  - Low

Criterion 2: Commitment

- Based on your assessment, how likely is the team to commit to this task? Please rate between 1-3 (half scores allowed):
  - 3: The task will help the team even to advance their own work, so they are likely to invest a lot of time in the task.
  - 2: The team has shown strong motivation, but their work is not directly linked to the shared task.
  - 1: The team's motivation is not clear/not well explained.
- Please justify/comment on your score:
- How confident are you on your assessment?
  - Very
  - Moderately
  - Low

Criterion 3: Details on Software Requirements

- How detailed are the requests made by the team in terms of software requirements (programming languages & versions, libraries & versions, language models)? Please rate between 1-3 (half scores allowed):
  - 3: The provided information are very detailed. One could set up everything the team has asked for, allowing the team to start working straight away.
  - 2: The provided information are adequate, but not complete. One could probably set up a working environment with many of the required languages/libraries/models, but clarifications would be needed on several aspects (e.g., on specific versions of libraries).
  - 1: The replies of the team are generic/missing. Clarifications are needed in almost all of the requirements.
- Please justify/comment on your score:
- How confident are you on your assessment?
  - Very
  - Moderately
  - Low

Final Question (not part of the assessment): For the isolated participants (i.e., those who are a team on their own: numMembers=1), who should we try to group together so that they form a single team? Try to reply based on their responses to the 5 questions.

Approximate Nearest Neighbour Extraction Techniques and Neural Networks for Suicide Risk Prediction in the CLPsych 2022 Shared Task

Gildo Fabregat[1] Ander Cejudo[3] Juan Martinez-Romo[1,2] Alicia Pérez[3]
Lourdes Araujo[1,2] Nuria Lebeña[3] Maite Oronoz[3] Arantza Casillas[3]
NLP & IR Group, Universidad Nacional de Educación a Distancia (UNED)[1]
Instituto Mixto de Investigación - Escuela Nacional de Sanidad (IMIENS)[2]
HiTZ Center - Ixa, University of the Basque Country (UPV/EHU)[3]

Abstract

This paper describes the participation of the groups NLP@UNED and IXA@EHU on the CLPsych 2022 shared task. For task A, which tries to capture changes in mood over time, we have applied an Approximate Nearest Neighbour (ANN) extraction technique with the aim of relabelling the user messages according to their proximity, based on the representation of these messages in a vector space. Regarding the subtask B, we have used the output of the subtask A to train a Recurrent Neural Network (RNN) to predict the risk of suicide at the user level. The results obtained are very competitive considering that our team was one of the few that made use of the organisers' proposed virtual environment and also made use of the Task A output to predict the Task B results.

1 Introduction

CLPsych 2022 Shared Task (Tsakalidis et al., 2022a) introduces the problem of assessing changes in a person's mood over time on the basis of their linguistic content (Tsakalidis et al., 2022b). The purpose of the organisers is to focus on posting activity in online social media platforms. In particular, given a user's posts over a certain period in time, the aim of the task is to capture those sub-periods during which a user's mood deviates from their baseline mood and to use this information to predict the suicide risk at user level. Thus, the CLPsych 2022 Shared Task consists of the two subtasks: (1) Identify mood changes in users' posts over time and; (2) Show how subtask A can help to assess the risk level of a user.

This paper presents our participation in the subtasks T1 and T2.

1.1 Dataset

The dataset (Tsakalidis et al., 2022b) provided by the organisers is composed of social media messages obtained from various sources (Losada and Crestani, 2016; Losada et al., 2020; Zirikly et al., 2019; Shing et al., 2018).

Specifically, the dataset is composed of 256 timelines from Reddit obtained from 186 users who at some point in time have written in subreddits related to mental health. In total, there are more than 6K posts obtained in a time range of about two months. In the annotation process, timelines were manually checked for content related to mood changes. Four annotators were employed for this task.

In terms of evaluation, three types of evaluation measures were used: traditional classification metrics, timeline-based classification metrics, and coverage-based metrics.

2 Methods

Our team's participation in the task has been based on a system for capturing changes in mood over time and the information generated by this system has been used by another system that allows the prediction of the level of suicide risk in social network users.

2.1 Task A: Capturing changes in mood over time

Given a user's timeline, the aim is to classify each post within it as belonging to a "Switch" (IS), an "Escalation" (IE), or "None" (O).

Taking into account that the source of information used to generate the dataset are messages from social networks, we have proposed the use of an Approximate Nearest Neighbour (ANN) extraction technique. In general terms, when this algorithm is applied to a small set of messages it tends to work similarly to a KNN (K-Nearest-Neighbor) algorithm. Specifically, we have used the NMSLIB (Non-Metric Space Library) (Boytsov and Naidan, 2013). This library, unlike other tree-based libraries such as Annoy, makes use of graph theory and a method called Hierarchical Navigable World

Proceedings of the Eighth Workshop on Computational Linguistics and Clinical Psychology, pages 199 - 204
July 15, 2022

graph (Malkov and Yashunin, 2016). In short, we have worked with the hypothesis that given a representation of the messages in a vector space V, those messages that share the same $label$ will be in an easily identifiable subspace of V.

In order to encode each of the messages in the same vector space, we have used the Universal Sentence Encoder (Cer et al., 2018). In similarity tasks, this encoder has proved to work efficiently, especially when it comes to encoding information present in the text and not inferred from it. In this way, this encoder has obtained a good performance for topic extraction, but not so much in tasks such as author profiling or sex gender identification. Two versions of this encoder are publicly available:

- Based on DAN or Deep Averaging Networks (Iyyer et al., 2015): As its name suggests, it calculates the average of all the components of a given text. That is, while aspects such as the frequency of similar terms are taken into account, other aspects such as the order of the different terms are not considered.
- Based on Transformers (Vaswani et al., 2017): A "novel" representation that includes aspects of seq2seq architectures but eliminating the presence of decoders in the last layers. These types of architectures also include the use of different attention mechanisms.

Although in this work we have prioritised the use of DAN over Transformers, we have explored both mechanisms, having generated two runs using DAN and one using Transformers. In total, the system consists of two parts: (1) the representation of the data; and (2) the generation of the structure on which to query the nearest neighbours. After the generation of the query index and for the processing of new instances, the following heuristics have been explored:

1. A new instance represented in V-space is considered to be of class O if it is at a distance greater than d from its nearest neighbour. If the instance to be classified is at a distance less than or equal to d from its nearest neighbour, it is assigned the $label$ of this neighbour.
2. A new instance represented in V-space is considered to belong to the class of the nearest neighbour retrieved in V.

The study of a d value for cases where the distance from its nearest neighbour is greater than d has been an approach we have considered in the last stages of experimentation. Although we have experimented with different values of d, this approach establishes a clear bias due to the preference of class O over the rest of the classes. Among other reasons, we discarded at the time a study of this parameter in order not to focus the conclusions obtained on aspects inherent to the corpus studied, e.g. the distribution of the classes. In future work, we will try to redefine d so that it does not consider aspects related to the distribution of classes in the corpus.

Heuristic 1 takes into account that the majority class is class O and tries to assume that isolated points in space V belong to that class, since no "reliable" information would be available. On the other hand, heuristic 2 removes the above restriction and considers any retrieved neighbour as informative, regardless of its distance from the instance to be classified.

2.2 Task B: Predicting the risk of suicide

The goal of Subtask B is to predict the suicide risk level, that is, it is a classification task at user level. The risk level is a label within $\mathcal{C}_{user} = \{No, Low, Moderate, Severe\}$ with the labels presented in increasing risk-level (meaning that $\mathcal{C}_{user}$ contains a finite-set of discrete, ordered values). However, the shared task aimed, specifically, to show how Subtask 1 could help to assess the risk level of a user. Accordingly we interpreted that Subtask 2 has to make use of meta-data from Subtask 1.

We characterized a user-timeline U_i by the sorted sequence of messages posted: $(P_{i1}, P_{i2}, \ldots, P_{il_i})$. Note that the number of posts is user dependent with l_i being the number of posts associated to U_i. From System 1, each post in the test set P_{ij} is associated with k-nearest posts from the training (labeled) set each of which with the corresponding similarity weight: $((P'_{ij1}, l_1, w_1), (P'_{ij2}, l_2, w_2), \ldots, (P'_{ijk}, l_k, w_k))$. Note that, in the triplet (P'_{ijn}, l_n, w_n) each component conveys the following information:

- P'_{ijn} is a post from the training set, indeed, the n-closest post, ranking the n-th position in terms of similarity with respect to P_{ij}
- w_n is the similarity score of P'_{ijn} with respect to P_{ij} as stated in Subtask 1, i.e.,

$sim(P_{ij}, P'_{ijn}) = w_n$ with w_n increasing with increasing similarity of P_{ij}, P'_{ijn}.

- l_n is the label with which the training post P'_{ijn} had been annotated. Note that the labels are bound to a finite set of labels stated in Subtask 1, i.e. $l_n \in \mathcal{C}_{post}$ with $\mathcal{C}_{post} = \{O, IS, IE\}$.

With this k neighbours we are able to summarize the essence of P_{ij} in each of the three states ($s \in \{O, IS, IE\}$) involving the k neighbours as in expression (1) with $\delta(s, l_n)$ being 1 if s is equal to l_n and 0 otherwise.

$$sim(P_{ij}, s) = 1/(\sum_{n=1}^{k} w_n \cdot \delta(l_n, s)) \quad (1)$$

Accordingly, P_{ij} is represented as in (2) with a triplet of similarities to each state s.

$$P_{ij} : (sim(P_{ij}, O), sim(P_{ij}, IS), sim(P_{ij}, IE)) \quad (2)$$

Recalling that a user-timeline U_i conveyed a series of posts as in (3).

$$U_i : (P_{i1}, P_{i2}, \ldots, P_{il_i}) \quad (3)$$

In brief, each user-timeline was described as a sequence of posts and each post as a triplet of similarities with respect to each mood. With this information, the aim was to assign a label within $\mathcal{C}_{user}$. Given that this process, intrinsically, has a sequential nature, we turned to a well known recurrent neural network able to learn from the context, that is, a BiLSTM (Schuster and Paliwal, 1997). In practice, the number of neighbours employed to get the tuple is set to 20. The number of neighbours to be retrieved and considered for class prediction of a given instance was studied using a validation set extracted from the training corpus. This partition was discarded in the test phase in order to ensure that the selected parameter was consistent with the previously conducted study. With regard to the practicalities of the implementation, we resorted to TensorFlow (Abadi et al., 2015). Having conceived this approach as a baseline, we simplified the architecture to the maximum and just incorporated 1 hidden layer and tested a batch size between 4 and 8.

At this point we should note that the number of messages posted by each user is variable (i.e. the number of posts l_i is not constant). Nevertheless, the implementation assumes a fixed-length input. Padding is a simple approach frequently used to address this issue. With this approach we forced the sequence of all users to a constant and predetermined length l. To address this restriction we distinguished two situations:

- For users with $l_i < l$, the user characterization was arbitrarily extended incorporating $l - l_i$ artificial tuples. The content of these tuples was fixed to as unknown or also called missing value (NaN).

- For users with $l_i > l$, the user characterization was arbitrarily restricted to the first l posts while discarding the latest $l_i - l$ posts. That is, for the user U_i with posts $(P_{i1}, P_{i2}, \ldots, P_{il_i})$ and $l_i > l$ we merely considered the posts $(P_{i1}, P_{i2}, \ldots, P_{il})$. Needless to say, this approach entails a loss of information, indeed, we are missing the latest or most recent information. Instead, we could have tried to discard the first posts.

In order to fix l, fine tuning was carried out in beam-search (not an exhaustive search) in a range between 2 and 30 and the optimum number of posts to keep was identified to be l = 10.

3 Results

Apart from the difficulty of the tasks themselves described in previous sections, another difficulty of the task was working in the environment that the organisers managed to access and work with the data. Instead of distributing the annotated dataset for training, the NORC Data Enclave environment was used. The NORC Data Enclave provides a confidential and protected environment in which only authorized participants could securely access and analyze remotely the data. However, due to the problems that some participants had in working in this environment, the datasets (training and test) were distributed to these groups. For this reason, a column in the results tables indicates the use of the Data Enclave environment to obtain these results.

3.1 Task A: Capturing changes in mood over time

In the section 2.1 two versions of the encoder used, were defined: Based on Deep Averaging Networks (DAN); and based on Transformers. In the same way, two types of heuristics (heuristic 1 and 2) were also defined. Thus, the configuration of the runs submitted to the subtask A is as follows:

- Run 1: DAN and Heuristic 1
- Run 2: DAN and Heuristic 2
- Run 3: Transformers and Heuristic 2

Table 1 shows the official results of task A at post level and for each of the participating teams. The organisers have selected the best run of results for each team. In the case of our team, the best run was "Run 1". Thus, the best configuration for this task has been the use of DAN and the Heuristic 1, in which a threshold was applied to select the maximum distance from nearest neighbours to be assigned the same label.

According to the results, our system leaves room for improvement in terms of accuracy and has an f1-measure comparable to the average of most systems. However, our system achieved recall scores that compensate the low scores of accuracy.

Although the DAN model is based on the unordered representation of the terms of a given text (applying the mean), this model has sufficient capacity to differentiate instances such as: "this is toy dog" Vs. "this is dog toy". The results obtained seem to indicate that under the same environment i.e., HNSW configuration and so on, the DAN-based model is better suited to the task than the Transformer-based model. Among the limitations of the Transformer-based model is the performance drop when processing excessively long texts. In the case of DAN, this limitation is also present but does not seem to be as important for the task at hand.

Task A - Post Level Macro-Average				
System	**DE**	**P**	**R**	**F1**
WResearch	**YES**	**0.62**	**0.58**	**0.60**
UArizona	**YES**	0.52	0.51	0.51
NLP-UNED	**YES**	0.49	0.52	0.50
UoS	NO	**0.69**	**0.62**	**0.65**
LAMA	NO	**0.55**	0.53	**0.52**
IIITH	NO	0.52	**0.60**	0.52
uOttawa-AI	NO	0.50	0.53	0.51
WWBP-SQT-lite	NO	0.51	0.51	0.51
BLUE	NO	0.50	0.49	0.50

Table 1: Official results of subtask A at post level and for each of the participating teams. DE: Use of the official shared task environment (Data Enclave); P: Precision; R: Recall, F1: F1 score.

Table 3 shows the official results of task A at coverage and for each of the participating teams. In the case of our team, the best run was "Run 1", as well as for the evaluation at the post level.

Task A - Coverage Macro-Average			
System	**DE**	**P**	**R**
WResearch	**YES**	**0.47**	**0.50**
UArizona	**YES**	0.42	0.42
NLP-UNED	**YES**	0.31	0.40
UoS	NO	**0.51**	**0.50**
LAMA	NO	0.38	**0.44**
IIITH	NO	0.35	0.41
uOttawa-AI	NO	0.35	0.43
WWBP-SQT-lite	NO	0.34	0.38
BLUE	NO	**0.50**	0.38

Table 2: Official results of subtask A at coverage and for each of the participating teams. DE: Use of the official shared task environment (Data Enclave); P: Precision; and R: Recall.

Table 3 shows the official results of task A Window-based and for each of the participating teams. The organisers have also selected the best run of each team. In our case the best run was the "Run 2". This means that in this case, heuristic 2 performs better when windows are taken into account compared to the post-level results, where heuristic 1 performed better. In both cases DAN performs better than the Transformer-based encoder.

According to the results, our system stands out in recall, especially for window sizes 2 and 3, where it obtains the best results among the systems that used Data Enclave, and in the case of window size 3 it obtains the best result taking into account all participating systems.

Task A - Window-based Macro-Average						
	Window 1		**Window 2**		**Window 3**	
System	**P**	**R**	**P**	**R**	**P**	**R**
WResearch	**.63**	**.62**	**.65**	**.65**	**.66**	.65
NLP-UNED	.53	.61	.55	**.65**	.58	**.69**
UArizona	**.58**	.56	**.60**	.58	**.62**	.60
UoS	**.68**	**.65**	**.69**	**.67**	**.71**	**.69**
uOttawa-AI	.53	.62	.56	.66	.60	**.69**
IIITH	.53	**.65**	.54	.66	.55	.67
LAMA	.57	.58	.59	.63	.61	.66
WWBP-SQT	.55	.57	.57	.60	.60	.62
BLUE	.54	.57	.56	.59	.58	.62

Table 3: Official results (rounded down) of subtask A at Window-based and for each of the participating teams. P: Precision; R: Recall.

3.2 Task B: Predicting the risk of suicide

Table 4 shows the results reported in Task B in two ways, either for all the teams or only for those teams that used the output of the Task A to cope with Task B. In general, the results achieved not using the output from Task A are better than when using it. However, our team decided to take up the organisers' challenge and use the output of task A to predict the risk of suicide in task B. Given that we perceived the use of Data Enclave (DE) as an added value, all our attempts are with DE by contrast to the majority of the systems involved.

Task B				
System	**DE**	**P**	**R**	**F1**
NLP-UNED	**YES**	0.36	0.39	0.37
WResearch	NO	0.47	**0.48**	**0.46**
UoS	NO	**0.62**	0.43	0.45
IIITH	NO	0.40	0.41	0.38
WWBP-SQT-lite	NO	0.35	0.37	0.35
uOttawa-AI	NO	0.33	0.36	0.34
LAMA	NO	0.31	0.42	0.30

Task B - With Task A Auxiliary				
System	**DE**	**P**	**R**	**F1**
NLP-UNED	**YES**	0.37	**0.39**	**0.36**
WResearch	NO	**0.37**	0.36	0.36

Table 4: Official results (rounded down) of subtask B: all the systems (top) and only those using the Task A for the prediction of task B (bottom). DE: using the official shared task environment (Data Enclave). P: Precision, R: Recall, F1: F1 score.

4 Conclusions

In this work, we introduce the Approximate Nearest Neighbour (ANN) extraction technique and the use of Recurrent Neural Networks (RNN) to automatically capture changes in mood over time and the prediction of the suicide risk at the user level.

The shared task had the added challenge of working in a virtual environment that the organisers had prepared to preserve the privacy of the real data with which we had to work. However, due to the problems of some participants in working in this environment, the data were distributed among these groups and could be processed outside the virtual environment. This fact, from our point of view, prevents a fair comparison among the systems that used the environment and those that did not. This is due to the fact that the virtual environment has no internet connection and therefore the resources available to process the data were only the libraries that previously had been installed at the beginning of the shared task.

Leaving this consideration aside, our system performed acceptably, having a high score in recall. The low precision we obtained is an aspect that we need to improve on, for future work.

As for the analysis of the organisers in terms of window size, it can be said that our system performed remarkably well for window sizes 2 and 3, obtaining the best recall scores in these cases.

Another challenge of the task was set by the organisers when planning the two sub-tasks. In this case, participants were encouraged to use the output of sub-task A as input for sub-task B to predict the suicide risk at the user level. In our case, we took up this challenge and together with just another team we were the only ones to use the output of subtask A to predict the suicide risk in subtask B. Moreover, by a very small margin with the other team, we obtained the best scores in F1-measure and recall.

Ethical Statement

Secure access to the shared task dataset was provided with IRB approval under University of Maryland, College Park protocol 1642625 and approval by the Biomedical and Scientific Research Ethics Committee (BSREC) at the University of Warwick (ethical application reference BSREC 40/19-20).

Acknowledgements

This work has been partially supported by the Spanish Ministry of Science and Innovation within the DOTT-HEALTH Project (MCI/AEI/FEDER, UE) under grant PID2019-106942RB-C32, as well as within the project RAICES (IMIENS 2022), the research network AEI RED2018-102312-T (IA-Biomed), the European Commission in a CHIST-ERA project (FEDER, ANTIDOTE PCI2020-120717-2) and the Basque Government (within IXA IT-1343-19 project as well as within Ikasiker grants published in the 03/02/2022 BOPV). The authors are particularly grateful to the anonymous users of Reddit whose data feature in this year's shared task dataset, to the annotators of the data for Task A, to the clinical experts from Bar-Ilan University who annotated the data for Task B, the American Association of Suicidology, to NORC for creating and administering the secure infrastructure and providing researcher support and to UKRI for providing funding to the CLPsych 2022 shared task organisers.

References

Martín Abadi, Ashish Agarwal, Paul Barham, Eugene Brevdo, Zhifeng Chen, Craig Citro, Greg S. Corrado, Andy Davis, Jeffrey Dean, Matthieu Devin, Sanjay Ghemawat, Ian Goodfellow, Andrew Harp, Geoffrey Irving, Michael Isard, Yangqing Jia, Rafal Jozefowicz, Lukasz Kaiser, Manjunath Kudlur, Josh Levenberg, Dandelion Mané, Rajat Monga, Sherry Moore, Derek Murray, Chris Olah, Mike Schuster, Jonathon Shlens, Benoit Steiner, Ilya Sutskever, Kunal Talwar, Paul Tucker, Vincent Vanhoucke, Vijay Vasudevan, Fernanda Viégas, Oriol Vinyals, Pete Warden, Martin Wattenberg, Martin Wicke, Yuan Yu, and Xiaoqiang Zheng. 2015. TensorFlow: Large-scale machine learning on heterogeneous systems. Software available from tensorflow.org.

Leonid Boytsov and Bilegsaikhan Naidan. 2013. Engineering efficient and effective non-metric space library. In *Similarity Search and Applications - 6th International Conference, SISAP 2013, A Coruña, Spain, October 2-4, 2013, Proceedings*, volume 8199 of *Lecture Notes in Computer Science*, pages 280–293. Springer.

Daniel Cer, Yinfei Yang, Sheng-yi Kong, Nan Hua, Nicole Limtiaco, Rhomni St. John, Noah Constant, Mario Guajardo-Cespedes, Steve Yuan, Chris Tar, Yun-Hsuan Sung, Brian Strope, and Ray Kurzweil. 2018. Universal sentence encoder. *CoRR*, abs/1803.11175.

Mohit Iyyer, Varun Manjunatha, Jordan L. Boyd-Graber, and Hal Daumé III. 2015. Deep unordered composition rivals syntactic methods for text classification. In *Proceedings of the 53rd Annual Meeting of the Association for Computational Linguistics and the 7th International Joint Conference on Natural Language Processing of the Asian Federation of Natural Language Processing, ACL 2015, July 26-31, 2015, Beijing, China, Volume 1: Long Papers*, pages 1681–1691. The Association for Computer Linguistics.

David E. Losada and Fabio Crestani. 2016. A test collection for research on depression and language use. In *Experimental IR Meets Multilinguality, Multimodality, and Interaction - 7th International Conference of the CLEF Association, CLEF 2016, Évora, Portugal, September 5-8, 2016, Proceedings*, volume 9822 of *Lecture Notes in Computer Science*, pages 28–39. Springer.

David E. Losada, Fabio Crestani, and Javier Parapar. 2020. Overview of erisk 2020: Early risk prediction on the internet. In *Experimental IR Meets Multilinguality, Multimodality, and Interaction - 11th International Conference of the CLEF Association, CLEF 2020, Thessaloniki, Greece, September 22-25, 2020, Proceedings*, volume 12260 of *Lecture Notes in Computer Science*, pages 272–287. Springer.

Yury A. Malkov and Dmitry A. Yashunin. 2016. Efficient and robust approximate nearest neighbor search using hierarchical navigable small world graphs. *CoRR*, abs/1603.09320.

M. Schuster and K.K. Paliwal. 1997. Bidirectional recurrent neural networks. *IEEE Transactions on Signal Processing*, 45(11):2673–2681.

Han-Chin Shing, Suraj Nair, Ayah Zirikly, Meir Friedenberg, Hal Daumé III, and Philip Resnik. 2018. Expert, crowdsourced, and machine assessment of suicide risk via online postings. In *Proceedings of the Fifth Workshop on Computational Linguistics and Clinical Psychology: From Keyboard to Clinic*, pages 25–36, New Orleans, LA. Association for Computational Linguistics.

Adam Tsakalidis, Jenny Chim, Iman Munire Bilal, Ayah Zirikly, Dana Atzil-Slonim, Federico Nanni, Philip Resnik, Manas Gaur, Kaushik Roy, Becky Inkster, Jeff Leintz, and Maria Liakata. 2022a. Overview of the CLPsych 2022 shared task: Capturing moments of change in longitudinal user posts. In *Proceedings of the Eighth Workshop on Computational Linguistics and Clinical Psychology: Mental Health in the Face of Change*.

Adam Tsakalidis, Federico Nanni, Anthony Hills, Jenny Chim, Jiayu Song, and Maria Liakata. 2022b. Identifying moments of change from longitudinal user text. In *Proceedings of the 60th Annual Meeting of the Association for Computational Linguistics (Volume 1: Long Papers)*, pages 4647–4660, Dublin, Ireland. Association for Computational Linguistics.

Ashish Vaswani, Noam Shazeer, Niki Parmar, Jakob Uszkoreit, Llion Jones, Aidan N. Gomez, Lukasz Kaiser, and Illia Polosukhin. 2017. Attention is all you need. *CoRR*, abs/1706.03762.

Ayah Zirikly, Philip Resnik, Özlem Uzuner, and Kristy Hollingshead. 2019. CLPsych 2019 shared task: Predicting the degree of suicide risk in Reddit posts. In *Proceedings of the Sixth Workshop on Computational Linguistics and Clinical Psychology*, pages 24–33, Minneapolis, Minnesota. Association for Computational Linguistics.

Capturing Changes in Mood Over Time in Longitudinal Data Using Ensemble Methodologies

Ana-Maria Bucur[1,2], **Hyewon Jang**[3], **Farhana Ferdousi Liza**[4]
[1]Interdisciplinary School of Doctoral Studies, University of Bucharest, Romania
[2]PRHLT Research Center, Universitat Politècnica de València, Spain
[3]Department of Linguistics, University of Konstanz, Germany
[4]School of Computing Sciences, University of East Anglia, UK
ana-maria.bucur@drd.unibuc.ro
hye-won.jang@uni-konstanz.de, f.liza@uea.ac.uk

Abstract

This paper presents the system description of team BLUE for Task A of the CLPsych 2022 Shared Task on identifying changes in mood and behaviour in longitudinal textual data. These moments of change are signals that can be used to screen and prevent suicide attempts. To detect these changes, we experimented with several text representation methods, such as TF-IDF, sentence embeddings, emotion-informed embeddings and several classical machine learning classifiers. We chose to submit three runs of ensemble systems based on maximum voting on the predictions from the best performing models. Of the nine participating teams in Task A, our team ranked second in the Precision-oriented Coverage-based Evaluation, with a score of 0.499. Our best system was an ensemble of Support Vector Machine, Logistic Regression, and Adaptive Boosting classifiers using emotion-informed embeddings as input representation that can model both the linguistic and emotional information found in users' posts.

1 Introduction

The changes in mood and behaviour in the social media discourse of users are markers that can be used for screening and prevention of future suicide attempts. The emotional signals expressed in language and switches to suicide ideation are used for assessing the suicide risk of online users. However, identifying a person's mood changes over time based on their linguistic content from the posting activity on online social media platforms is a challenging task. Challenges come from different perspectives, including methodological challenges of noisy natural language understanding (Farzindar and Inkpen, 2017), ethical implications of research and deployment (Benton et al., 2017; Chancellor et al., 2019; Resnik et al., 2021) and challenges associated with longitudinal data analysis. Despite different challenges, the potential role of Artificial Intelligence (AI) based language technologies in mental health is gaining increasing attention (Lee et al., 2021). For example, some social media domains started implementing auto-detection tools to prevent suicide (Ji et al., 2020). In this paper, we present the methodology and the results of the machine learning models developed using the 2022 CLPsych Shared Task dataset (Tsakalidis et al., 2022a). We experiment with machine learning algorithms for the classification task using as input text representations based on statistical TF-IDF, pre-trained GloVe embeddings (Pennington et al., 2014) and embeddings extracted from pre-trained transformer models. After that, we develop a majority voting scheme over the predictions to report the final labels for a user timeline. Our best strategy is based on majority voting of Logistic Regression (LR), Support Vector Machine (SVM) and Adaptive Boosting (AdaBoost) classifiers using as input the embeddings extracted from the pre-trained transformer models fine-tuned for emotion detection. Our team BLUE ranked second in terms of Precision-oriented Coverage-based Evaluation (macro-avg) metric with an overall score of 0.499, whereas the top score in this evaluation metric is 0.506.

2 Related Work

With the rise in social media use, more people started discussing their mental health problems and seeking support online. This allowed Natural Language Processing and Psychology researchers to use social media data to search for cues of mental illnesses. The frequently used social media platforms for studying these issues are Twitter (Sawhney et al., 2020b; Coppersmith et al., 2016) and Reddit (Zirikly et al., 2019a; Losada et al., 2020).

For suicide detection, there are two methodologies for screening the online content: at the user level or post level. For user-level classification, the aim is to detect from the whole history of the user

Proceedings of the Eighth Workshop on Computational Linguistics and Clinical Psychology, pages 205 - 212
July 15, 2022

if they are at risk of suicide or if they show suicide ideation prior to the attempt, for an intervention to be made and for trying to save their life (Coppersmith et al., 2018; Zirikly et al., 2019b; Sawhney et al., 2020a).

Post-level classification is performed by screening one post at a time, searching for posts that are indicative of a user being at risk of suicide (O'dea et al., 2015; Sawhney et al., 2018; Tadesse et al., 2019). O'dea et al. (2015) collect suicide-related tweets and annotate them as *strongly concerning*, *possibly concerning* or *safe to ignore*. Afterwards, the authors train machine learning classifiers (SVM, LR) to distinguish the concern level for these tweets containing suicide-related words.

Coppersmith et al. (2016) explore the language of Twitter users prior to a suicide attempt to find quantifiable signals that can be used for screening and prevention. Their article reveals that users have more posts expressing *anger* and *sadness* before trying to commit suicide. However, these emotions get to the same level as control users after the attempt. Furthermore, people who attempt suicide have a higher proportion of emotional posts, increasing after the incident. In line with these findings, several works are modelling the emotional information found in the online discourse of users for classifying the suicide risk (Ji et al., 2021; Sawhney et al., 2021; Bitew et al., 2019; Chen et al., 2019).

Regarding longitudinal approaches for suicide detection, De Choudhury et al. (2016) extract markers of shifts to suicide ideation from users engaged in the online discourse revolving around mental illnesses, such as hopelessness, high self-attention focus, anxiety, impulsiveness and others. Using these markers, the authors can predict which individuals are more prone to express suicide ideation in future posts. Through a time-aware approach, Sawhney et al. (2021) propose a framework that uses people's historical and emotional spectrum when assessing the risk of a specific post.

Tsakalidis et al. (2022b) propose to take the temporal information into account by identifying the changes in people's behaviour and mood on social media. The changes considered are switches (sudden mood changes) and escalation (gradual mood progression). These changes in mood or emotion found in the online discourse can be used for assessing the suicide risk of users.

Although the potential role of language technology in mental health using information from social media datasets is gaining increasing attention, continued progress on NLP for mental health is hampered by obstacles to shared, community-level access to relevant data. The 2021 CLPsych Shared Task was introduced to address this problem by conducting a shared task using sensitive data in a secure environment (MacAvaney et al., 2021) and continued in the 2022 CLPsych Shared Task (Tsakalidis et al., 2022a). The goal of the tasks from the previous year was to assess the suicide risk of a user from posts 30 days or 6 months prior to a suicide attempt. The best-performing models used approaches such as weighted ensemble of different machine learning classifiers (LR, Naive Bayes classifiers, linear SVM) (Bayram and Benhiba, 2021), LSTM architecture with topic modelling and dictionary-based features (Gollapalli et al., 2021) and Bayesian modelling of features from Linguistic Inquiry and Word Count (LIWC) (Pennebaker et al., 2001), behavioural information or other features derived from already available or custom dictionaries (Gamoran et al., 2021).

3 Data and Task A

We participate in Task A in the 2022 CLPsych Shared Task, intending to capture the mood changes of individuals in a given time window based on their Reddit posts. The dataset for this task was collected in Tsakalidis et al. (2022b). The posts from Reddit's mental health-related subreddits in a given time window (timeline) (Losada et al., 2020; Losada and Crestani, 2016; Zirikly et al., 2019a; Shing et al., 2018) were annotated by four annotators on the basis of three labels hinting at moments of change (Tsakalidis et al., 2022b): none (O), escalation (IE), and switch (IS). A total of 256 timelines and 6,205 posts are available for Task A. Thus, given a user's timeline, the aim is to classify each post as either a 'switch' (IS), or an 'escalation' (IE) or 'none' (O).

Three metrics are used for evaluating the performance of the models in Task A (Tsakalidis et al., 2022b). *Post-level* evaluation calculates the traditional Precision, Recall, and F1 scores per post and class, with the macro-average to get the final score. Apart from the traditional post-level metric, timeline-based scores are also used for the evaluation, given the sequential nature of Task A. In the *window-based* evaluation, Precision and Recall scores are calculated based on whether correct labels are in a certain time window. In the *coverage-*

based evaluation, Precision and Recall scores are calculated based on the models' ability to capture *regions of change*.

4 Method

4.1 Text Representation

We experiment with several methods for encoding the textual data, such as TF-IDF, GloVe embeddings and transformer-based representations.

Term Frequency–Inverse Document Frequency (TF-IDF) As a baseline approach, we use TF-IDF vectorization to model our data. We experiment with different N-gram sizes and find that converting text into TF-IDF matrix using unigrams only (N=1) produces the best results.

Sentence Embeddings We experiment with pre-trained models from the Sentence Transformers library (Reimers and Gurevych, 2019) that are not specifically fine-tuned on emotion data: *paraphrase-MiniLM-L6-v2* (Wang et al., 2020), *distilbert-base-uncased* (Sanh et al., 2019), and *average_word_embeddings_glove.6B.300d* (Pennington et al., 2014). We chose these models based on the small model size and computational efficiency.

Emotion-Informed Embeddings Given the nature of the task and the presence of different positive and negative emotions in the users' timelines, we posit that models fine-tuned on the emotion detection task could provide better textual representations for our data, by modelling both the linguistic and emotion information found in users' posts. We experiment with various text representations extracted using pre-trained transformer models fine-tuned on several datasets for emotion detection (Saravia et al., 2018; Mohammad et al., 2018; Busso et al., 2008; Poria et al., 2019) provided by Hugging face[1]. The models used in this work, that were compatible with the Sentence Transformers library, are: *bertweet-emotion-base* [2] (fine-tuned version of BERTweet (Nguyen et al., 2020) for emotion detection), *distilbert-base-uncased-emotion* [3] (fine-tuned version of DistilBERT (Sanh et al., 2019)), *emoberta-base* [4] (Kim and Vossen, 2021), *twitter_emotions* [5] (fine-tuned version of MiniLM (Wang et al., 2020)), *albert-base-v2-emotion* [6] (ALBERT (Lan et al., 2019) fine-tuned), *roberta-base-emotion* [7] and *twitter-roberta-base-emotion* [8] (RoBERTa (Liu et al., 2019) models fine-tuned for emotion detection).

4.2 Models

For classifying the data using the different text representation methods, we train several classical machine learning models for detecting the escalation (IE) and switch (IS) in the dataset, including Logistic Regression (LR), Decision Trees (DT), Random Forest (RF), Support Vector Machine (SVM), the Adaptive Boosting (AdaBoost). We develop a majority voting scheme over the predictions to report the final labels for a user timeline. In order to choose which machine learning classifier to use, we experiment with multiple models trained on 70% of the data and evaluate them using the remaining held-out 30% of the data (the validation data). Our final submissions were the top-performing models evaluated on the validation data.

We perform a hyperparameter grid search for the classification models that use the emotion-informed embeddings to find the best hyperparameters for these models. The search space used for grid search can be found in Appendix A. We choose the best performing classification model and the best hyperparameters for each method of representing the input (based on the fine-tuned models for emotion detection).

4.3 Submitted Runs

We submitted three runs for Task A using the following models:

Run 1: *ensemble_without_emotion_features:* We use an ensemble method based on maximum voting on the classification results obtained from the Adaptive Boosting Ensemble classifier using non-emotion embeddings (TF-IDF and sentence embeddings).

Run 2: *ensemble_with_all_models:* We experiment with the same ensemble method based on maximum voting on the classification results obtained from all our models (Run 1 and Run 3).

Run 3: *ensemble_with_emotion_features:* For the third run, we use the ensemble method based on

[1] https://huggingface.co/

[2] https://huggingface.co/Emanuel/bertweet-emotion-base

[3] https://huggingface.co/bhadresh-savani/distilbert-base-uncased-emotion

[4] https://huggingface.co/tae898/emoberta-base

[5] https://huggingface.co/trnt/twitter_emotions

[6] https://huggingface.co/bhadresh-savani/albert-base-v2-emotion

[7] https://huggingface.co/bhadresh-savani/roberta-base-emotion

[8] https://huggingface.co/cardiffnlp/twitter-roberta-base-emotion

maximum voting on the predictions obtained from the classifiers using as input the emotion-informed embeddings. The ensemble was comprised of predictions from LR, SVM and AdaBoost classifiers (the best performing models).

5 Results and Discussion

At the time of writing the paper, we do not have access to the test data ground truth labels. Therefore, we present the performance of our three ensemble systems on the validation data and the official results from the task organisers on the test data. In addition, we perform an error analysis by exploring in more detail at the predictions of the models on the validation data.

	Post-Level			Window-based		Coverage-based	
Run	**P**	**R**	**F1**	**P**	**R**	**P**	**R**
Run 1	0.52	0.55	0.53	0.55	0.61	0.39	0.49
Run 2	0.67	0.55	0.59	0.67	0.56	0.55	0.44
Run 3	0.64	0.55	0.58	0.67	0.58	0.49	0.45

Table 1: Macro Average of Validation Scores. Precision (P), Recall (R), F1 score (F1) for post-level, window-based (window=1), and coverage-based metrics.

	Post-Level			Window-based		Coverage-based	
Run	**P**	**R**	**F1**	**P**	**R**	**P**	**R**
Run 1	0.50	0.50	0.50	0.54	0.57	0.38	0.45
Run 2	0.48	0.46	0.46	0.51	0.51	0.33	0.38
Run 3	0.63	0.46	0.46	0.62	0.50	0.50	0.38
Baseline 1	0.55	0.50	0.49	0.38	0.42	0.50	0.54
Baseline 2	0.52	0.39	0.38	0.26	0.20	0.58	0.39

Table 2: Macro Average of Official Test Scores. Precision (P), Recall (R), F1 score (F1) for post-level, window-based (window=1), and coverage-based metrics. Baseline 1 is a LR approach on TF-IDF features, Baseline 2 is a BERT model trained on Talklife data using focal loss.

5.1 Results

Nine teams participated in Task A of the 2022 CLPsych Shared Task. Our team ranked second in the Precision-oriented Coverage-based Evaluation, with a score of 0.499, whereas the score of the top-ranking system was 0.506.

In Table 1, we present the results on the validation data for the identification of moments of change. We report the macro-average of the scores for the post-level, window-based and coverage-based evaluation methods. Table 2 shows the official results for the three runs and two baselines provided by the organisers. Baseline 1 is an LR model trained on TF-IDF features, and Baseline 2 is a BERT model trained on Talklife data (Tsakalidis et al., 2022b) using focal loss (Lin et al., 2017). All our runs surpass the baseline methods in the window-based evaluation. The ensemble model using as input the emotion-informed embeddings (Run 3) has the highest Precision for the three evaluation metrics, post-level, window-based and coverage-based. In contrast, the ensemble from Run 1 performs best in terms of Recall. Even if the system from Run 2 is the best performing model on the validation data, its performance is the lowest when predicting on test data.

5.2 Error analysis

We perform a brief error analysis on the predictions of our systems on the validation data. There are cases when the user has a large number of posts in a row labelled as escalations, and the model can identify most of them successfully. However, in some cases, the model failed to identify the escalations. Furthermore, in some cases, the model can recognise the mood changes, but it fails to distinguish whether the changes are escalations or switches.

The system also predicts false positives (IS or IE) when the users mentions about someone close who has suicide ideation or has depression in their posts and do not talk about themselves (e.g., "my friend talks about taking their own life with me", "you suffer from depression", "I despise seeing you suffer.[9]). To address this, we plan to incorporate anaphora resolution techniques into the modelling in the future.

There is a specific case when the system cannot recognise a moment of change because it seems a neutral text. However, it contains a mention of *klonopin*[10], a drug from the class of *benzodiazepines*, used for treating different physical and mental health problems. This drug can cause addiction and lead to overdose when combined with other drugs or alcohol. To improve the identification of mood changes in these special cases, additional knowledge related to specific medications for mental health problems can be added to the modelling.

It is worth mentioning that some of the errors may stem from the difficulty associated with the longitudinal labelling of data. It is generally hard to determine what is an escalation of a mood and

[9]not actual examples from the dataset, but equivalent sentences in order to maintain anonymity

[10]https://drugabuse.com/benzodiazepines/klonopin/overdose/

what is a sudden switch. In one example of our error analysis, our system (Run 2) classified several posts in a row as IE (escalation) when the ground truth labels were mostly O (no mood change) with occasional IS (switch). This example shows that a model performance can exponentially degrade due to the connectivity of each data point to the adjacent ones; IS (switch) is less likely to appear if the preceding texts are not O (no mood change). It would mean that if a model makes a mistake for one post, the following predictions are likely to be wrong accordingly (*domino effect*).

Moreover, there are instances where we agreed more with the classification labels produced by our system than the ground truth labels. For instance, *I've messed up a lot of stuff. (...) I am sorry. (...) I am so sorry. (...)*[11] showed obvious signs of emotional turbulence and can facilitate prominently in understanding of the emotional underpinnings of depressive symptoms (Kim et al., 2011); however, the ground truth label was O (our system predicted IE). As such, difficulty associated with the annotation of longitudinal data could be addressed in future research.

6 Conclusion

In this paper, we presented the system description and results of team BLUE for the task of identifying moments of change from the CLPSych 2022 Shared Task. We experimented with several text representation methods, such as TF-IDF, sentence embeddings (from pre-trained transformer models, GloVe) and emotion-informed embeddings (extracted from the pre-trained transformer models fine-tuned for emotion detection). To identify the mood changes, we trained several classical machine learning classifiers. We chose to submit three ensemble systems based on maximum voting on the best performing models (SVM, LR, AdaBoost) with different inputs. Of the nine participating teams in Task A, our team ranked second in the Precision-oriented Coverage-based Evaluation, with a score of 0.499 (the top team had a score of 0.506). Our best run was an ensemble method of SVM, LR, and AdaBoost classifiers using as input emotion-informed embeddings that can model both the linguistic and emotional information found in users' posts. Due to the Enclave data system's technical difficulties, we have developed systems in three working days after getting the data in our local system. For future work, we plan to investigate the dataset in detail and develop improved models for identifying mood changes in longitudinal textual data and assess the suicide risk of social media users.

[11]not actual examples from the dataset, but equivalent sentences in order to maintain anonymity

Ethical Statement

Secure access to the shared task dataset was provided with IRB approval under University of Maryland, College Park protocol 1642625 and approval by the Biomedical and Scientific Research Ethics Committee (BSREC) at the University of Warwick (ethical application reference BSREC 40/19-20).

Acknowledgements

The authors are particularly grateful to the anonymous users of Reddit whose data feature in this year's shared task dataset, to the annotators of the data for Task A, to the clinical experts from Bar-Ilan University who annotated the data for Task B, the American Association of Suicidology, to NORC for creating and administering the secure infrastructure and providing researcher support and to UKRI for providing funding to the CLPsych 2022 shared task organisers.

References

Ulya Bayram and Lamia Benhiba. 2021. Determining a person's suicide risk by voting on the short-term history of tweets for the clpsych 2021 shared task. In *Proceedings of the Seventh Workshop on Computational Linguistics and Clinical Psychology: Improving Access*, pages 81–86.

Adrian Benton, Glen Coppersmith, and Mark Dredze. 2017. Ethical research protocols for social media health research. In *Proceedings of the first ACL workshop on ethics in natural language processing*, pages 94–102.

Semere Kiros Bitew, Giannis Bekoulis, Johannes Deleu, Lucas Sterckx, Klim Zaporojets, Thomas Demeester, and Chris Develder. 2019. Predicting suicide risk from online postings in Reddit the UGent-IDLab submission to the CLPysch 2019 shared task a. In *Proceedings of the Sixth Workshop on Computational Linguistics and Clinical Psychology*, pages 158–161, Minneapolis, Minnesota. Association for Computational Linguistics.

Carlos Busso, Murtaza Bulut, Chi-Chun Lee, Abe Kazemzadeh, Emily Mower, Samuel Kim, Jeannette N Chang, Sungbok Lee, and Shrikanth S Narayanan. 2008. Iemocap: Interactive emotional dyadic motion capture database. *Language resources and evaluation*, 42(4):335–359.

Stevie Chancellor, Michael L Birnbaum, Eric D Caine, Vincent MB Silenzio, and Munmun De Choudhury. 2019. A taxonomy of ethical tensions in inferring mental health states from social media. In *Proceedings of the conference on fairness, accountability, and transparency*, pages 79–88.

Lushi Chen, Abeer Aldayel, Nikolay Bogoychev, and Tao Gong. 2019. Similar minds post alike: Assessment of suicide risk using a hybrid model. In *Proceedings of the Sixth Workshop on Computational Linguistics and Clinical Psychology*, pages 152–157.

Glen Coppersmith, Ryan Leary, Patrick Crutchley, and Alex Fine. 2018. Natural language processing of social media as screening for suicide risk. *Biomedical informatics insights*, 10:1178222618792860.

Glen Coppersmith, Kim Ngo, Ryan Leary, and Anthony Wood. 2016. Exploratory analysis of social media prior to a suicide attempt. In *Proceedings of the third workshop on computational linguistics and clinical psychology*, pages 106–117.

Munmun De Choudhury, Emre Kiciman, Mark Dredze, Glen Coppersmith, and Mrinal Kumar. 2016. Discovering shifts to suicidal ideation from mental health content in social media. In *Proceedings of the 2016 CHI conference on human factors in computing systems*, pages 2098–2110.

Atefeh Farzindar and Diana Inkpen. 2017. Natural language processing for social media. *Synthesis Lectures on Human Language Technologies*, 10(2):1–195.

Avi Gamoran, Yonatan Kaplan, Almog Simchon, and Michael Gilead. 2021. Using psychologically-informed priors for suicide prediction in the clpsych 2021 shared task. In *Proceedings of the Seventh Workshop on Computational Linguistics and Clinical Psychology: Improving Access*, pages 103–109.

Sujatha Das Gollapalli, Guilherme Augusto Zagatti, and See Kiong Ng. 2021. Suicide risk prediction by tracking self-harm aspects in tweets: Nus-ids at the clpsych 2021 shared task. In *Proceedings of the Seventh Workshop on Computational Linguistics and Clinical Psychology: Improving Access*, pages 93–98.

Shaoxiong Ji, Xue Li, Zi Huang, and Erik Cambria. 2021. Suicidal ideation and mental disorder detection with attentive relation networks. *Neural Computing and Applications*, pages 1–11.

Shaoxiong Ji, Shirui Pan, Xue Li, Erik Cambria, Guodong Long, and Zi Huang. 2020. Suicidal ideation detection: A review of machine learning methods and applications. *IEEE Transactions on Computational Social Systems*, 8(1):214–226.

Sangmoon Kim, Ryan Thibodeau, and Randall S Jorgensen. 2011. Shame, guilt, and depressive symptoms: a meta-analytic review. *Psychological bulletin*, 137(1):68.

Taewoon Kim and Piek Vossen. 2021. Emoberta: Speaker-aware emotion recognition in conversation with roberta. *arXiv preprint arXiv:2108.12009*.

Zhenzhong Lan, Mingda Chen, Sebastian Goodman, Kevin Gimpel, Piyush Sharma, and Radu Soricut. 2019. Albert: A lite bert for self-supervised learning of language representations. In *International Conference on Learning Representations*.

Ellen E Lee, John Torous, Munmun De Choudhury, Colin A Depp, Sarah A Graham, Ho-Cheol Kim, Martin P Paulus, John H Krystal, and Dilip V Jeste. 2021. Artificial intelligence for mental health care: clinical applications, barriers, facilitators, and artificial wisdom. *Biological Psychiatry: Cognitive Neuroscience and Neuroimaging*, 6(9):856–864.

Tsung-Yi Lin, Priya Goyal, Ross Girshick, Kaiming He, and Piotr Dollár. 2017. Focal loss for dense object detection. In *Proceedings of the IEEE international conference on computer vision*, pages 2980–2988.

Yinhan Liu, Myle Ott, Naman Goyal, Jingfei Du, Mandar Joshi, Danqi Chen, Omer Levy, Mike Lewis, Luke Zettlemoyer, and Veselin Stoyanov. 2019. Roberta: A robustly optimized bert pretraining approach. *arXiv preprint arXiv:1907.11692*.

David E. Losada and Fabio Crestani. 2016. A test collection for research on depression and language use. In *Experimental IR Meets Multilinguality, Multimodality, and Interaction - 7th International Conference of the CLEF Association, CLEF 2016, Évora, Portugal, September 5-8, 2016, Proceedings*, volume 9822 of *Lecture Notes in Computer Science*, pages 28–39. Springer.

David E. Losada, Fabio Crestani, and Javier Parapar. 2020. Overview of erisk 2020: Early risk prediction on the internet. In *Experimental IR Meets Multilinguality, Multimodality, and Interaction - 11th International Conference of the CLEF Association, CLEF 2020, Thessaloniki, Greece, September 22-25, 2020, Proceedings*, volume 12260 of *Lecture Notes in Computer Science*, pages 272–287. Springer.

Sean MacAvaney, Anjali Mittu, Glen Coppersmith, Jeff Leintz, and Philip Resnik. 2021. Community-level research on suicidality prediction in a secure environment: Overview of the CLPsych 2021 shared task. In *Proceedings of the Seventh Workshop on Computational Linguistics and Clinical Psychology: Improving Access*, pages 70–80, Online. Association for Computational Linguistics.

Saif Mohammad, Felipe Bravo-Marquez, Mohammad Salameh, and Svetlana Kiritchenko. 2018. SemEval-2018 task 1: Affect in tweets. In *Proceedings of The 12th International Workshop on Semantic Evaluation,* pages 1–17, New Orleans, Louisiana. Association for Computational Linguistics.

Dat Quoc Nguyen, Thanh Vu, and Anh Tuan Nguyen. 2020. BERTweet: A pre-trained language model for English Tweets. In *Proceedings of the 2020 Conference on Empirical Methods in Natural Language Processing: System Demonstrations*, pages 9–14.

Bridianne O'dea, Stephen Wan, Philip J Batterham, Alison L Calear, Cecile Paris, and Helen Christensen. 2015. Detecting suicidality on twitter. *Internet Interventions*, 2(2):183–188.

James W Pennebaker, Martha E Francis, and Roger J Booth. 2001. Linguistic inquiry and word count: Liwc 2001. *Mahway: Lawrence Erlbaum Associates*, 71(2001):2001.

Jeffrey Pennington, Richard Socher, and Christopher D Manning. 2014. Glove: Global vectors for word representation. In *Proceedings of the 2014 conference on empirical methods in natural language processing (EMNLP)*, pages 1532–1543.

Soujanya Poria, Devamanyu Hazarika, Navonil Majumder, Gautam Naik, Erik Cambria, and Rada Mihalcea. 2019. MELD: A multimodal multi-party dataset for emotion recognition in conversations. In *Proceedings of the 57th Annual Meeting of the Association for Computational Linguistics*, pages 527–536, Florence, Italy. Association for Computational Linguistics.

Nils Reimers and Iryna Gurevych. 2019. Sentence-bert: Sentence embeddings using siamese bert-networks. In *Proceedings of the 2019 Conference on Empirical Methods in Natural Language Processing*. Association for Computational Linguistics.

Philip Resnik, April Foreman, Michelle Kuchuk, Katherine Musacchio Schafer, and Beau Pinkham. 2021. Naturally occurring language as a source of evidence in suicide prevention. *Suicide and Life-Threatening Behavior*, 51(1):88–96.

Victor Sanh, Lysandre Debut, Julien Chaumond, and Thomas Wolf. 2019. Distilbert, a distilled version of bert: smaller, faster, cheaper and lighter. *ArXiv*, abs/1910.01108.

Elvis Saravia, Hsien-Chi Toby Liu, Yen-Hao Huang, Junlin Wu, and Yi-Shin Chen. 2018. CARER: Contextualized affect representations for emotion recognition. In *Proceedings of the 2018 Conference on Empirical Methods in Natural Language Processing*, pages 3687–3697, Brussels, Belgium. Association for Computational Linguistics.

Ramit Sawhney, Harshit Joshi, Lucie Flek, and Rajiv Shah. 2021. Phase: Learning emotional phase-aware representations for suicide ideation detection on social media. In *Proceedings of the 16th Conference of the European Chapter of the Association for Computational Linguistics: Main Volume*, pages 2415–2428.

Ramit Sawhney, Harshit Joshi, Saumya Gandhi, and Rajiv Shah. 2020a. A time-aware transformer based model for suicide ideation detection on social media. In *Proceedings of the 2020 conference on empirical methods in natural language processing (EMNLP)*, pages 7685–7697.

Ramit Sawhney, Harshit Joshi, Saumya Gandhi, and Rajiv Ratn Shah. 2020b. A time-aware transformer based model for suicide ideation detection on social media. In *Proceedings of the 2020 Conference on Empirical Methods in Natural Language Processing (EMNLP)*, pages 7685–7697, Online. Association for Computational Linguistics.

Ramit Sawhney, Prachi Manchanda, Puneet Mathur, Rajiv Shah, and Raj Singh. 2018. Exploring and learning suicidal ideation connotations on social media with deep learning. In *Proceedings of the 9th workshop on computational approaches to subjectivity, sentiment and social media analysis*, pages 167–175.

Han-Chin Shing, Suraj Nair, Ayah Zirikly, Meir Friedenberg, Hal Daumé III, and Philip Resnik. 2018. Expert, crowdsourced, and machine assessment of suicide risk via online postings. In *Proceedings of the Fifth Workshop on Computational Linguistics and Clinical Psychology: From Keyboard to Clinic*, pages 25–36, New Orleans, LA. Association for Computational Linguistics.

Michael Mesfin Tadesse, Hongfei Lin, Bo Xu, and Liang Yang. 2019. Detection of suicide ideation in social media forums using deep learning. *Algorithms*, 13(1):7.

Adam Tsakalidis, Jenny Chim, Iman Munire Bilal, Ayah Zirikly, Dana Atzil-Slonim, Federico Nanni, Philip Resnik, Manas Gaur, Kaushik Roy, Becky Inkster, Jeff Leintz, and Maria Liakata. 2022a. Overview of the CLPsych 2022 shared task: Capturing moments of change in longitudinal user posts. In *Proceedings of the Eighth Workshop on Computational Linguistics and Clinical Psychology: Mental Health in the Face of Change*.

Adam Tsakalidis, Federico Nanni, Anthony Hills, Jenny Chim, Jiayu Song, and Maria Liakata. 2022b. Identifying moments of change from longitudinal user text. In *Proceedings of the 60th Annual Meeting of the Association for Computational Linguistics (Volume 1: Long Papers)*, pages 4647–4660, Dublin, Ireland. Association for Computational Linguistics.

Wenhui Wang, Furu Wei, Li Dong, Hangbo Bao, Nan Yang, and Ming Zhou. 2020. Minilm: Deep self-attention distillation for task-agnostic compression of pre-trained transformers. *Advances in Neural Information Processing Systems*, 33:5776–5788.

Ayah Zirikly, Philip Resnik, Özlem Uzuner, and Kristy Hollingshead. 2019a. CLPsych 2019 shared task: Predicting the degree of suicide risk in Reddit posts.

In *Proceedings of the Sixth Workshop on Computational Linguistics and Clinical Psychology*, pages 24–33, Minneapolis, Minnesota. Association for Computational Linguistics.

Ayah Zirikly, Philip Resnik, Ozlem Uzuner, and Kristy Hollingshead. 2019b. Clpsych 2019 shared task: Predicting the degree of suicide risk in reddit posts. In *Proceedings of the sixth workshop on computational linguistics and clinical psychology*, pages 24–33.

Detecting Moments of Change and Suicidal Risks in Longitudinal User Texts Using Multi-task Learning

Tayyaba Azim*, Loitongbam Gyanendro Singh*, Stuart E. Middleton
School of Electronics and Computer Science,
University of Southampton, Southampton, UK
{ta7g21, gsl1r22, sem03}@soton.ac.uk

Abstract

This work describes the classification system proposed for the Computational Linguistics and Clinical Psychology (CLPsych) Shared Task 2022. We propose the use of multitask learning approach with a bidirectional long-short term memory (Bi-LSTM) model for predicting changes in user's mood (Task A) and their suicidal risk level (Task B). The two classification tasks have been solved independently or in an augmented way previously, where the output of one task is leveraged for learning another task, however this work proposes an 'all-in-one' framework that jointly learns the related mental health tasks. Our experimental results (ranked top for task A) suggest that the proposed multi-task framework outperforms the alternative single-task frameworks submitted to the challenge and evaluated via the timeline based and coverage based performance metrics shared by the organisers. We also assess the potential of using various types of feature embedding schemes that could prove useful in initialising the Bi-LSTM model for better multitask learning in the mental health domain.

1 Introduction

Mental illness has greatly affected a vast majority of world's population due to COVID-19 and its resulting economic recession. According to the world health organisation (WHO), global prevalence of anxiety and depression has increased by a massive 25% raising concerns about providing mental health and psychosocial support to the population as a COVID-19 response plan[1]. Many social media platforms have risen to the challenge by offering space to online users to self report their mental health issues, receive counselling support and resolve their mental health issues. This activity has

Table 1: Statistics of the training data set provided for the CLPSych Shared Task 2022.

Moments of Change (Task A)				
Data Set Attributes	None (O)	Escalation (IE)	Switch (IS)	Total
No. of Users	147	87	118	352
No. of Posts	4043	773	327	5143
Avg. No. of Users per post	27.50	8.88	2.77	–
Avg. No. of Words Per Post	75.33	231.82	214.085	–

Suicidal Risk Levels (Task B)				
Data Set Attributes	Low	Moderate	Severe	Total
No. of Users	14	87	103	204
Avg No. of Timelines	1.42	2.17	1.60	–

resulted in two research trends: (1) the surge in development of machine learning algorithms that can automatically detect mental health issues from the language used in social media platforms and (2) the development of new and better diagnostic measures and mental health monitoring tools suitable for the clinical community. Most of the research tasks revolve around classifying individuals on the basis of suicide risk or having a mental health condition (Chancellor and De Choudhury, 2020), however a few have thought of monitoring individual's mood and mental health in real time (Tsakalidis et al., 2022a,b). Despite the growing interest in this interdisciplinary space, there are challenges regarding the availability, use and validity of mental health data gathered from social media platforms and decisions drawn from it.

This paper describes our work identifying moments of change in user's mood (Task A) and suicidal risk level (Task B) in the CLPsych Shared Task 2022. We have experimented with several different sentence and word embedding techniques to draw semantically meaningful features for initialising the multitask sequential model. The model utilised for sequential representation of data is Bidirectional Long Short-Term Memory (Bi-LSTM) (Balikas et al., 2017), trained jointly for multiple tasks (Task A and Task B). The multi task outputs determine

*Equal contributions.

[1]https://www.who.int/news/item/02-03-2022-covid-19-pandemic-triggers-25-increase-in-prevalence-of-anxiety-and-depression-worldwide (Accessed on 25.5.2022)

Proceedings of the Eighth Workshop on Computational Linguistics and Clinical Psychology, pages 213 - 218
July 15, 2022

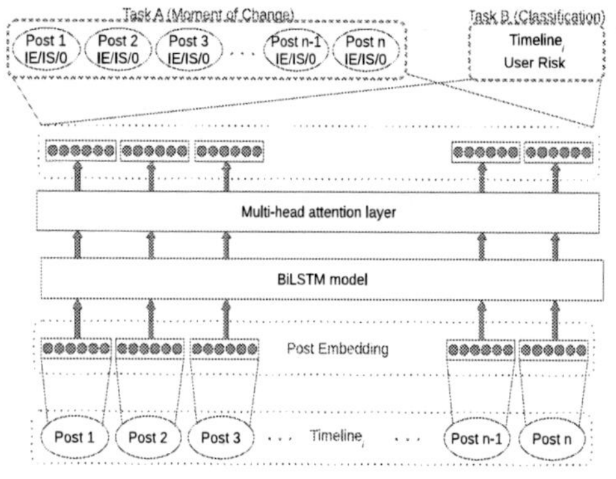

Figure 1: A high-level architecture of the proposed multi-task model for determining moments of change and suicidal risk of users in a particular timeline.

the moments of change in user's mood as well as assess the level of suicidal risk from their posts.

2 Shared Task and Data Set

We have participated in two tasks introduced by the organisers: The first task (*Task A*) is to predict the changes in user's mood over time based on the linguistic content gathered from their posting activity shared on online social media platforms. This is a post-level sequential classification task that aims to detect those sub-periods where a user's mood deviates from their baseline mood. Sequence of an individual's posts over a time span of two months is collected for this shared task (Losada and Crestani, 2016; Losada et al.). The progression in user's mood is categorised as follows: (1) Switch *(IS)*, which signifies a sudden change in user's mood, (2) Escalation *(IE)*, which denotes a gradual shift in user's mood and (3) None *(O)*, denoting no change in user's mood over time. The mood shifting is graded on a scale from positive to negative. This information is further used for *Task B* where user's suicidal risk level is predicted as *Low*, *Moderate* and *Severe* based on the longitudinal mood changes of the user(Shing et al., 2018; Zirikly et al., 2019). The class distribution of the data for each of these labels is shown in Table 1. In order to tackle data imbalance issues, the 'No Risk' and 'Low Risk' label instances were merged and represented as 'Low Risk' examples in the data set for Task B. The task participants were required to sign data use agreements and abide by ethical practice during the competition.

3 Methodology

This section demonstrates the stages involved in developing the proposed multi-task model for determining moments of change in mood and user's suicidal risk determined through a sequence of posts in user's timeline. Figure 1 shows the high-level model architecture for both the tasks.

3.1 Text Preprocessing

The content of the user posts go through several preprocessing steps, including removing stopwords and normalizing keywords (converting to lowercase, removing URL links). Furthermore, the username[2] present in the post is replaced with *@user* to anonymize the mentioned user.

3.2 Semantic Embedding of User Posts

After preprocessing, the user posts are represented using off-the-shelf pre-trained embedding methods to capture the user post's semantics. The pre-trained embedding methods represent the semantics of the posts using fastText word embedding (Bojanowski et al., 2016) and Sentence-BERT (SBERT) (Reimers and Gurevych, 2019). Each post P_i with n tokens can be represented using the pre-trained fastText (FT) word embedding[3] by simply averaging the semantic embeddings of the words present in the post, i.e.,

$$P_i^{FT} = \frac{1}{n} \sum_{pi=1}^{n} \mathbf{w}_{pi}, \mathbf{w}_{pi} \in \mathbb{R}^{300} \tag{1}$$

Recently, the RoBERTa model has yielded much better results in recognizing emotions than other transformer variants such as BERT, XLNet, DistillBERT, and ELECTRA (Cortiz, 2021). Therefore, the RoBERTa-based natural language inference pre-trained model (*'nlirobertalarge'*) is used in addition to fastText embedding to represent the post representation P_i, i.e., $P_i^{SBERT} \in \mathbb{R}^{1024}$.

In order to understand the emotional expressions in text, user's posts are further classified using pre-trained RoBERTa-base model[4] trained on $\widetilde{58}$ million tweets from the TweetEval benchmark (Barbieri et al., 2020) for six different tasks: *emoji*,

[2]Tokens starting with @ symbol.

[3]Pre-trained word embeddings obtained from the Wikipedia corpus `https://dl.fbaipublicfiles.com/fastText/vectors-english/wiki-news-300d-1M.vec.zip`

[4]`https://huggingface.co/cardiffnlp/twitter-roberta-base-sentiment`

emotion, *hate*, *irony*, *offensive*, and *sentiment*. The emoji classification task has 20 categories; emotion classification has four; hate, irony, and offensive classification tasks each have two categories; and sentiment classification tasks have three categories. The task-specific scores therefore represent an additional 33 dimensional feature vector to differentiate each user's posts based on the task-specific scores. The post P_i can be represented by aggregating the scores of task-specific pre-trained model ($Scores_t$):

$$P_i^{Score} = \forall_{t \in T} Concat(Scores_t(P_i)), P_i^{Score} \in \mathbb{R}^{33} \quad (2)$$

where T is the set of six tasks, i.e., *emoji*, *emotion*, *hate*, *irony*, *offensive*, and *sentiment* and $Concat$ represents the score concatenation for all six tasks.

3.3 Multi-Task Model

A user can post *n* number of posts in a particular timeline t_{ij} ranging from time i to j. The objective of the proposed multi-task model is to predict the moments of change (either IE, IS, O) in the user's posts (Task A) and also classify the suicidal risk of the users (Task B) given the sequence of posts in a particular timeline t_{ij}.

3.3.1 Moments of Change Classification

The problem of predicting the moments of change in the user's mood can be viewed as a sequence tagging problem. The learning model predicts the changes in user's mood for each post sequentially, given the sequence of posts in a timeline. This study proposes to use the bidirectional LSTM (Bi-LSTM) (Zhang et al., 2015) model to capture the sequential information of the user posts in a timeline. The Bi-LSTM model generates dense representation for each post, encoding the sequential information of neighbouring posts in both directions, i.e., the user's previous and subsequent posts. Specifically, the Bi-LSTM model encodes the post sequence representation by concatenating the outputs of two LSTMs, namely LSTM-forward ($LSTM_f$) and LSTM-backward ($LSTM_b$) models. $LSTM_f$ processes the post sequence from left to right, i.e., $P_1, P_2, \ldots, P_n$, whereas $LSTM_b$ process the post sequence from right to left, i.e., $P_n, P_{n-1}, \ldots, P_1$. Each LSTM model consists of a repeating unit called memory cell, which takes current post, previous hidden state, previous cell state ($\mathbf{x}_t$, $\mathbf{h}_{t-1}$, $\mathbf{c}_{t-1}$) as input and produces current hidden state and cell state information i.e. $(\mathbf{h}_t, \mathbf{c}_t) = LSTM(\mathbf{x}_t, \mathbf{h}_{t-1}, \mathbf{c}_{t-1})$. Therefore, the encoded representation of post P_t is generated by concatenating the hidden state information obtained by $LSTM_f$ and $LSTM_b$ outputs, i.e., $\mathbf{h}_t = (\mathbf{h}_t^{(f)} \oplus \mathbf{h}_t^{(b)})$. The whole timeline t_{ij} can be represented as $\mathbf{H_{ij}} \in \mathbb{R}^{n \times d}$ where $\mathbf{H_{ij}}$ is a matrix of the encoded representation of n posts of d dimension[5]. The encoded representation of the posts is then fed to the softmax classifier to predict the user's moment of change, i.e.,

$$\mathbf{Task_a} = Softmax(\mathbf{H_{ij}} \mathbf{W_a}^T + \mathbf{B}) \quad (3)$$

where $\mathbf{W_a} \in \mathbb{R}^{c \times d}$ is the neural weight parameters, c is the three classes of the moment of change categories (i.e., IE, IS, O), and $\mathbf{B} \in \mathbb{R}^{n \times c}$ being the neural network biases.

3.3.2 User Suicidal Risk classification

Using the same encoded representation, the user's risk can be classified for the timeline t_{ij} by flattening the matrix $\mathbf{H_{ij}}$, i.e.,

$$\mathbf{Task_b} = Softmax(flatten(\mathbf{H_{ij}}) \mathbf{W_b}^T + \mathbf{B}) \quad (4)$$

where $\mathbf{W_b} \in \mathbb{R}^{r \times nd}$ is the neural weight parameters, r being the number of user risk categories in Task B, and $\mathbf{B} \in \mathbb{R}^r$ being the neural network biases. Further, the user risk can also be classified by embedding an attention layer over the encoded representation $\mathbf{H_{ij}}$ before $flattening$ to give more attention to the user's post that influences the user risk classification decision. The output of the multi-head attention[6] layers generate an attention weighted encoded representation $\mathbf{H_{ij}^a}$ of the same dimension as $\mathbf{H_{ij}}$. The impact of adding an attention layer could be seen in the tables discussed in the results section.

The current model classifies the user's suicidal risk for a particular timeline t_{ij}. However, a user can have multiple timelines $\{t_{ab}, t_{cd}, ..., t_{ij}\}$, hence the user risk must be classified considering all the timelines. Since the model classifies the user risk for each timeline, i.e., $\{\mathbf{Task_{b}}_{ab}, \mathbf{Task_{b}}_{cd}, ..., \mathbf{Task_{b}}_{ij}\}$, the final user risk $\mathbf{Task_b}$ is classified using a simple heuristic approach. The user risk is classified based on the prediction of the user's risk severity level across the timelines, i.e., if the model has predicted *Severe* in one of the timeline then the user is considered to be at *Severe* risk; followed by

[5] 100 LSTM units

[6] 8 heads

Moderate-level and *Low*-level risks. We can also consider a voting method to classify the user risk based on the output of all timelines. This study consider evaluating the user risk classification based on the heuristics of risk severity level.

4 Experiment and Results

In this work we have used two different combinations of feature embeddings for the user posts. For the ease of reference, we consider naming them as P_{emb} which is the concatenation of fastText and SBERT embeddings ($P^{FT} \oplus P^{SBERT}$) and $P_{task-emb}$ which is the concatenation of fastText, SBERT, and task-specific scores of the post ($P^{FT} \oplus P^{SBERT} \oplus P^{Score}$).

Models: The efficacy of the proposed model is evaluated on two types of post embeddings (P_{emb}, $P_{task-emb}$), with and without the attention layers. This, eventually leads us to four different types of models for evaluation: (i) Multitask: model using P_{emb}, (ii) Multitask-score: model using $P_{task-emb}$, (iii) Multitask-attn: model with attention layer using P_{emb}, and (iv) Multitask-attn-score: model with attention layer using $P_{task-emb}$.

Evaluation Metrics: The performance of the proposed model is evaluated using metrics Precision, Recall and F1 Score on the validation set. We also show window-based and coverage-based evaluation metrics (Tsakalidis et al., 2022b) used by the CLPsych organisers to assess the models' performance on the test set.[7]

Implementation Details: The train data set is initially divided into train, validation and test sets using the ratio: 60:20:20, to optimise the Bi-LSTM parameters. Once the parameters are fine tuned using the validation set, we retrain the model again with 80% of the train data and test it on 20% of the unseen test data. After fine tuning, the Bi-LSTM model is trained for 50 epochs with 64 batch size. The maximum sequence length for Bi-LSTM is set to the maximum number of posts in a timeline, i.e., 122 (see Appendix). Categorical cross-entropy loss and Adam optimizer are used to train the model on both the tasks. The implementation was done using Keras API and is available at `https://github.com/stuartemiddleton/uos_clpsych`.

Table 2 shows the results of our model on the validation set using the standard evaluation metrics. Here, the precision, recall and F1 score values obtained for each class (see Table 5 in the appendix) have been macro-averaged by calculating the arithmetic mean of individual classes' precision, recall and F1 scores. We have used the macro-averaging score to treat all the classes equally for evaluating the overall performance of the classifier regardless of their support values (i.e the actual occurrences of the class in the data set). Here, we observe that **Multitask-attn-score** model gives more promising results as compared to other enlisted models on both tasks. This behaviour is reflected in the classification results on test data too (Table 3), where **Multitask-attn-score** has outperformed the remaining feature embeddings with the Bi-LSTM model as well as the baseline state of the art results (Tsakalidis et al., 2022a). From the model outcomes in Table 2 and 3, one could also see the impact of introducing attention layers in the Bi-LSTM model. Adding attention layers in Bi-LSTM model has helped accuracy for both the tasks.

Table 2: Performance of the proposed models on Task A and Task B using the validation set.

Model	Moments of Change			Suicidal Risk Levels		
	P	R	F1	P	R	F1
Multitask-attn-score	0.674	**0.800**	**0.724**	**0.415**	**0.397**	0.382
Multitask-score	**0.680**	0.760	0.713	0.355	0.331	0.334
Multitask	0.582	0.717	0.629	0.352	0.327	0.335
Multitask-attn	0.663	0.697	0.676	0.408	0.378	**0.388**

Given the class imbalance in the data set with majority of post instances belonging to the *None(0)* class and minority instances to *Escalation (IE)* and *Switch (IS)* classes, we see the performance is compromised and biased towards the majority class, i.e. the classifier is more sensitive to detecting the majority class (*None(0)*) patterns precisely but less sensitive to detecting the minority class patterns $\{IE, IS\}$. See Table 5 in the Appendix to observe the precision, recall and F1 score of the models for each individual class in task A. The data distribution is skewed for task B too, thus influencing its results for majority and minority classes shown in Table 6. Overall, on the validation set, the proposed models have shown better recall rate than precision, revealing low false negatives than the false positives.

Table 3 and Table 4 show the performance of our proposed approach with variable feature encoding schemes and attention layers in Bi-LSTM on the test set provided by the CLPsych Shared Task 2022. The entire train set comprising of 5143 posts is used to train the proposed model with the

[7]Please note that the data set is imbalanced and therefore intuitions just drawn from only accuracy are not correct.

Table 3: Performance of the proposed models on Task A using the test set. The traditional post-level, coverage-based, and timeline-based evaluation metrics based on precision (P), recall(R) and F1 score are shown for comparison and analysis with the baseline results (Tsakalidis et al., 2022a).

	Post-level Metrics (Macro average)			Coverage-based Metrics (Macro average)		Timeline-level Metrics (Macro average) Window-1		Window-2		Window-3	
	P	R	F1	P	R	P	R	P	R	P	R
Multitask-attn-score	**0.689**	**0.625**	**0.649**	0.506	**0.503**	**0.676**	**0.652**	0.693	**0.670**	0.708	**0.686**
Multitask-score	0.677	0.595	0.625	0.492	0.467	0.662	0.605	0.681	0.622	0.695	0.632
Multitask	0.680	0.579	0.607	**0.521**	0.441	0.674	0.592	**0.695**	0.608	**0.723**	0.623
Majority	NaN	0.333	0.280	NaN	0.141	NaN	0.333	NaN	0.333	NaN	0.333
TFIDF-LR	0.545	0.495	0.492	0.377	0.424	0.496	0.539	0.505	0.550	0.506	0.551
BERT-TalkLife-Focal	0.522	0.386	0.380	0.260	0.204	0.582	0.392	0.608	0.405	0.608	0.405

Table 4: Performance of the proposed model on Task B using the test set. The precision (P), recall (R), and F1-scores (F1) shown are macro-averaged over the user's risk categories and compared to the baseline results (Tsakalidis et al., 2022a).

	(Macro average)			(Micro average)		
	P	R	F1	P	R	F1
Multitask-attn-score	**0.618**	**0.427**	**0.451**	**0.482**	**0.469**	**0.438**
Majority	0.156	0.333	0.212	0.219	0.468	0.299
TFIDF-LR	0.302	0.338	0.295	0.412	0.468	0.406

optimal parameters defined above and then its efficacy is assessed on the given test set comprising of 1052 posts. On the test set, the proposed models have shown higher precision than recall. When compared to the baseline results, our submission on task A has topped the ranking results on the test set, whereas for task B we stood second in the shared task based on the timeline based and coverage based metrics.

5 Conclusion and Future Work

This work demonstrates the power of using various feature embeddings for multi task learning with Bi-LSTM on the CLPsych Shared Task 2022 data set. We have tried several different textual embeddings to represent the content of user's posts. These embeddings are passed on to the Bi-LSTM which is trained to learn two labels jointly. The model has shown to give promising results on the test set when attention layer is incorporated and complete set of feature embeddings (fastText+SBERT+TaskScore) is utilised. On Task A, our team topped the post-level classification problem based on the window based and coverage based statistics, whereas for Task B, we showed second best results in the competition.

In future, we would like to compare our proposed model with other single task learning models trained using separate loss functions. Given the correlation between the shared tasks, multi-task learning is expected to yield good results as shown in this paper, however it will be interesting to explore the underlying user information (e.g. age, gender, etc) that could be explicitly added to support tasks for mental health and suicidal risk prediction. Also in order to mitigate the effects of imbalanced classes, we would like to improve our developed pipeline using resampling techniques.

Ethical Statement

Secure access to the shared task dataset was provided with IRB approval under University of Maryland, College Park protocol 1642625 and approval by the Biomedical and Scientific Research Ethics Committee (BSREC) at the University of Warwick (ethical application reference BSREC 40/19-20).

Acknowledgements

The authors are particularly grateful to the anonymous users of Reddit whose data feature in this year's shared task dataset, to the annotators of the data for Task A, to the clinical experts from Bar-Ilan University who annotated the data for Task B, the American Association of Suicidology, to NORC for creating and administering the secure infrastructure and providing researcher support and to UKRI for providing funding to the CLPsych 2022 shared task organisers. This work was supported by the Engineering and Physical Sciences Research Council (EP/V00784X/1).

References

Rie Kubota Ando and Tong Zhang. 2005. A Framework for Learning Predictive Structures from Multiple Tasks and Unlabeled Data. *Journal of Machine Learning Research*, 6:1817–1853.

Georgios Balikas, Simon Moura, and Massih-Reza Amini. 2017. Multitask Learning for Fine-grained

Twitter Sentiment Analysis. In *Proceedings of the 40th International ACM SIGIR Conference on Research and Development in Information Retrieval*, pages 1005–1008.

Francesco Barbieri, Jose Camacho-Collados, Leonardo Neves, and Luis Espinosa-Anke. 2020. Tweeteval: Unified Benchmark and Comparative Evaluation for Tweet Classification. *arXiv preprint arXiv:2010.12421*.

Piotr Bojanowski, Edouard Grave, Armand Joulin, and Tomas Mikolov. 2016. Enriching Word Vectors with Subword Information. *arXiv preprint arXiv:1607.04606*.

Stevie Chancellor and Munmun De Choudhury. 2020. Methods in Predictive Techniques for Mental Health Status on Social Media: A Critical Review. *NPJ Digital Medicine*, 3(1):1–11.

Diogo Cortiz. 2021. Exploring Transformers in Emotion recognition: A Comparison of BERT, Distillbert, Roberta, Xlnet and Electra. *arXiv preprint arXiv:2104.02041*.

David E. Losada and Fabio Crestani. 2016. A test Collection for Research on Depression and Language Use. In *Experimental IR Meets Multilinguality, Multimodality, and Interaction - 7th International Conference of the CLEF Association, CLEF 2016, Évora, Portugal, September 5-8, 2016, Proceedings*, volume 9822 of *Lecture Notes in Computer Science*, pages 28–39. Springer.

David E. Losada, Fabio Crestani, and Javier Parapar. Overview of erisk 2020: Early Risk Prediction on the Internet. In *Experimental IR Meets Multilinguality, Multimodality, and Interaction - 11th International Conference of the CLEF Association, CLEF 2020, Thessaloniki, Greece, September 22-25, 2020, Proceedings*, Lecture.

Nils Reimers and Iryna Gurevych. 2019. Sentence-BERT: Sentence Embeddings using Siamese BERT-Networks.

Han-Chin Shing, Suraj Nair, Ayah Zirikly, Meir Friedenberg, Hal Daumé III, and Philip Resnik. 2018. Expert, Crowdsourced, and Machine Assessment of Suicide Risk via Online Postings. In *Proceedings of the Fifth Workshop on Computational Linguistics and Clinical Psychology: From Keyboard to Clinic*, pages 25–36, New Orleans, LA. Association for Computational Linguistics.

Adam Tsakalidis, Jenny Chim, Iman Munire Bilal, Ayah Zirikly, Dana Atzil-Slonim, Federico Nanni, Philip Resnik, Manas Gaur, Kaushik Roy, Becky Inkster, Jeff Leintz, and Maria Liakata. 2022a. Overview of the CLPsych 2022 Shared Task: Capturing Moments of Change in Longitudinal User Posts. In *Proceedings of the Eighth Workshop on Computational Linguistics and Clinical Psychology: Mental Health in the Face of Change*.

Adam Tsakalidis, Federico Nanni, Anthony Hills, Jenny Chim, Jiayu Song, and Maria Liakata. 2022b. Identifying Moments of Change from Longitudinal User Text. *arXiv preprint arXiv:2205.05593*.

Shu Zhang, Dequan Zheng, Xinchen Hu, and Ming Yang. 2015. Bidirectional Long Short-term Memory Networks for Relation Classification. In *Proceedings of the 29th Pacific Asia Conference on Language, Information and Computation*, pages 73–78.

Ayah Zirikly, Philip Resnik, Özlem Uzuner, and Kristy Hollingshead. 2019. CLPsych 2019 Shared Task: Predicting the Degree of Suicide Risk in Reddit posts. In *Proceedings of the Sixth Workshop on Computational Linguistics and Clinical Psychology*, pages 24–33, Minneapolis, Minnesota. Association for Computational Linguistics.

Appendices

Tables 5 and 6 show the evaluation metrics by class in Task A and Task B. Figure 2 shows the post distribution in the training set.

Table 5: Performance of the proposed models on Task A using the validation set. The traditional post-level evaluation metrics based on precision (P), recall (R) and F1 score are shown for comparison and analysis.

	Precision			Recall			F1 Score		
	IE	IS	0	IE	IS	0	IE	IS	0
Multitask-attn-score	0.539	0.512	0.971	0.739	0.75	0.909	0.623	0.608	0.939
Multitask-score	0.614	0.485	0.938	0.712	0.68	0.887	0.660	0.566	0.912
Multitask	0.429	0.346	0.970	0.710	0.566	0.873	0.535	0.430	0.919
Multitask-attn	0.677	0.414	0.897	0.630	0.566	0.893	0.653	0.478	0.895

Table 6: Performance of the proposed models on Task B using the validation set. The traditional post-level evaluation metrics based on precision (P), recall (R) and F1 score are shown for comparison and analysis.

	Precision			Recall			F1 Score		
	Severe	Moderate	Low	Severe	Moderate	Low	Severe	Moderate	Low
Multitask-attn-score	0.555	0.500	0.00	0.625	0.357	0.00	0.588	0.416	0.00
Multitask-score	0.588	0.636	0.00	0.666	0.466	0.00	0.625	0.538	0.00
Multitask	0.764	0.300	0.00	0.565	0.428	0.00	0.650	0.352	0.00
Multitask-attn	0.846	0.400	0.000	0.523	0.666	0.00	0.647	0.500	0.000

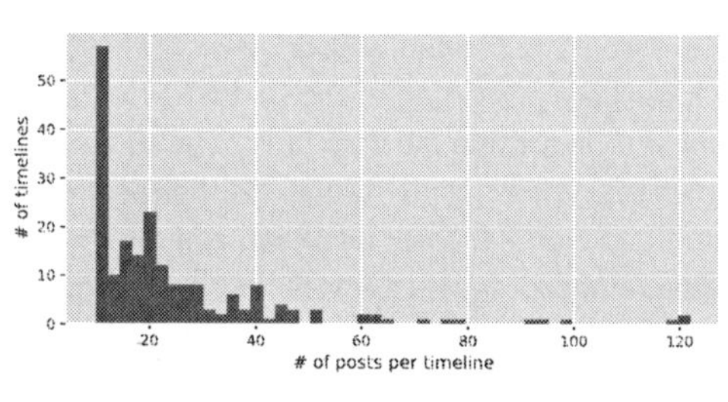

Figure 2: Distribution of number of posts per timeline in the training dataset. The *x-axis* represents the number of posts per timeline and *y-axis* represents the number of timelines having that number of posts. The maximum number of posts in a timeline is 122.

Emotionally-Informed Models for Detecting Moments of Change and Suicide Risk Levels in Longitudinal Social Media Data

Ulya Bayram
Dept. of Electrical & Electronics Engineering,
Çanakkale Onsekiz Mart University
Çanakkale, Turkey
ulya.bayram@comu.edu.tr

Lamia Benhiba
IAD Department, ENSIAS,
Mohammed V University in Rabat
Rabat, Morocco
lamia.benhiba@um5.ac.ma

Abstract

In this shared task, we focus on detecting mental health signals in Reddit users' posts through two main challenges: A) capturing mood changes (anomalies) from the longitudinal set of posts (called timelines), and B) assessing the users' suicide risk-levels. Our approaches leverage emotion recognition on linguistic content by computing emotion/sentiment scores using pre-trained BERTs on users' posts and feeding them to machine learning models, including XGBoost, Bi-LSTM, and logistic regression. For Task-A, we detect longitudinal anomalies using a sequence-to-sequence (seq2seq) autoencoder and capture regions of mood deviations. For Task-B, our two models utilize the BERT emotion/sentiment scores. The first computes emotion bandwidths and merges them with n-gram features, and employs logistic regression to detect users' suicide risk levels. The second model predicts suicide risk on the timeline level using a Bi-LSTM on Task-A results and sentiment scores. Our results outperformed most participating teams and ranked in the top three in Task-A. In Task-B, our methods surpass all others and return the best macro and micro F1 scores.

1 Introduction

Tracking and identifying moments of change in a user's social media longitudinal data could be a possible identifier of their mental health deterioration and be especially useful for those with suicidal ideation (Tsakalidis et al., 2022b). In this 2022 CLPsych shared task, the goal is to tackle two challenges. Task-A aims to identify mood shifts and gradual mood progressions from users' timelines, where each timeline has a list of longitudinal posts from a close time range. Meantime, Task-B aims to detect suicide risk levels of the users. We were allowed to provide three submissions for Task-A and two for Task-B. The second Task-B submission was expected to use the results from Task-A.

The dataset of this shared task is a mixture of three separate datasets: UMD from 2019 CLPsych (Shing et al., 2018; Zirikly et al., 2019), E-Risk with some additional data (Losada and Crestani, 2016; Losada et al., 2020), and a new collection called Reddit-New (Tsakalidis et al., 2022a). The dataset has 255 timelines: 204 in training/51 in the unlabeled test set.

Our team (called WResearch for "Women in Research") decided to use emotionally-informed features for their ability to capture mood changes. In Task-A, we combine a seq2seq autoencoder and machine learning (ML) models to capture moments of change in a user's timeline. Meanwhile, in Task-B, we were partially influenced by the 2021 CLPsych results, which showed that merging long-term posts of a user could capture long-term suicidal ideation (Bayram and Benhiba, 2021; Macavaney et al., 2021). We used the post-level features extracted in Task-A to compute user-level emotion-bandwidth features and concatenated them with statistical n-gram features to detect suicidal risk levels. Additionally, we experimented with a timeline-level prediction model using Bi-LSTM. The success of our results compared to the other teams and the baselines suggest that our emotionally-informed models are advantageous for dealing with the tasks at hand.

2 Methods

The training set in this challenge includes data on users with three suicide risk levels (Severe/Moderate/Low). A user can have multiple timelines, where a timeline is a chronologically ordered sequence of posts. Each post is labeled as IS for switches in mood (sudden mood shifts from positive to negative, or vice versa), IE for mood escalations (gradual mood changes from neutral or positive to a higher positive, or neutral, or negative to a higher negative), or O to represent the baseline (neutral) mood (Tsakalidis et al., 2022b). In

Proceedings of the Eighth Workshop on Computational Linguistics and Clinical Psychology, pages 219 - 225
July 15, 2022 ©2022 Association for Computational Linguistics

the implementations, for machine learning models, Scikit-learn (version 1.0.2) (Pedregosa et al., 2011), for deep learning models, PyTorch (version 1.11.0+cu102) and Keras (version 2.7.0) libraries (Paszke et al., 2019) are used.

2.1 Task A

Feature Extraction: The main set of features used in Task-A is obtained from three pre-trained BERT models. The first model is Bertweet-base-sentiment, trained with SemEval 2017 corpus (around 40k tweets) using a RoBERTa (Pérez et al., 2021). It returns three sentiments $\{Positive, Negative, Neutral\}$ per text. The second model is EmoRoBERTa, trained with 58,000 Reddit comments and returns 28 emotion scores per text (Ghoshal, 2021). The third model is Twitter-roberta-base-emotion (CardiffNLP, 2021), trained on $\tilde{5}8$M tweets and fine-tuned for emotion recognition with the TweetEval benchmark (Barbieri et al., 2020). As shown in Figure 1, we concatenate the sentiment and emotion scores into an emotionally-informed feature vector of length 35 for each post in the data collection.

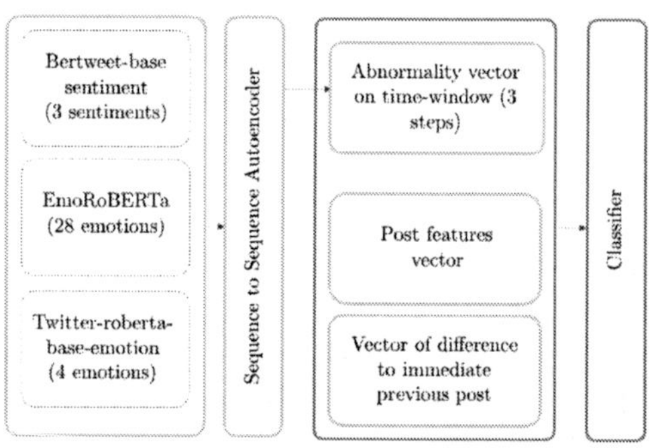

Figure 1: Task A Learning model

Mood Anomaly detection: Before feeding the emotionally-informed features to classifiers, we compute a feature vector that reflects abnormalities in the user-expressed mood based on past behavior. To compute the abnormality vector, we use a seq2seq learning model for multivariate time-series forecasting (Provotar et al., 2019). We generate a series of (t-n) feature vectors for each post at time t, where n is the length of the look-back time window. This input is fed to the autoencoder. We aim to predict the emotionally-informed feature vector of the next step, i.e., the feature vector of the post at t+1. The error margin is thereafter calculated based on the outputs of the autoencoder and the actual emotionally-informed feature vectors. We follow the same methodology as Tran et al. (Tran et al., 2019) to compute the irregularities vector and use it as a proxy for identifying mood anomalies. Upon experimentation, we found that, while the abnormality vector helps detect escalations, it did not succeed for switches. We thus concatenated the emotionally-informed features, window-based abnormality vectors, and a feature vector denoting the emotional difference between a post and the previous one. We implement the seq2seq learning model in Keras with two LSTMs with 100 neurons and a final dense layer with 35 neurons. We use a Learning Rate Scheduler that decreases the learning rate (lr) with a factor of 1e-3 * 0.90 ** lr when the learning stagnates. We train using the Adam optimizer and Huber loss function with a batch size of 16 and early stopping (patience=3).

Classification: We pass the output of the previous step as an input to ML classifiers to predict the label of a post (O, IE, IS). We experiment with three models: a Logistic Regression (LR) [class_weight="balanced", multi_class="multinomial", solver="saga"], XGBoost, and a stacked Ensemble of four classifiers: LR, Random Forest, XGBoost, and Extremely Randomized Trees. Being mindful of the data imbalance, we choose to assign a higher class weight to the minority classes (IE, IS) while reducing the weight of the majority class (O). We apply stratified 10-folds cross-validation and grid-search on the tree-based models (n_estimators=[400, 700, 1000], colsample_bytree=[0.7,0.8], max_depth=[15,20,25], subsample=[0.7,0.8,0.9]) to optimize the hyperparameters and avoid overfitting.

2.2 Task B

In this task, we eliminate all users with suicide risk label N/A from the labeled set, thus work on a three-class classification problem: Low, Moderate, Severe suicide risk detection.

Feature Extraction: For the first submission, we use two types of features. The first feature, n-grams, is selected due to their success in previous suicide risk detection research (Bayram and Benhiba, 2021; Pestian et al., 2020). Our n-gram features consist of unigrams and bigrams ($n \in \{1, 2\}$). To extract them, we perform lowercase conversion and punctuation removal, then use a spaCy library (en_core_web_lg) (Honnibal and Montani, 2017).

As the goal is to obtain user-level suicide risk, we perform the detection on the merged posts per user. However, the leave-one-out cross-validation experiments returned low results on the labeled set, so we decided to use/merge only the posts with "IE" or "IS" labels in training since they contain strong emotions that might be associated with suicidal ideation. In the test set, we merge all posts per person (since they lack IE and IS labels) and obtain the user's suicide risk-level prediction.

The training set provides 5,808 n-gram features. Next, we train an LR to collect feature importance scores for performing feature elimination. Upon applying a leave-one-out cross-validation on the labeled set, also using LR, we exploit classification performance scores from the top features to find the optimal feature subset. Figure 2 shows a peak at top 900 n-gram features, corresponding to 300 top features per class. We save these features and use them as the final features on the test set.

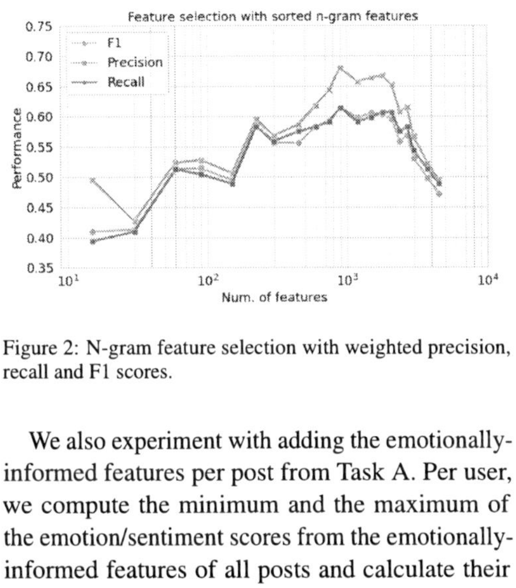

Figure 2: N-gram feature selection with weighted precision, recall and F1 scores.

We also experiment with adding the emotionally-informed features per post from Task A. Per user, we compute the minimum and the maximum of the emotion/sentiment scores from the emotionally-informed features of all posts and calculate their absolute difference. Thus, in the new feature vector, each element reflects the range (bandwidth) of emotions/sentiments of that user. We hypothesize that these bandwidths of emotions/sentiments could help identify suicide risk. Next, we concatenate the n-gram feature vector and the obtained emotion bandwidth vector per user for classification.

Classification: In the first submission of Task-B, we use simple methods that do not require a lot of training data and that can perform multiclass classification: LR (lbfgs, sag, saga, newton-cg solvers), non-linear support vector machines (SVM) (rbf, poly, and sigmoid kernels), random forest (RF), and XGBoost. We obtain leave-one-out results on the training set, where LR with lbfgs solver (weighted F1=0.718) and SVM with the sigmoid kernel (weighted F1=0.710) achieve the best performance, possibly due to their success in handling small datasets (RF's weighted F1=0.433, XGBoost's weighted F1=0.278). Thus, we select LR as the ML model to be used with ngrams+emotional bandwidth features (class_weight="balanced", multi_class = "multinomial", solver="lbfgs", random_state=7, remaining parameters are kept at default values (Pedregosa et al., 2011)).

Timeline-level risk prediction: The second submission for Task-B leverages Task-A's mood change predictions and the emotionally-informed features to predict a user's suicide risk level. Since timelines (longitudinal posts) were obtained around a user's mood change-points during data collection (Tsakalidis et al., 2022b), we predict the suicide risk on the timeline level. As was the case in the first model, we only include posts with IS or IE labels in our training set while also including O labels in the validation and test data. We use a Bi-LSTM to classify the suicide risk in the timeline by exploiting past and future emotional contexts of posts. To aggregate predictions on the user level, we experiment with computing average, majority voting, and argmax on the timeline-level results and select argmax due to its accuracy. The Bi-LSTM model is a gated recurrent unit (GRU) wrapped in a Bi-LSTM, followed by a dropout layer and two dense layers (Dropout_rate=0.1, Dense layer 1: 50-neurons with Relu, Dense layer 2: 3-neurons with softmax, batch_size=16, Rmsprop optimizer, categorical cross-entropy loss, and early-stopping with patience=3).

3 Results

In Tables 1, 2, and 3, we present the test set results of Task-A obtained from three different evaluation techniques. Each table summarizes the results obtained on the three submissions: seq2seq + one of the selected classifiers (i.e., 1=LR, 2=XGBoost, and 3=the Ensemble method). Table 1 shows results at the post-level, while Table 2 and 3 report results on a timeline basis using the coverage metric and the window-based evaluation metric with window size = 3 (more details on the evaluation methods can be found in (Tsakalidis et al., 2022b)).

Table 4 shows results for Task-B where the first

model (1) is the n-grams + emotion bandwidth features with LR classifier, and the second (2) is the Bi-LSTM model.

Table 1: Task-A post-level evaluation for seq2seq+classifier (resp. (1) Logistic Regression (LR), (2) XGBoost, (3) Ensemble). (B1) tf-idf LR and (B2) BERT are baselines. Max & Min results from all CLPsych'22 submissions are also included.

	Sub.	Precision	Recall	F1
IS	1	0.204	**0.512**	0.292
	2	0.362	0.256	**0.300**
	3	**0.478**	0.134	0.209
	B1	0.222	0.024	0.044
	B2	0.091	0.012	0.021
	Max	**0.500**	**0.585**	**0.376**
	Min	0	0	0
IE	1	0.500	**0.625**	0.556
	2	0.646	0.553	**0.596**
	3	0.644	0.505	0.566
	B1	0.569	0.514	0.540
	B2	**0.723**	0.163	0.267
	Max	**0.748**	**0.630**	**0.662**
	Min	0.273	0.029	0.052
O	1	**0.944**	0.726	0.820
	2	0.868	0.929	**0.897**
	3	0.838	**0.953**	0.892
	B1	0.844	0.947	0.893
	B2	0.753	**0.983**	0.853
	Max	**0.954**	**0.968**	**0.910**
	Min	0.729	0.647	0.771
Macro avg	1	0.549	**0.621**	0.556
	2	0.625	0.579	**0.598**
	3	**0.654**	0.531	0.556
	B1	0.545	0.495	0.492
	B2	0.523	0.386	0.380
	Max	**0.689**	**0.625**	**0.649**
	Min	0.354	0.337	0.305

The shared task provided two baselines from the mood change study (Tsakalidis et al., 2022b). The first baseline (B1 in the tables) uses tf-idf features with LR. The second baseline (B2) uses BERT trained with Talklife website posts, treats each post as an instance (i.e., completely ignoring the timeline sequence), and is trained using the alpha-weighted focal loss. We also include the best (Max) and worst (Min) values for each metric obtained by competing submissions to allow better readability of the results. We add an asterisk (*) next to the results when the best performance is achieved by our models.

4 Discussion

In comparison to the submissions of other teams that participated in this shared task (Tsakalidis et al., 2022a), our models achieved the top three macro average F1 scores for Task-A on all three evaluation techniques. Meanwhile, in Task-B, the first model returns the highest micro and macro average F1 scores in Clpysch'22.

Table 2: Task-A coverage evaluation for seq2seq+classifier (resp. (1) Logistic Regression (LR), (2) XGBoost, (3) Ensemble). (B1) tf-idf LR and (B2) BERT are baselines. Max & Min results from all CLPsych'22 submissions are also included.

	Sub.	Precision	Recall	F1
IS	1	0.211	**0.563**	0.307
	2	0.406	0.318	**0.357**
	3	**0.511**	0.199	0.287
	B1	0.111	0.008	0.0148
	B2	0.025	0.007	0.011
	Max	**0.517**	**0.575**	**0.390**
	Min	0	0	0
IE	1	0.198	0.406	0.266
	2	**0.307**	**0.467***	**0.370**
	3	0.302	0.452	0.362
	B1	0.284	**0.504**	0.363
	B2	0.226	0.094	0.132
	Max	**0.369**	**0.467***	**0.406**
	Min	0.070	0.050	0.073
O	1	0.520	0.537	0.528
	2	0.703	0.725	0.713
	3	0.675	0.700	0.687
	B1	**0.738**	**0.762**	**0.750**
	B2	0.529	0.513	0.521
	Max	0.720	0.737	0.728
	Min	0.510	0.486	0.503
Macro avg	1	0.310	0.502	0.383
	2	0.472	**0.503***	**0.487**
	3	**0.496**	0.450	0.472
	B1	0.378	0.425	0.400
	B2	0.260	0.204	0.229
	Max	**0.521**	**0.503***	**0.504**
	Min	0.220	0.186	0.202

Task-A: In the post-level, the seq2seq + XGBoost achieves robust performance by balancing between precision and recall. It outperforms the baseline methods on all macro-average evaluation metrics and achieves second best F1 scores in all categories (e.g., IE, IS, O, average). At the timeline level, the coverage metric demonstrates the ability of a model to capture regions of change. In this respect, the seq2seq + XGBoost strikes a balance between precision and recall again, and performs second best on the macro-average F1. In the window-based evaluation the seq2seq + LR achieves the third highest F1 performance overall and renders the best macro-average recall. The ensemble method achieves the best precision on the IS class but tends to over-predict, as demonstrated by its low coverage recall. Experimenting with various look-back time windows can provide more insight on the rationale behind the results.

Table 3: Task-A window-based (window size = 3) evaluation for seq2seq+classifier (resp. (1) Logistic Regression (LR), (2) XGBoost, (3) Ensemble). (B1) tf-idf LR and (B2) BERT are baselines. Max & Min results from all CLPsych'22 submissions are also included.

	Sub.	Precision	Recall	F1
IS	1	0.368	**0.814**	**0.507**
	2	0.525	0.372	0.435
	3	**0.711***	0.224	0.341
	B1	0.167	0.008	0.015
	B2	0.450	0.065	0.113
	Max	**0.711***	**0.872**	**0.512**
	Min	0.200	0.004	0.008
IE	1	0.429	**0.748**	0.545
	2	0.566	0.620	0.592
	3	0.570	0.622	0.595
	B1	0.477	0.675	0.559
	B2	0.612	0.158	0.251
	Max	**0.630**	**0.773**	**0.637**
	Min	0.371	0.010	0.168
O	1	**0.956***	0.755	0.844
	2	0.881	0.968	**0.923***
	3	0.854	**0.992**	0.918
	B1	0.875	0.973	**0.922**
	B2	0.762	**0.995**	0.863
	Max	**0.956***	**0.996**	**0.923***
	Min	0.769	0.610	0.742
Macro avg	1	0.584	**0.773***	**0.665**
	2	0.657	0.653	0.655
	3	**0.712**	0.613	0.658
	B1	0.506	0.552	0.528
	B2	0.608	0.406	0.487
	Max	**0.723**	**0.773***	**0.697**
	Min	0.523	0.399	0.455

Task-B: In Task-B, we wanted to contrast the user suicide risk prediction performance when obtained at the user level in the n-grams+emotion bandwidth+LR model and at the timeline level using the Bi-LSTM model. The latter leverages Task A's moments-of-change results to help predict the user's suicide risk level.

The n-grams+emotion bandwidth+LR model returns the best F1 scores in CLPsych'22 based on micro and macro average metrics in Table 4, showing the viability of our approach. This outcome is also a good inspiration for future suicide risk detection studies in which mood change labels are available or obtainable.

The Bi-LSTM model was built on the premise that emotional context from past and future posts, including the moments of change, would allow better inference of the timeline's suicide risk level. While the model is slightly better than the baseline, we suppose that it might have rendered better results had it been trained on timeline-level rather than user-level labels. In an attempt to err on the side of safety, we chose argmax for aggregation. However, it biased the model in favor of moderate and severe risk levels. Other aggregation methods will be explored in the future to help address the prediction of low-level suicide risk.

Table 4: Task-B evaluation for the models (1) n-gram+emotion bandwith+Logistic Regression (LR), and (2) Bi-LSTM. A baseline (B1) tf-idf LR, and Max & Min results from all CLPsych'22 submissions are also included.

Level	Sub.	Precision	Recall	F1
Low	1	**0.200**	**0.333**	**0.250**
	2	0	0	0
	B1	0	0	0
	Max	**1**	**0.667**	**0.500**
	Min	0	0	0
Moderate	1	0.533	0.571	**0.552**
	2	**0.545**	0.429	0.480
	B1	0.429	0.214	0.286
	Max	**0.625**	**0.714**	**0.588**
	Min	0.250	0.071	0.111
Severe	1	**0.667***	0.533	0.593
	2	0.556	0.667	**0.606**
	B1	0.480	**0.800**	**0.600**
	Max	**0.667***	**0.867**	**0.684**
	Min	0.478	0.467	0.500
Macro avg	1	0.467	**0.479***	**0.465***
	2	0.367	0.365	0.362
	B1	0.303	0.338	0.295
	Max	**0.618**	**0.479***	**0.465***
	Min	0.306	0.365	0.298
Micro avg	1	**0.565***	0.531	**0.543***
	2	0.499	0.500	0.494
	B1	0.412	0.469	0.406
	Max	**0.565***	**0.562**	**0.543***
	Min	0.359	0.344	0.315

5 Conclusion

In this shared task, we tackled two problems: capturing mood changes from timelines of posts of Reddit users and detecting their suicide risk levels. The results reveal that our methods performed the highest macro and micro F1 scores in suicide risk-level detection and performed in the top three in mood-change detection. Our models can inspire future research for accurately detecting abrupt mood changes among social media users. These models also might shed light on users' suicide risk levels, thus enabling early mental-health intervention to prevent suicidal events.

Ethical Statement

Secure access to the shared task dataset was provided with IRB approval under University of Maryland, College Park protocol 1642625 and approval by the Biomedical and Scientific Research Ethics

Committee (BSREC) at the University of Warwick (ethical application reference BSREC 40/19-20). Before being granted access, we signed a Non-Disclosure Agreement (NDA) and a Data Enclave Use Agreement (DUA).

Acknowledgements

The authors are particularly grateful to the anonymous users of Reddit whose data feature in this year's shared task dataset, to the annotators of the data for Task A, to the clinical experts from Bar-Ilan University who annotated the data for Task B, the American Association of Suicidology, to NORC for creating and administering the secure infrastructure and providing researcher support and to UKRI for providing funding to the CLPsych 2022 shared task organisers.

References

Francesco Barbieri, Jose Camacho-Collados, Leonardo Neves, and Luis Espinosa-Anke. 2020. Tweeteval: Unified benchmark and comparative evaluation for tweet classification. *arXiv preprint arXiv:2010.12421*.

Ulya Bayram and Lamia Benhiba. 2021. Determining a person's suicide risk by voting on the short-term history of tweets for the clpsych 2021 shared task. In *Proceedings of the Seventh Workshop on Computational Linguistics and Clinical Psychology: Improving Access*, pages 81–86.

CardiffNLP. 2021. Twitter-roBERTa-base for emotion recognition.

Arpan Ghoshal. 2021. EmoRoBERTa.

Matthew Honnibal and Ines Montani. 2017. spaCy 2: Natural language understanding with Bloom embeddings, convolutional neural networks and incremental parsing. To appear.

David E. Losada and Fabio Crestani. 2016. A test collection for research on depression and language use. In *Experimental IR Meets Multilinguality, Multimodality, and Interaction - 7th International Conference of the CLEF Association, CLEF 2016, Évora, Portugal, September 5-8, 2016, Proceedings*, volume 9822 of *Lecture Notes in Computer Science*, pages 28–39. Springer.

David E. Losada, Fabio Crestani, and Javier Parapar. 2020. Overview of erisk 2020: Early risk prediction on the internet. In *Experimental IR Meets Multilinguality, Multimodality, and Interaction - 11th International Conference of the CLEF Association, CLEF 2020, Thessaloniki, Greece, September 22-25, 2020, Proceedings*, volume 12260 of *Lecture Notes in Computer Science*, pages 272–287. Springer.

Sean Macavaney, Anjali Mittu, Glen Coppersmith, Jeff Leintz, and Philip Resnik. 2021. Community-level research on suicidality prediction in a secure environment: Overview of the clpsych 2021 shared task. In *Proceedings of the Seventh Workshop on Computational Linguistics and Clinical Psychology: Improving Access*, pages 70–80.

Adam Paszke, Sam Gross, Francisco Massa, Adam Lerer, James Bradbury, Gregory Chanan, Trevor Killeen, Zeming Lin, Natalia Gimelshein, Luca Antiga, Alban Desmaison, Andreas Kopf, Edward Yang, Zachary DeVito, Martin Raison, Alykhan Tejani, Sasank Chilamkurthy, Benoit Steiner, Lu Fang, Junjie Bai, and Soumith Chintala. 2019. Pytorch: An imperative style, high-performance deep learning library. In *Advances in Neural Information Processing Systems 32*, pages 8024–8035. Curran Associates, Inc.

F. Pedregosa, G. Varoquaux, A. Gramfort, V. Michel, B. Thirion, O. Grisel, M. Blondel, P. Prettenhofer, R. Weiss, V. Dubourg, J. Vanderplas, A. Passos, D. Cournapeau, M. Brucher, M. Perrot, and E. Duchesnay. 2011. Scikit-learn: Machine learning in Python. *Journal of Machine Learning Research*, 12:2825–2830.

John Pestian, Daniel Santel, Michael Sorter, Ulya Bayram, Brian Connolly, Tracy Glauser, Melissa DelBello, Suzanne Tamang, and Kevin Cohen. 2020. A machine learning approach to identifying changes in suicidal language. *Suicide and Life-Threatening Behavior*, 50(5):939–947.

Oleksandr I Provotar, Yaroslav M Linder, and Maksym M Veres. 2019. Unsupervised anomaly detection in time series using lstm-based autoencoders. In *2019 IEEE International Conference on Advanced Trends in Information Theory (ATIT)*, pages 513–517. IEEE.

Juan Manuel Pérez, Juan Carlos Giudici, and Franco Luque. 2021. pysentimiento: A python toolkit for sentiment analysis and socialnlp tasks.

Han-Chin Shing, Suraj Nair, Ayah Zirikly, Meir Friedenberg, Hal Daumé III, and Philip Resnik. 2018. Expert, crowdsourced, and machine assessment of suicide risk via online postings. In *Proceedings of the fifth workshop on computational linguistics and clinical psychology: from keyboard to clinic*, pages 25–36.

Kim Phuc Tran, Huu Du Nguyen, and Sébastien Thomassey. 2019. Anomaly detection using long short term memory networks and its applications in supply chain management. *IFAC-PapersOnLine*, 52(13):2408–2412.

Adam Tsakalidis, Jenny Chim, Iman Munire Bilal, Ayah Zirikly, Dana Atzil-Slonim, Federico Nanni, Philip Resnik, Manas Gaur, Kaushik Roy, Becky Inkster, Jeff Leintz, and Maria Liakata. 2022a. Overview of the CLPsych 2022 shared task: Capturing moments of change in longitudinal user posts. In *Proceedings of the Eighth Workshop on Computational Linguistics and Clinical Psychology: Mental Health in the Face of Change*.

Adam Tsakalidis, Federico Nanni, Anthony Hills, Jenny Chim, Jiayu Song, and Maria Liakata. 2022b. Identifying moments of change from longitudinal user text. In *Proceedings of the 60th Annual Meeting of the Association for Computational Linguistics (Volume 1: Long Papers)*, pages 4647–4660, Dublin, Ireland. Association for Computational Linguistics.

Ayah Zirikly, Philip Resnik, Ozlem Uzuner, and Kristy Hollingshead. 2019. Clpsych 2019 shared task: Predicting the degree of suicide risk in reddit posts. In *Proceedings of the sixth workshop on computational linguistics and clinical psychology*, pages 24–33.

Exploring transformers and time lag features for predicting changes in mood over time

John Culnan, Damian Y. Romero Diaz, Steven Bethard
University of Arizona
Tucson, AZ, USA
{jmculnan, damianiji, bethard}@email.arizona.edu

Abstract

This paper presents transformer-based models created for the CLPsych 2022 shared task. Using posts from Reddit users over a period of time, we aim to predict changes in mood from post to post. We test models that preserve timeline information through explicit ordering of posts as well as those that do not order posts but preserve features on the length of time between a user's posts. We find that a model with temporal information may provide slight benefits over the same model without such information, although a RoBERTa transformer model provides enough information to make similar predictions without custom-encoded time information.

1 Introduction

With the ubiquity of data online come opportunities for studying and providing support to individuals and communities. For example, a user's posts on Reddit fora may reveal information about that user's emotional state over time (Tsakalidis et al., 2022b). Additionally, these tasks may seek to make early predictions about mental states, allowing for prompt intervention when needed (Losada et al., 2020). This work represents one such attempt as part of the 2022 CLPsych shared task (Tsakalidis et al., 2022a),[1] using a transformer-based architecture to make predictions about changes in Reddit user moods over time. We demonstrate how state-of-the-art transformer models like RoBERTa (Liu et al., 2019) provide predictions of changes in mood that are difficult to improve upon with custom features or sequential architectures.

2 Related work

Previous work has used social media to examine the ability of neural networks to make predictions about depression (Losada and Crestani, 2016), suicidality (Benton et al., 2017), and related mental health disorders (Wongkoblap et al., 2017). Losada et al. (2020) introduce a task where participants attempt to make early identifications of depression from social media, finding that further improvements needed to be made before such models could successfully be used in a clinical setting.

Work on predicting temporal shifts in language use has frequently focused on lexical-semantic changes over time, with only recent research focusing on the impacts of temporally-aware language models on downstream tasks (Dhingra et al., 2022; Rosin et al., 2022). For example, in a span prediction task, Dhingra et al. (2022) used a simple string representation of the year when texts were first created to finetune T5 language generation models. They found that adding the year as a prefix to the input aided learning of seen facts, improving performance on predictions of future events.

Tsakalidis et al. (2022b) identify individuals' changes in mental health over time. This temporal dimension can be helpful in monitoring clinical outcomes and it can also help online platform moderators prioritize interventions depending on an individual's vulnerability at a certain moment in time. They provide strong baseline models for this task, including both timeline-based models and timeline-agnostic models, finding that BERT-based models outperform their remaining systems. Thus, it is reasonable to assume that finetuning existing language models using the time information available in social media posts can help detect changes in mental health.

3 Approach

We examine both timeline-agnostic models, which accept single data points in random order and timeline-preserving models, which require the order of posts in each timeline to be maintained. Timeline-preserving models are expected to be most successful, as the dataset includes labels such as *switch in mood* (IS) that require information

[1] https://clpsych.org/sharedtask2022/

Proceedings of the Eighth Workshop on Computational Linguistics and Clinical Psychology, pages 226 - 231
July 15, 2022 ©2022 Association for Computational Linguistics

from past data points to predict the label of the present data point. We incorporate such information both through sequence models such as LSTMs (Hochreiter and Schmidhuber, 1997) that encode and preserve information from previous data points to make predictions, as well as through explicit custom features representing the time between data points, which we refer to as time lag features. We choose RoBERTa as a base for our models, as (Tsakalidis et al., 2022b) find BERT-based models perform well on this task, and RoBERTa models frequently outperform BERT in practice (Liu et al., 2019).

3.1 Time lag features

To get the time lags between posts, we calculate the time difference (in seconds) between the current post and the previous post. Formally, for each post i we define:

$$lag(i) = time(i) - time(i-1)$$

For the first post in every timeline, we use the absolute mean time for that timeline:

$$lag(0) = \frac{1}{N} \sum_{i}^{N} lag(i)$$

If the time stamp of post i or $i-1$ is missing from the data, we define $lag(i)$ as one day in seconds.

3.2 Timeline-agnostic models

For timeline-agnostic models, we consider three ways to represent posts:

RoBERTa Feed the tokens of the post through RoBERTa (Liu et al., 2019) and produce the contextualized embedding of the first token in the post, the pseudo-token `[CLS]`.

RoBERTa-lin Obtain the RoBERTa representation as above, and feed it through linear layers to reduce its dimensionality to 50, then increase it to 100.

RoBERTa-lin-lag Obtain the RoBERTa-lin representation as above, feed it through a linear layer to reduce its dimensionality to 50, concatenate it with a single item representing the amount of time between the user's previous post and current post, then feed it through a linear layer to increase its dimensionality to 100.

Post representations were fed into a final linear layer to reduce dimensionality to 3, the number of labels in the task. All of the models above examine points in isolation, although the time lag feature adds information about the previous data point.

3.3 Timeline-preserving models

For our timeline-preserving models, we consider two approaches. Due to the memory constraints of the computing system, we restricted the amount of context considered to three posts: the post of interest plus the previous two posts. We consider two ways to represent timelines.

RoBERTa-pre2-lin Concatenate the three posts, with posts represented as in the timeline-agnostic RoBERTa-lin-lag, and feed this concatenated vector through a linear layer to reduce its dimensionality to 100.

RoBERTa-pre2-lstm Feed the three posts through an LSTM, with posts represented as in the timeline-agnostic RoBERTa-lin-lag, and take the final LSTM state as the representation.

Timeline representations were fed into a final linear layer to reduce dimensionality to 3, the number of labels in the task. These models examine whether the explicit inclusion of information from previous posts increases prediction accuracy, as might be expected since the task requires knowledge of a user's previous moods to correctly predict labels like *switch in mood* (IS).

4 Data

The data used in this work are those selected for the CLPsych 2022 shared task (Tsakalidis et al., 2022a) and drawn from the UMD Reddit Suicidality Dataset Version 2 (Shing et al., 2018; Zirikly et al., 2019) with Queen Mary University of London annotations, Reddit-New, a new dataset created from posts by Reddit users who posted on mental-health related subreddits and annotated for suicidality and moments of change (Tsakalidis et al., 2022a,b), and the eRisk Dataset (Losada and Crestani, 2016; Losada et al., 2020). These data consist of timelines of Reddit posts by a series of users, selected based on individuals who participated in subreddit fora related to mental health. Data points are labeled for moments of change–changes in mood over time–and individual users'

Class	Train	Dev	Test
IS	178	41	82
IE	323	177	208
O	2012	991	762
Total	2513	1209	1053

Table 1: Number of items in each partition of the dataset

overall suicide risk; here, we focus solely on predictions of changes in mood over time. In order to access the data, each member of this team signed a data usage agreement and an NDA due to the sensitive nature of this data.

The data consists of a total of 4775 posts, broken down as shown in table 1. Each post in the dataset was labeled for one of three mood classes: an *escalation in mood* (IE), a *switch in mood* (IS), or *no change from the baseline* (O). An escalation label may refer to a change from positive to more positive or from negative to more negative. A switch may likewise refer to either a change from negative to positive or from positive to negative. These labels indicate changes from previous posts, which suggests that information about timelines may be crucial for making successful predictions.

5 Implementation details

RoBERTa (Liu et al., 2019) models were based on Hugging Face's `roberta-base`[2] and were trained via the pytorch (Paszke et al., 2017) version of the `RobertaForSequenceClassification` class using cross-entropy loss. RoBERTa is not frozen for any of the architectures; linear layers, LSTMs, etc. were trained alongside the RoBERTa weights.

For timeline-agnostic models, we randomized the order of all posts in the training data. For timeline-preserving models, we randomized the order of the timelines in the training data but preserved the order of individual items within each timeline. For timeline-preserving models, when fewer than two previous posts were available (e.g., at the beginning of a timeline), padded masked posts were fed instead but were not used to update model parameters.

6 Model selection on the development set

We used the development data to experiment with the various architectures we considered, with the goal of selecting the best models to evaluate on the test set. Each of the models described in section 3 was evaluated using the development partition.

Table 2 presents the performance of each model at the post level and at the timeline level. This table shows that adding linear or sequential structure on top of RoBERTa does not improve performance. The baseline timeline-agnostic RoBERTa model outperforms all other models overall and in most individual evaluation metrics, with the second-best performance belonging to RoBERTa-lin-lag, the timeline-agnostic RoBERTa model with the time lag feature concatenated to the RoBERTa representation.

The timeline-preserving models (RoBERTa-pre2-lin and RoBERTa-pre2-lstm) showed much worse performance than the timeline-agnostic models, although the RoBERTa-pre2-lin model that concatenated the three posts and fed them through linear layers did perform best for precision in the switch class and recall in the no-change class. Still, its overall performance as measured by macro F1 was much worse than the timeline-agnostic models. The timeline-sensitive model using LSTM layers performed even worse, making predictions only for the no-change majority class.

Based on these overall trends, two models were selected to make predictions on the test set: the RoBERTa baseline model and RoBERTa-lin-lag. We engaged in small-scale focused parameter tuning using the development set, selecting the best dropout and learning rate for each model from among a limited set of items. For the RoBERTa baseline model, tuning selected a hidden dropout rate of 0.2, a learning rate of 3e-5, and a minibatch size of 8. For the RoBERTa-lin-lag model, tuning selected a hidden dropout rate of 0.2, a learning rate of 5e-6, and a minibatch size of 8. Other parameters used the default values from `roberta-base`.

7 Results on the test set

The two selected models were used to make predictions on the held-out test set. The results in table 3 demonstrate that the models perform similarly. Macro-average at both the post-level and coverage-based evaluations are within .003 of each other. The main tradeoff is that the baseline RoBERTa model is better at *escalation in mood* (IE), while RoBERTa-lin-lag is better at *switch in mood* (IS). This is reasonable, given that only RoBERTa-lin-lag knows anything about the timeline, and the IS

[2] `https://huggingface.co/roberta-base`

Model	post-level evaluation												coverage-based metrics							
	IS			IE			O			macro-avg			IS		IE		O		macro-avg	
	P	R	F1	P	R	F1	P	R	F1	P	R	F1	CP	CR	CP	CR	CP	CR	CP	CR
RoBERTa	.099	**.293**	.148	**.667**	**.587**	**.624**	**.943**	.887	.914	**.570**	**.589**	**.579**	**.234**	**.257**	**.357**	.418	**.674**	**.708**	**.422**	**.461**
R-lin	—	.000	.000	.522	.542	.532	.896	.925	.910	.473	.489	.481	—	.000	.304	**.492**	.656	.697	.320	.396
R-lin-lag	.127	.220	**.161**	.552	.452	.497	.918	.920	**.919**	.532	.531	.531	.207	.201	.296	.376	.653	.703	.385	.427
R-pre2-lin	**.154**	.049	.074	.247	.102	.144	.826	**.936**	.878	.409	.362	.384	.107	.014	.166	.051	.501	.451	.258	.172
R-pre2-lstm	—	.000	.000	—	.000	.000	.820	1.00	.901	.273	.333	.300	—	.000	—	.000	.523	.481	.174	.160

Table 2: Performance of trained models on development partition comprising 30% of training dataset. Models are as defined in section 3 except that 'RoBERTa' is abbreviated as 'R' for space. The best performance on each metric is shown in **bold**.

Model	post-level evaluation												coverage-based metrics							
	IS			IE			O			macro-avg			IS		IE		O		macro-avg	
	P	R	F1	P	R	F1	P	R	F1	P	R	F1	CP	CR	CP	CR	CP	CR	CP	CR
Majority	—	.000	.000	—	.000	.000	.724	**1.000**	.840	—	.333	.280	—	.000	—	.000	.489	. 426	—	.142
LogReg	.222	.024	.044	.569	**.514**	**.540**	.844	.948	**.893**	**.545**	.495	.492	.111	.008	**.284**	**.504**	**.738**	**.762**	.378	**.425**
BERT (f)	.091	.012	.022	**.723**	.163	.267	.754	.983	.853	.523	.386	.380	.025	.007	.226	.094	.529	.513	.260	.204
RoBERTa	.142	**.220**	.172	.561	.423	.482	**.872**	.879	.876	.525	**.507**	**.510**	.158	.211	.230	.332	.657	.695	.348	.413
R-lin-lag	**.267**	.195	**.225**	.476	.375	.419	.841	.913	.875	.527	.495	.507	**.368**	**.248**	.202	.285	.682	.716	**.418**	.416

Table 3: Results of our best models on the test partition (RoBERTa, R-lin-lag), with a majority class classifier (Majority), logistic regression model with TF-IDF features (LogReg), and BERT with focal loss (BERT (f)), all from Tsakalidis et al. (2022b). The best performing model on each evaluation metric is shown in **bold**.

label requires knowledge of past mood.

These models were compared to baseline models from Tsakalidis et al. (2022b) whose results were provided to participants in the shared task. These models are **Majority**, where only the majority (O) class is selected, **LogReg**, where a logistic regression model is trained on TF-IDF features, and **BERT (f)**, a BERT model trained on focal loss.

Compared to the baseline models, our models show mixed results. Both of our models outperform the baselines on recall and F1 for the IS class, with our R-lin-lag also outperforming all baselines on precision for the IS class. For the IE class, however, they are beaten by the logistic regression model. Our RoBERTa model outperforms the baseline for precision on the O class, though not recall or F1. Overall, our models have the best macro average F1 at the post level. For coverage-based metrics, our models again perform best for the IS class, although the logistic regression baseline again outperforms our models for the IE class, as well as for the O class and macro average recall. Our model with time lag features performs the best for macro-average precision.

8 Qualitative error analysis

To better understand the types of posts that prove problematic for our models, we examine a small subset of the prediction errors produced on the development partition of the dataset. We specifically focus on times when our model produced a *no-change* (O) label while the gold label was IS or IE, as well as the reverse. Due to the sensitive nature of this data, we do not provide specific examples, but rather describe trends in the data.

The following are situations in which our models tends to predict a change in mood but no change should be predicted:

1. The user discusses difficult situations from the past but is not in a current state of distress.
2. The user comments on another person's depression, anxiety or desperation.
3. The user worries about potential scenarios that would cause him or her significant mental anguish but that have not come to pass.

Our models tend to predict IE or IS labels whenever a post discusses unhealthy or dangerous scenarios, such as traumatic experiences, or when someone expresses desperation. However, as seen in items 1 to 3, this does not always provide accurate results. This type of error accounted for the majority of incorrect predictions in the sample of the development set examined.

Additionally, our models occasionally predict that a post does not show a change in mood when it is an example of a IS or IE. In these cases, errors are typically due to:

4. Largely neutral texts containing one strong indicator of distress.

5. Posts with a title but no content.

6. Short posts containing both positive and distressed content.

With these items, errors are typically caused by posts where there are both positive and negative elements, or where there is one very negative element that is limited to a minority of the post. Additionally, in cases where there is no content in the post, our models always make a prediction of no change; however, there are cases where the post title alone reveals that an IS or IE label is more appropriate.

9 Conclusion

We examined the ability of timeline-agnostic and timeline-preserving transformer-based models to make predictions about changes in mood over time, finding that more complex models do not necessarily improve predictions. We furthermore experiment with a custom feature representing the length of time between one post and another, demonstrating that this may provide some support to more complex models. Overall, we see that this remains a difficult task, suggesting that further improvements need to be made to methods of longitudinal mood modeling.

Ethics Statement

This work was completed following the ACL code of ethics. Secure access to the shared task dataset was provided with IRB approval under University of Maryland, College Park protocol 1642625 and approval by the Biomedical and Scientific Research Ethics Committee (BSREC) at the University of Warwick (ethical application reference BSREC 40/19-20). Each member of the team completed an NDA and a data usage agreement, ensuring that the data used for this work would not be misused, distributed, or otherwise compromised. Due to the sensitive nature of this data, dataset creators and the shared task organizers were de-identified, and each team member agreed to make no attempt to identify the individuals whose data was used for the task. We completed our analyses using the secure NORC Data Enclave to further protect the data.[3]

Acknowledgements

We are particularly grateful to the anonymous users of Reddit whose data feature in this year's shared task dataset, the annotators of the data for the post-level annotations, the American Association of Suicidology, NORC, who created and administered the secure infrastructure and provided researcher support, and UKRI, who provided funding to the CLPsych 2022 shared task organisers. We would also like to thank the organizers of the CLPsych 2022 shared task, who made this work possible, and the two anonymous reviewers who helped improve the quality of this paper.

References

Adrian Benton, Margaret Mitchell, and Dirk Hovy. 2017. Multitask learning for mental health conditions with limited social media data. In *Proceedings of the 15th Conference of the European Chapter of the Association for Computational Linguistics: Volume 1, Long Papers*, pages 152–162, Valencia, Spain. Association for Computational Linguistics.

Bhuwan Dhingra, Jeremy R. Cole, Julian Martin Eisenschlos, Daniel Gillick, Jacob Eisenstein, and William W. Cohen. 2022. Time-Aware Language Models as Temporal Knowledge Bases. *Transactions of the Association for Computational Linguistics*, 10:257–273.

Sepp Hochreiter and Jürgen Schmidhuber. 1997. Long short-term memory. *Neural computation*, 9(8):1735–1780.

Yinhan Liu, Myle Ott, Naman Goyal, Jingfei Du, Mandar Joshi, Danqi Chen, Omer Levy, Mike Lewis, Luke Zettlemoyer, and Veselin Stoyanov. 2019. Roberta: A robustly optimized bert pretraining approach. *arXiv preprint arXiv:1907.11692*.

David E Losada and Fabio Crestani. 2016. A test collection for research on depression and language use. In *International Conference of the Cross-Language Evaluation Forum for European Languages*, pages 28–39. Springer.

David E Losada, Fabio Crestani, and Javier Parapar. 2020. Overview of erisk 2020: Early risk prediction on the internet. *CLEF (Working Notes)*.

[3]https://www.norc.org/Research/Capabilities/Pages/data-enclave.aspx.

Adam Paszke, Sam Gross, Soumith Chintala, Gregory Chanan, Edward Yang, Zachary DeVito, Zeming Lin, Alban Desmaison, Luca Antiga, and Adam Lerer. 2017. Automatic differentiation in pytorch.

Guy D. Rosin, Ido Guy, and Kira Radinsky. 2022. Time masking for temporal language models. In *Proceedings of the Fifteenth ACM International Conference on Web Search and Data Mining*, WSDM '22, page 833–841, New York, NY, USA. Association for Computing Machinery.

Han-Chin Shing, Suraj Nair, Ayah Zirikly, Meir Friedenberg, Hal Daumé III, and Philip Resnik. 2018. Expert, crowdsourced, and machine assessment of suicide risk via online postings. In *Proceedings of the Fifth Workshop on Computational Linguistics and Clinical Psychology: From Keyboard to Clinic*, pages 25–36, New Orleans, LA. Association for Computational Linguistics.

Adam Tsakalidis, Jenny Chim, Iman Munire Bilal, Ayah Zirikly, Dana Atzil-Slonim, Federico Nanni, Philip Resnik, Manas Gaur, Kaushik Roy, Becky Inkster, Jeff Leintz, and Maria Liakata. 2022a. Overview of the CLPsych 2022 shared task: Capturing moments of change in longitudinal user posts. In *Proceedings of the Eighth Workshop on Computational Linguistics and Clinical Psychology: Mental Health in the Face of Change*.

Adam Tsakalidis, Federico Nanni, Anthony Hills, Jenny Chim, Jiayu Song, and Maria Liakata. 2022b. Identifying moments of change from longitudinal user text. In *Proceedings of the 60th Annual Meeting of the Association for Computational Linguistics (Volume 1: Long Papers)*, pages 4647–4660, Dublin, Ireland. Association for Computational Linguistics.

Akkapon Wongkoblap, Miguel A Vadillo, and Vasa Curcin. 2017. Researching mental health disorders in the era of social media: systematic review. *Journal of medical Internet research*, 19(6):e228.

Ayah Zirikly, Philip Resnik, Özlem Uzuner, and Kristy Hollingshead. 2019. CLPsych 2019 shared task: Predicting the degree of suicide risk in Reddit posts. In *Proceedings of the Sixth Workshop on Computational Linguistics and Clinical Psychology*, pages 24–33, Minneapolis, Minnesota. Association for Computational Linguistics.

Multi-Task Learning to Capture Changes in Mood Over Time

Prasadith Buddhitha, Ahmed Husseini Orabi, Mahmoud Husseini Orabi, Diana Inkpen
School of Electrical Engineering and Computer Science
University of Ottawa, Ottawa, ON, K1N 6N5, Canada
{pgamaara, ahuss045, mhuss092, diana.inkpen}@uottawa.ca

Abstract

This paper investigates the impact of using Multi-Task Learning (MTL) to predict mood changes over time for each individual (social media user). The presented models were developed as a part of the Computational Linguistics and Clinical Psychology (CLPsych) 2022 shared task. Given the limited number of Reddit social media users, as well as their posts, we decided to experiment with different multi-task learning architectures to identify to what extent knowledge can be shared among similar tasks. Due to class imbalance at both post and user levels and to accommodate task alignment, we randomly sampled an equal number of instances from the respective classes and performed ensemble learning to reduce prediction variance. Faced with several constraints, we managed to produce competitive results that could provide insights into the use of multi-task learning to identify mood changes over time and suicide ideation risk.

1 Introduction

For many countries, suicide has been a formidable challenge, where 1.3% of world deaths in 2019 were due to suicides. Of the committed suicides, most of them were by individuals before reaching their fifties and in countries with low to middle income (World Health Organization, 2021). Considering these factors, it is of utmost importance for any institution responsible for the mental health of the population to early detect users susceptible to suicide ideation and mental disorders. In recent years, social media has become an integral part of the everyday life of many. According to Schimmele et al. (2021), more than 25% of users aged between 15 and 64 have shared their personal information (e.g., pictures, videos, text-based posts) publicly. This data rich with personal information opens the pathway for many research opportunities.

The importance of using social media data to detect users susceptible to suicide ideation (MacAvaney et al., 2021; Zirikly et al., 2019) and mental disorders (Coppersmith et al., 2015b; Milne et al., 2016) was demonstrated throughout the CLPsych workshop series. When analyzing research, including publications in the CLPsych workshop series, we could see that in comparison to traditional machine learning methods (Cohan et al., 2016; Coppersmith et al., 2015a; Jamil et al., 2017; Schwartz et al., 2014), recent research has focused more on using deep learning architectures (Husseini Orabi et al., 2018; Kshirsagar et al., 2017; Mohammadi et al., 2019) that considerably reduce the time and effort required for feature engineering. However, researchers have continued using traditional machine learning methods to predict individuals susceptible to mental disorders and suicide ideation, which could be due to the lack of large sets of annotated data (e.g., Hauser et al. (2019) or to the requirement of explainability (e.g., Saha et al. (2022)).

In this paper, we describe the experiments conducted using deep learning methods, specifically with multi-task learning, to predict a user's mood change over time (i.e., either a switch or an escalation in the mood) and also the suicide ideation risk level where a selected user can be categorized into one of the following risk categories: low, moderate, or severe. The main reason for selecting multi-task learning is to leverage its capabilities of sharing knowledge between related but different tasks that could potentially alleviate the negative impact of having a small number of training instances. For example, we identified the negative impact of having a limited number of data points during model training, specifically when using deep learning architectures where different regularization methods were used to reduce model overfitting and increase the model's generalizability. When predicting suicide ideation risk level, we used an additional dataset from Cohan et al. (2018), named the Self-Reported Mental Health Diagnoses (SMHD) dataset, which

Proceedings of the Eighth Workshop on Computational Linguistics and Clinical Psychology, pages 232 - 238
July 15, 2022

consists of users who have self-declared mental disorders. Similar to Gamaarachchige (2021), which demonstrates the impact different mental disorders have on suicide ideation detection (i.e., whether an individual is susceptible to suicide ideation or not), we investigate the impact mental disorders have on different suicide ideation risk levels (i.e., low, moderate, or severe).

2 Task and Data

The CLPsych 2022 shared task consisted of two subtasks (Tsakalidis et al., 2022a). The first task was to identify a user's mood change over time (Tsakalidis et al., 2022b), and the second task was to predict the level of suicidality risk for an individual (Shing et al., 2018; Zirikly et al., 2019). Then, when predicting the suicidality risk, the participants were encouraged to discover if there is any relationship between the mood change over time and the risk of suicidality. The dataset provided to the task participants consisted of users and their posts extracted from the Reddit social media platform. Apart from 3,089 posts distributed across 139 timelines posted by 83 users, the rest of the users were sampled from the University of Maryland Reddit Suicidality Dataset (Shing et al., 2018; Zirikly et al., 2019) and the eRisk dataset (Losada and Crestani, 2016; Losada et al., 2020). The combined dataset statistics are shown in table 1.

# Timelines	Users	Posts
204	149	5,063

Table 1: CLPsych 2022 training data.

For both tasks, we combined the text fields "title" and "content", and after several preliminary preprocessing steps, we identified 5,143 posts where the majority of the posts were categorized as "None". The distribution of the classes in the training dataset is shown in table 2, for Task A.

Label	Count	Percentage
None (O)	4,043	79%
Escalation (IE)	773	15%
Switch (IS)	327	6%

Table 2: Post-level class distribution.

For "Task B", we grouped all the posts per user and trained our proposed deep learning model on a dataset that contained 127 users distributed among three classes as shown in table 3.

Label (risk level)	Count	Percentage
Low	11	9%
Moderate	55	43%
Severe	61	48%

Table 3: Suicide ideation risk level class distribution.

A considerable class imbalance can be identified when analyzing the class distribution for both tasks. Such imbalance could adversely impact model training and its generalizability, which we will discuss more in the following sections.

For "Task B" only, we used an external dataset from Cohan et al. (2018), that contains users who have self-declared single or multiple mental disorders. Based on the conclusions derived by Gamaarachchige (2021), we sampled users who have self-declared Post-Traumatic Stress Disorder (PTSD), Anxiety, and Bipolar Disorder as the input for the mental illness detection task within the MTL environment. However, we did not include any users who have self-declared other mental illnesses due to time constraints.

Macro-averaged precision, recall, and F1-score were used as evaluation metrics at the post, timeline, and coverage levels.

To generalize and reduce input noise, we performed the following preprocessing steps: lowercased the texts, kept only a selected set of stop words, removed most of the non-alphanumeric characters, removed numbers and URLs, and expanded contractions.

3 Methodology

As mentioned before, we based our experiments on multi-task learning and specifically an architecture using a combination of soft and hard parameter sharing. Multi-task learning allows related tasks to share representations (Caruana, 1997), and based on how parameters are being shared, can be categorized into two types of architectures, which are hard parameter sharing and soft parameter sharing (Ruder, 2017). Each task will share model weights in hard parameter sharing, and features unique to individual tasks will be extracted through the task-specific layers. Even though model weights are not shared between layers in soft parameter sharing, the parameters are regularized between the layers to discover similarities. We used a custom loss function that combines "categorical cross-entropy", "mean squared error", and "cosine similarity" to regularize layer weights.

When using MTL for “Task A” (i.e., according to figure 1), the two tasks were to predict whether the post is a “Switch” or “None” (i.e., “IS” or “O”) or whether it is an “Escalation” or “None” (i.e., “IE” or “O”). To prepare the training and validation input for each task, we sampled an equal number of instances from each class where the number of instances to sample is based on the minority class. Selecting an equal number of instances for each class made it possible to align the tasks so that similar tasks could potentially share a common feature space. We kept aside a sample with a class distribution to be the same as the original dataset for testing.

For “Task A”, the task-specific layers consist of a multi-channel Convolutional Neural Network (CNN) (Kim, 2014) where each channel was responsible for filtering features constituting bigrams and trigrams. To reduce the number of learnable parameters, the output from the CNN layers was further transformed using Global Maximum Pooling and then sent through a feedforward neural network. The output from each channel was then merged to form vectors that represent the task-specific features. These vectors were submitted to a loss function to regularize the network weights further. The merged outputs from each task-specific layer were concatenated to form the shared representation where each task will learn from a common feature space. It was identified that the model started to overfit the training data within a few epochs and consequently generated poor results during inference. To overcome model overfitting, we used several regularization techniques such as dropout (Srivastava et al., 2014) (i.e., a probability of 0.4 for "Task A" and 0.2 for "Task B")and L1 and L2 regularization to penalize larger weights in the multi-channel CNN. Further experiments discovered that making the model more or less complex reduced prediction accuracies due to either overfitting or underfitting, respectively.

We adopted an ensemble learning approach to reduce the variance in the results, which could be due to noise and random sampling. Model training and evaluation were done on three stratified training and validation splits where the final output is generated using an ensemble strategy on the combined predictions. We used the model averaging ensemble (Brownlee, 2018) strategy to generate the output.

For “Task B”, we used the same methods as for "Task A", except that we used an additional dataset to enhance the shared feature space between users susceptible to suicide ideation and mental disorders. Therefore, we selected a random sample of users similar to the number of users in the suicide ideation detection dataset. For example, to extract shared hidden features between users with severe suicide ideation risk and PTSD, we randomly selected 61 users who have self-declared PTSD from the SMHD dataset. The number of users is identified from the training dataset, where 61 users are categorized with severe suicide ideation risk.

The output of the suicide ideation detection task predicts three classes, that is, whether the user has a “Low”, “Moderate”, or “Severe” suicide ideation risk. For the second task, we conducted experiments using a different combination of mental disorders by predicting whether a given user has PTSD, Anxiety, or Bipolar Disorder. The final predictions are based on a model where users with “Moderate” and “Severe” suicide risks were aligned (i.e., sharing a common feature space) with users who have self-declared PTSD, and users with “Low” risk were aligned with users who have self-declared anxiety.

We used randomly initialized and trainable embedding layers with a dimension of 300 units for both subtasks. For task-specific layers, we used Rectified Linear Unit (ReLU) (2010) activation function and Adam optimizer (2015) with a learning rate 0.001 to update network weights.

4 Experiments and Results

We trained our models for fifteen epochs and reduced the learning rate by a factor of 0.1 if the validation loss did not improve. If the validation loss did not continuously improve, we stopped training and returned the model weights that produced the minimum loss. For both tasks, we trained our models using a mini-batch of size 16. Finally, we selected the label with the highest probability from the output generated using the model averaging ensemble.

We submitted three results for “Task A” and one for “Task B”. The difference between our two submissions, “uOttawa-AI(2)” and “uOttawa-AI(3)”, is based on regularization, where with more optimized regularization hyperparameters (i.e., on the

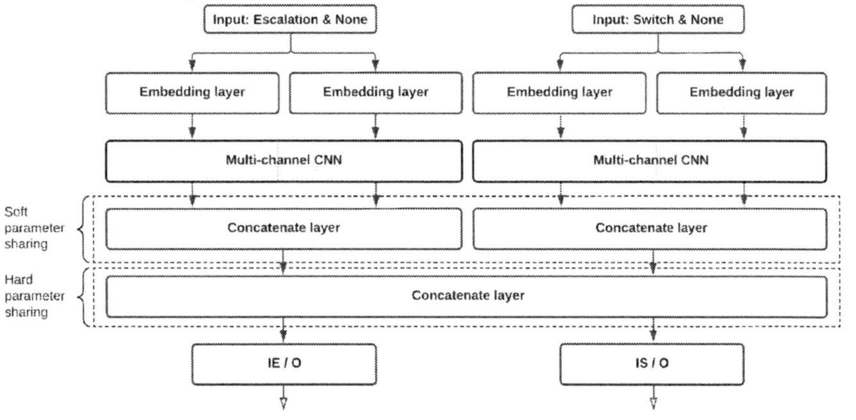

Figure 1: The Proposed multi-task learning architecture with hard and soft parameter sharing. The mentioned architecture is used mainly for "Task A". For "Task B", instead of IE/O and IS/O, we use suicide ideation risk levels as one output and the selected mental disorders as the second (i.e., PTSD/Anxiety).

submission uOttawa-AI(2)), we managed to train our model for more epochs and as a result produced a more generalized model. The "uOttawa-AI(1)" submission results are from a model with fewer learnable parameters.

Our results, compared to a majority class baseline and two preliminary experiments conducted by the task organizers (Tsakalidis et al., 2022b), are mentioned in tables 4, 5, 6, and 7. The results are macro averaged at the post level, window-based, and coverage-based (please refer Tsakalidis et al. (2022a) for more details on the evaluation metrics).

	Precision	Recall	F1
uOttawa-AI(2)	0.504	0.529	0.511
Majority	nan	0.333	0.280
TFIDF	0.545	0.495	0.492
BERT	0.522	0.386	0.380

Table 4: Post-level macro averaged results.

	Precision	Recall
uOttawa-AI(2)	0.347	0.453
Majority	nan	0.141
TFIDF	0.377	0.424
BERT	0.260	0.204

Table 5: Coverage-based macro averaged results.

5 Discussion

When analyzing the results of "Task A", we could see that our proposed architecture has produced competitive results when compared against the baseline and two of the preliminary experiments that use TF-IDF features with logistic regression and the BERT (Bidirectional Encoder Representations from Transformers) language model trained using the Talklife dataset (Tsakalidis et al., 2022b). We also identified that our submission "uOttawa-AI(2)" has produced better coverage and window-based (refer to table 6) predictions.

Even though the test results for the "Task B" model have produced better outcomes than the majority class baseline and the preliminary models trained by the task organizers (refer to table to 7, our model has not performed well in comparison to the best results. One of the critical reasons for the low results is class imbalance. During training, there were only 11 instances for the "Low" risk class compared to 55 and 61 for "Moderate" and "Severe" risk (refer to table 3). During inference, our model has not predicted "Low" risk labels but only "Moderate" and "Severe" labels. Another reason that we identified is the use of mental illness data as a complementary task. Even though the mental illness detection task has shared a common feature space with the suicide ideation detection task (i.e., suicide ideation or not) in Gamaarachchige (2021), when it comes to a more granular level (i.e., level of risk), mental ill-

	Window 1		Window 2		Window 3	
	P	R	P	R	P	R
uOttawa-AI(2)	0.529	0.621	0.559	0.662	0.596	0.691
Majority	nan	0.333	nan	0.333	nan	0.333
TFIDF	0.496	0.539	0.505	0.550	0.506	0.551
BERT	0.582	0.392	0.608	0.405	0.608	0.405

Table 6: Window-based macro averaged results.

	Precision	Recall	F1
uOttawa-AI	0.329	0.365	0.344
Majority	0.156	0.333	0.212
TFIDF	0.302	0.338	0.295

Table 7: Task B macro averaged results.

ness detection task has not managed to share features with suicide ideation risk levels. Even though we could not derive a conclusion on the suicide risk level and its correlation with a particular mental disorder, it could be assumed that more data representing different risk categories could derive a stronger relationship with certain mental disorders.

6 Conclusion and Future Work

We have investigated the applicability of multi-task learning to predict the change in mood of a social media user over time. With limited experiments, we managed to identify that MTL can be effectively applied to predict whether a post contains a mood shift, an escalation, or no change. Using different MTL architectures, which adopted different forms of parameter sharing strategies, it was identified that a combination of both the parameter sharing strategies (i.e., hard and soft parameter sharing) managed to produce better results. The main drawbacks we faced when using deep learning methods for classification are the class imbalance and the limited number of data points. For both tasks, we adopted a sampling strategy that facilitates task alignment. For "Task B", we introduced a complementary task intending to enrich the hidden features space so that we could, to a certain extent, eliminate the negative impact of having a smaller dataset with class imbalance. When analyzing the prediction outcomes, we could assume that features shared by certain mental disorders are not sufficient to define a decision boundary over suicide ideation risk levels.

In future research, we will look into the possibilities of improving the prediction accuracies by making changes to the current architecture (e.g., by changing the constructs of task-specific and shared layers) and also by adding contextual (e.g., ELMo[1] (Peters et al., 2018), BERT (Devlin et al., 2019)) and non-contextual embeddings (e.g., word2vec (Mikolov et al., 2013), fastText (Joulin et al., 2017)).

Ethics Statement

Secure access to the shared task dataset was provided with IRB approval under the University of Maryland, College Park protocol 1642625 and approval by the Biomedical and Scientific Research Ethics Committee (BSREC) at the University of Warwick (ethical application reference BSREC 40/19-20).

Acknowledgements

The authors are particularly grateful to the anonymous users of Reddit whose data feature in this year's shared task dataset, to the annotators of the data for Task A, to the clinical experts from Bar-Ilan University who annotated the data for Task B, and the American Association of Suicidology, to NORC for creating and administering the secure infrastructure and providing researcher support and to UKRI for providing funding to the CLPsych 2022 shared task organizers.

We would also like to thank the anonymous reviewers for their in-depth feedback and guidance, and the CLPsych shared task organizers for organizing the task and especially Adam Tsakalidis for continuously supporting us through the challenging times. Finally, we would like to convey our appreciation to all who have made it possible for the research domain to move forward by collecting, annotating and distributing data and conducting research to discover valuable insights.

[1]Embeddings from Language Models

References

Jason Brownlee. 2018. *Better Deep Learning: Train Faster, Reduce Overfitting, and Make Better Predictions.*, 1 edition. Machine Learning Mastery.

Rich Caruana. 1997. Multitask Learning. *Machine Learning*, 28(1):41–75.

Arman Cohan, Bart Desmet, Andrew Yates, Luca Soldaini, Sean MacAvaney, and Nazli Goharian. 2018. SMHD: A Large-Scale Resource for Exploring Online Language Usage for Multiple Mental Health Conditions. In *Proceedings of the 27th International Conference on Computational Linguistics*, pages 1485–1497, Santa Fe, New Mexico, USA. Association for Computational Linguistics.

Arman Cohan, Sydney Young, and Nazli Goharian. 2016. Triaging Mental Health Forum Posts. In *Proceedings of the 3rd Workshop on Computational Linguistics and Clinical Psychology: From Linguistic Signal to Clinical Reality*, pages 143–147, San Diego, CA, USA. Association for Computational Linguistics.

Glen Coppersmith, Mark Dredze, Craig Harman, and Kristy Hollingshead. 2015a. From ADHD to SAD: Analyzing the Language of Mental Health on Twitter through Self-Reported Diagnoses. In *Proceedings of the 2nd Workshop on Computational Linguistics and Clinical Psychology: From Linguistic Signal to Clinical Reality*, pages 1–10, Denver, Colorado. Association for Computational Linguistics.

Glen Coppersmith, Mark Dredze, Craig Harman, Hollingshead Kristy, and Margaret Mitchell. 2015b. CLPsych 2015 Shared Task: Depression and PTSD on Twitter. In *Proceedings of the 2nd Workshop on Computational Linguistics and Clinical Psychology: From Linguistic Signal to Clinical Reality*, pages 31–39, Denver, Colorado. Association for Computational Linguistics.

Jacob Devlin, Ming-Wei Chang, Kenton Lee, and Kristina Toutanova. 2019. BERT: Pre-training of Deep Bidirectional Transformers for Language Understanding. In *Proceedings of the 2019 Conference of the North American Chapter of the Association for Computational Linguistics: Human Language Technologies*, pages 4171–4186, Minneapolis, MN, USA. Association for Computational Linguistics.

Prasadith Buddhitha Kirinde Gamaarachchige. 2021. *Mental Illness and Suicide Ideation Detection Using Social Media Data.* Ph.D. thesis, University of Ottawa.

Michael Hauser, Evangelos Sariyanidi, Birkan Tunc, Casey Zampella, Edward Brodkin, Robert Schultz, and Julia Parish-Morris. 2019. Using natural conversations to classify autism with limited data: Age matters. In *Proceedings of the Sixth Workshop on Computational Linguistics and Clinical Psychology*, pages 45–54, Minneapolis, Minnesota. Association for Computational Linguistics.

Ahmed Husseini Orabi, Prasadith Buddhitha, Mahmoud Husseini Orabi, and Diana Inkpen. 2018. Deep Learning for Depression Detection of Twitter Users. In *Proceedings of the Fifth Workshop on Computational Linguistics and Clinical Psychology: From Keyboard to Clinic*, pages 88–97, New Orleans, LA. Association for Computational Linguistics.

Zunaira Jamil, Diana Inkpen, Prasadith Buddhitha, and Kenton White. 2017. Monitoring Tweets for Depression to Detect At-risk Users. In *Proceedings of the Fourth Workshop on Computational Linguistics and Clinical Psychology*, pages 32–40, Vancouver. Association for Computational Linguistics.

Armand Joulin, Edouard Grave, Piotr Bojanowski, and Tomas Mikolov. 2017. Bag of Tricks for Efficient Text Classification. In *Proceedings of the 15th Conference of the European Chapter of the Association for Computational Linguistics*, pages 427–431, Valencia, Spain. Association for Computational Linguistics.

Yoon Kim. 2014. Convolutional neural networks for sentence classification. In *Proceedings of the 2014 Conference on Empirical Methods in Natural Language Processing (EMNLP)*, pages 1746–1751, Doha, Qatar. Association for Computational Linguistics.

Diederik P. Kingma and Jimmy Lei Ba. 2015. Adam: A Method for Stochastic Optimization. In *Proceedings of the 3rd International Conference on Learning Representations 2015*, pages 1–15, San Diego, CA, USA.

Rohan Kshirsagar, Robert Morris, and Samuel Bowman. 2017. Detecting and Explaining Crisis. In *Proceedings of the Fourth Workshop on Computational Linguistics and Clinical Psychology — From Linguistic Signal to Clinical Reality*, pages 66–73, Vancouver, BC. Association for Computational Linguistics.

David E. Losada and Fabio Crestani. 2016. A test collection for research on depression and language use. In *Experimental IR Meets Multilinguality, Multimodality, and Interaction, 7th International Conference of the {CLEF} Association*, volume 9822 LNCS, Evora, Portugal.

David E. Losada, Fabio Crestani, and Javier Parapar. 2020. Overview of eRisk 2020: Early Risk Prediction on the Internet. *Lecture Notes in Computer Science (including subseries Lecture Notes in Artificial Intelligence and Lecture Notes in Bioinformatics)*, 12260 LNCS:272–287.

Sean MacAvaney, Anjali Mittu, Glen Coppersmith, Jeff Leintz, and Philip Resnik. 2021. Community-level Research on Suicidality Prediction in a Secure Environment: Overview of the CLPsych 2021 Shared Task. In *Proceedings of the Seventh Workshop on Computational Linguistics and Clinical Psychology*, pages 70–80, Online. Association for Computational Linguistics.

Tomas Mikolov, Kai Chen, Greg Corrado, and Jeffrey Dean. 2013. Distributed Representations of Words and Phrases and their Compositionality. In *Proceedings of the 26th International Conference on Neural Information Processing Systems*, page 3111–3119, Red Hook, NY, USA. Curran Associates Inc.

David N Milne, Glen Pink, Ben Hachey, and Rafael A Calvo. 2016. CLPsych 2016 Shared Task : Triaging content in online peer-support forums. In *Proceedings of the Third Workshop on Computational Linguistics and Clinical Psycholog*, pages 118–127, San Diego, CA, USA. Association for Computational Linguistics.

Elham Mohammadi, Hessam Amini, and Leila Kosseim. 2019. CLaC at CLPsych 2019: Fusion of Neural Features and Predicted Class Probabilities for Suicide Risk Assessment Based on Online Posts. In *Proceedings of the Sixth Workshop on Computational Linguistics and Clinical Psychology*, pages 34–38. Association for Computational Linguistics.

Vinod Nair and Geoffrey E Hinton. 2010. Rectified Linear Units Improve Restricted Boltzmann Machines. In *Proceedings of the 27th International Conference on International Conference on Machine Learning*, ICML'10, page 807–814, Madison, WI, USA. Omnipress.

Matthew E. Peters, Mark Neumann, Mohit Iyyer, Matt Gardner, Christopher Clark, Kenton Lee, and Luke Zettlemoyer. 2018. Deep contextualized word representations. In *Proceedings of the 2018 Conference of the North American Chapter of the Association for Computational Linguistics: Human Language Technologies*, pages 2227–2237, New Orleans, Louisiana. Association for Computational Linguistics.

Sebastian Ruder. 2017. An Overview of Multi-Task Learning in Deep Neural Networks. *CoRR*, abs/1706.0.

Koustuv Saha, Asra Yousuf, Ryan L. Boyd, James W. Pennebaker, and Munmun De Choudhury. 2022. Social Media Discussions Predict Mental Health Consultations on College Campuses. *Scientific Reports*, 12(1):1–11.

Christoph Schimmele, Jonathan Fonberg, and Grant Schellenberg. 2021. Canadians' assessments of social media in their lives. Technical report, Statistics Canada.

H Andrew Schwartz, Johannes Eichstaedt, Margaret L Kern, Gregory Park, Maarten Sap, David Stillwell, Michal Kosinski, and Lyle Ungar. 2014. Towards Assessing Changes in Degree of Depression through Facebook. In *Proceedings of the Workshop on Computational Linguistics and Clinical Psychology: From Linguistic Signal to Clinical Reality*, pages 118–125, Baltimore, Maryland, USA. Association for Computational Linguistics.

Han-chin Shing, Suraj Nair, Ayah Zirikly, Meir Friedenberg, Hal Daumé III, and Philip Resnik. 2018. Expert, Crowdsourced, and Machine Assessment of Suicide Risk via Online Postings. In *Proceedings of the Fifth Workshop on Computational Linguistics and Clinical Psychology: From Keyboard to Clinic*, pages 25–36, New Orleans, LA. Association for Computational Linguistics.

Nitish Srivastava, Geoffrey Hinton, Alex Krizhevsky, Ilya Sutskever, and Ruslan Salakhutdinov. 2014. Dropout: A Simple Way to Prevent Neural Networks from Overfitting. *Journal of Machine Learning Research*, 15(2):1929–1958.

Adam Tsakalidis, Jenny Chim, Iman Munire Bilal, Ayah Zirikly, Dana Atzil-Slonim, Federico Nanni, Philip Resnik, Manas Gaur, Kaushik Roy, Becky Inkster, Jeff Leintz, and Maria Liakata. 2022a. Overview of the CLPsych 2022 Shared Task: Capturing Moments of Change in Longitudinal User Posts. In *Proceedings of the Eighth Workshop on Computational Linguistics and Clinical Psychology: Mental Health in the Face of Change*.

Adam Tsakalidis, Federico Nanni, Anthony Hills, Jenny Chim, Jiayu Song, and Maria Liakata. 2022b. Identifying Moments of Change from Longitudinal User Text. In *Proceedings of the 60th Annual Meeting of the Association for Computational Linguistics*, volume 1, pages 4647–4660, Dublin, Ireland. Association for Computational Linguistics.

World Health Organization. 2021. Suicide worldwide in 2019: global health estimates. Technical report, World Health Organization, Geneva.

Ayah Zirikly, Philip Resnik, Ozlem Uzuner, and Kristy Hollingshead. 2019. CLPsych 2019 Shared Task: Predicting the Degree of Suicide Risk in Reddit Posts. In *Proceedings of the Sixth Workshop on Computational Linguistics and Clinical Psychology*, pages 24–33, Minneapolis, Minnesota. Association for Computational Linguistics.

Predicting Moments of Mood Changes Overtime from Imbalanced Social Media Data

Falwah AlHamed[1,3], **Julia Ive**[2], and **Lucia Specia**[1]

[1]Department of Computing, Imperial College London, London, UK
[1]{f.alhamed20,l.specia}@imperial.ac.uk
[2]Queen Mary University of London, London, UK
[2]j.ive@qmul.ac.uk
[3]King Abdulaziz City for Science and Technology(KACST), Riyadh, Saudi Arabia

Abstract

Social media data have been used in research for many years to understand users' mental health. In this paper, using user-generated content we aim to achieve two goals: the first is detecting moments of mood change over time using timelines of users from Reddit. The second is predicting the degree of suicide risk as a user-level classification task. We used different approaches to address longitudinal modelling as well as the problem of a severely imbalanced dataset. For the first task, using BERT with undersampling techniques performed the best among models tested, including LSTM and random forests models. For the second task, extracting features related to suicide from posts' text contributed to the overall performance improvement. Specifically, a feature representing of a number of suicide-related words in a post improved accuracy by 17%.

1 Introduction

Social media platforms are widely used nowadays. The nature of these platforms allows people to be open and express themselves and share daily details about their activities and thoughts. As a result, social media data have been used in research for many years to understand users' mental health. A number of techniques have been proposed in the recent literature on monitoring mental health state over time. For example, a study by (Sawhney et al., 2020) was conducted to investigate suicidal risks from Twitter. The authors used a time-aware transformer model with a pre-collected data set for suicide ideation and applied their model on 34,306 tweets from 32,558 users. The main goal was to classify if the person is at risk based on their sequence of tweets. Another study was conducted for detecting mood change by (Pruksachatkun et al., 2019). They proposed a predictive model to determine if a post is associated with a moment of cognitive change.

In this paper, we explain our approach to the CLPsych (Tsakalidis et al., 2022a) shared task, which consists of two subtasks, as follows:

Subtask A: Subtask A tries to capture those moments when a user's mood deviates from their baseline mood based on a user's postings throughout a specific time period — this is a post-level sequential classification task. The full task description can be found in (Tsakalidis et al., 2022b).

Subtask B: A user-level classification task on predicting the degree of suicide risk. An individual/user is considered to belong to one of four categories: no, low, medium or severe risk based on their posts on Reddit "r/SuicideWatch". The full task description can be found in (Zirikly et al., 2019).

2 Dataset

Data used for this shared task was pulled from Reddit. This well-known social media platform contains communities known as "subreddits", each of which covers a different topic.

For Subtask A, subreddits relating to mental health were used in this task. A total of 186 users were included in this study, with 256 timelines and a total of 6205 posts. The average time span for each user is 2 months. Data annotation was carried out by four annotators with multiple training rounds and mediation. Timelines were manually checked to ensure that they contain content indicating mood. Each post was labelled with one of three labels: IS for Switches i.e (mood shifts from positive to negative, or vice versa), IE for Escalations – gradual mood progression from negative (positive) to very negative (very positive), and 0 for no change. Subtask A data can be found on (Losada and Crestani, 2016; Losada et al., 2020; Shing et al., 2018). The data for this task was severely imbalanced. The values distribution were 79% for 0, 15% for IE, and only 6% labelled as IS.

For subtask B, four clinical experts annotated the user based on data from the *SuicideWatch* subreddit to one relative suicide risk severity. SubTask B data

Proceedings of the Eighth Workshop on Computational Linguistics and Clinical Psychology, pages 239 - 244
July 15, 2022 ©2022 Association for Computational Linguistics

can be found on (Shing et al., 2018).Each user was labelled with one of four labels: "None", "Low", "Moderate", "Severe" representing their suicidal risk level. The classes "Low" and "None" were merged together to address the class sparsity issue. The resulting class set is composed of the "Severe", "Moderate" or "Low" classes for 127 users with the frequencies of 48%, 43% and 9% respectively.

All authors have signed a Data User Agreement (DUA) and Non-Disclosure Agreement (NDA) to have access to the dataset.

3 Methods

In this section, we will describe the methods we developed to address these two shared subtasks.

3.1 Subtask A

We looked at various strategies to address the problem of data imbalance and also to consider longitudinal modelling.

3.1.1 Pre-processing

Different preprocessing techniques were applied on the posts in sequential manner using regular expressions operations. This includes cleaning for special characters and words such as users' mentions (special character '@'). Some characters were defined to be word boundaries characters which include comma, period, colon, question mark and semicolon. All these characters are replaced with a white space. Also, all URL hyperlinks were removed from posts with Regex.

3.1.2 Undersampling

We used undersampling to address the severe class imbalance. For this, we inspect sentiments in texts posts using TextBlob. [1] We found that most posts labelled with "0" have a positive sentiment with polarity greater than 0.2 (polarity ranges between -1 to 1), while "IS" and "IE" posts are connected with negative sentiment. This allowed us to remove 649 (out of 5143) samples labelled with "0" (polarity >= 0.2) and improve the dataset balance. We note that oversampling could be an alternative technique to avoid reducing sample size, which we leave for future work.

3.1.3 BERT

Models built by fine-tuning BERT (Devlin et al., 2019) or related pre-trained language models achieve state-of-the-art performance in a number of NLP tasks. This approach has been shown to give good results in multiple classification tasks, outperforming other algorithms (Acheampong et al., 2021; Al-Garadi et al., 2021). We used BERT with sequence length of 512 for post-level classification. In other words, *the predictions are performed per post without taking the preceding sequence of posts into account.* We experimented with the following different hyperparameters: `batch size:4,8,16,32; epochs: 8,16,32,64`. We reported the best parameters in Section 4.2.

3.1.4 LSTM

LSTMs are widely used for predicting sequential and temporal events, for example in (Chiu et al., 2021; Mirheidari and Christensen, 2019; Sawhney et al., 2020). We used LSTM for monitoring and predicting mood changes over time *taking into account the previous sequence of posts.* Since the baseline model for this task uses LSTM with BERT embeddings, we tried different embedding types, namely GloVe [2] and SpaCy Tok2Vec.[3] We tuned different hyperparameters to improve accuracy of the model. `batch size = [16,32,64,128] epochs=[16,32,40,64] learning rate=[0.01,0.02,0.05,0.1,0.2,0.5]`. We reported the best parameters in Section 4.2.

3.2 Subtask B

The aim of this subtask is to classify users to the correspondent suicide risk level. It is clear that the "Low" class is the least represented, which we take into account in our models.

3.2.1 Extra Features

To improve models performance and to account for the class imbalance, we extracted extra features that could positively affect the models' results. Since data size is small for this task (only 127 user), we used all data without undersampling.

Sentiment: Using TextBlob,[4] we extracted the sentiment of each post in user's data, then we sum the sentiment and based on the total we assign to each user a value of "Positive" if the total is greater than zero or "Negative" if the total is less than zero.

Polarity: We extracted the polarity of each post as a value between -1 and 1 (where -1 is severe

[1] https://textblob.readthedocs.io/en/dev/

[2] https://nlp.stanford.edu/projects/glove/

[3] https://spacy.io/api/tok2vec

[4] https://textblob.readthedocs.io/en/dev/

Table 1: List of suicidal words for Task B

Suicidal Words	
kill	die
knife	survive
dead	end my life
I'm gone	live anymore
I'm done	taking my life
killing	overdose
jump	suicide
wrist	hang
burn	self-harm
self harm	pesticide
death	take my life
call for help	

negative and 1 is extreme positive) using TextBlob, then we calculated the sum for all posts to get the polarity feature as a numeric value. Most users with severe risk level received a negative value, and most users with low risk levels received a positive value. The polarity was chosen as an indicator of the sentiment intensity.

Number of Suicidal Words: We inspected the posts of the three classes and found that the Severe class contains many words related to suicide attempts and ideas. Thus, we created a list of suicidal words by combining words from (Yang et al., 2022) and other words inferred from manual posts inspection. The word list is shown in Table 1. Then for each user, we calculated the number of words from the suicidal list that occurred in their posts. We added the total frequency of suicidal words as a feature. Related research has shown that combining lexical features besides machine learning models can improve the prediction results (AlHamed and AlGwaiz, 2020; Carvalho and Plastino, 2021).

3.2.2 Random Forests

Random forest is an ensemble machine learning model that relies on constructing multiple decision trees, then comparing the output of trees to predict the class. The class selected by random forests is the class that was selected by most of the trees via majority voting. Random forest was chosen as a non-neural algorithm as it has been shown to achieve higher accuracy in text classification tasks compared to other traditional machine learning algorithms such as KNNs (Biau and Scornet, 2016; Pranckevivius and Marcinkevicius, 2017). We used three random forest models in this task. The first with only word embeddings as features (RF1). The second with word embeddings and the additional extracted featured (RF2). The third with only the extracted features without word embeddings (RF3). We performed random grid search with the following hyperparameters: `no. of estimators = [200,300,400,500... 2000]; max features = ['auto', 'sqrt']; max depth = [10,20,30,...110]; min samples split = [2, 5, 10]; min samples leaf = [1, 2, 4]; bootstrap = [True, False].` Best performing parameters are reported in section 4.3.

4 Results and Evaluation

Results from the all models in both tasks on the blind test set are shown in Table 2. Baseline models (as reported by the shared task organisers) are Majority, TFIDF-LR, and BERT-Talklife-focal.

4.1 Evaluation metrics

As per (Tsakalidis et al., 2022b), the evaluation is carried out using two types of metrics. The first one is post-level metrics, which assesses the model's performance using precision, recall, and F1 score. The second type is coverage-level, these are the same metrics (precision, recall, and F1 score) but assessing the performance at the timeline level to assure that the model captures the sequence of mood changes overtime.

4.2 Subtask A Results

For this task we used three models, LSTM with SpaCy embeddings (LSTM-SpaCy), LSTM with GloVe embeddings (LSTM-GloVe), and BERT. All models are trained on data after undersampling. BERT performed the best in all the evaluation metrics, we think the reason behind that is BERT was fine-tuned on the dataset while LSTM models used pre-trained embeddings. Results for "IS" are the lowest as the class is underrepresented. For LSTM, the best hyperparameters are as follows: `batch size = 32, epochs=40, learning rate=0.05, optimiser = Adam`. For BERT, the best results were obtained for a model with batch size = 8 and number of epochs = 8.

4.3 Subtask B Results

For this task, we tried three types of random forests models.

The best results were obtained with the following settings: max depth=60, max features='sqrt', min samples leaf=2, min samples split=10, no. of estimators=600, random state=3. Surprisingly, RF3 where we used only the extra features as input (without using embeddings) outperforms the other RF models by 17% in accuracy. The reason behind this could be that the high similarity of words presented in all classes negatively affected prediction, and using only suicidal words and sentiments provided better context inference. This indicates that extracting additional meaningful features from text can enhance classification results.

5 Discussion

For subTask A, as shown in table 2, our BERT model outperformed the baseline BERT model - where LSTM over BERT embeddings is used - for macro-average results in both coverage based metrics and post-level metrics evaluation. Our proposed model with undersampling also scored the highest in precision and recall for the least presented class "IS". The reason behind this could be that the model was able to learn the features of this class after undersampling. On the other hand, the model performed less well in detecting "0" class. It could be an effect of undersampling, or that because other models were trained on the severely imbalanced dataset, they were biased toward predicting "0", and thus scoring higher precision and recall values.

When it comes to all participants in this year's CLPsych shared task, our BERT model ranked the third best performing model for post-level metrics evaluation. This emphasizes the feasibility and usefulness of the undersampling technique used.

For subTask B, compared to baseline models, RF3 was the only model able to predict the class "low". A possible explanation is that using the suicidal words count feature helped in identifying "low" suicidal risk users. On the other hand, results for "Moderate" and "Severe" classes were less compared to baseline models, this might be because we did not normalize the number of suicidal words to the number of posts per user and thus the model was inflated for users with more posts.

It is essential that the limitations of this study are considered in future studies. Firstly, the suicidal words list is collected from different sources and from analysis and manual inspection of the dataset. It could be expanded and validated to include more suicide related words. Another limitation is that we did not fine-tune any embedding model to our dataset (except for BERT). We used general pre-trained embeddings such as GloVe. Also, we aimed to try oversampling techniques to address data imbalance but we could not achieve that due to time constraints.

6 Conclusions and Future Work

In this paper, we presented our system description for CLPsych shared task (Tsakalidis et al., 2022a). The task consists of two subtasks. Subtask A aims to detect moments of mood change for posts in a timeline. For this, first we undersample the dataset to address the severe imbalance in dataset by filtering out the posts with positive sentiment irrelevant to mood changes. BERT without explicit modelling of the post sequence outperforms other models. Subtask B aims to classify a user to correspondent suicide risk-level. For this task, we extracted additional features and performed random forests. The proposed model succeeded in detecting the least represented class. In future, we aim to perform oversampling using GPT-3 to balance the dataset. We also aim to expand the suicidal words list and extract additional features from the text that could enhance obtained results.

Ethics Statement

Secure access to the shared task dataset was provided with IRB approval under University of Maryland, College Park protocol 1642625 and approval by the Biomedical and Scientific Research Ethics Committee (BSREC) at the University of Warwick (ethical application reference BSREC 40/19-20).

Acknowledgements

The authors are particularly grateful to the anonymous users of Reddit whose data feature in this year's shared task dataset, to the annotators of the data for Task A, to the clinical experts from Bar-Ilan University who annotated the data for Task B, the American Association of Suicidology, to NORC for creating and administering the secure infrastructure and providing researcher support and to UKRI for providing funding to the CLPsych 2022 shared task organisers.

Table 2: Results of all models for subTask A and the best variant of RF for subTask B on the official test set. We boldface the best results.

SubTask A												
1- Coverage Based Metrics												
	Macro- Average			IS			IE			0		
	P	**R**	**F1**	**P**	**R**	**F1**	**P**	**R**	**F1**	**P**	**R**	**F1**
Majority	nan	0.141	-	nan	0	-	nan	0	-	0.489	0.426	-
TFIDF-LR	**0.378**	0.424	-	0.111	0.008	-	**0.284**	**0.504**	-	**0.738**	**0.762**	-
BERT-Talklife-focal	0.260	0.204	-	0.025	0.007	-	0.226	0.093	-	0.530	0.513	-
LSTM-SpaCy	0.220	0.186	0.202	0.016	0.013	0.014	0.134	0.049	0.072	0.509	0.496	0.503
LSTM-GloVe	0.260	0.205	0.229	0.123	0.053	0.074	0.138	0.071	0.094	0.518	0.492	0.505
BERT	0.375	**0.440**	**0.405**	**0.253**	**0.372**	**0.301**	0.193	0.243	**0.215**	0.680	0.705	**0.692**
2- Post-level Metrics												
Majority	nan	0.333	0.280	nan	0	0	nan	0	0	0.724	1	0.840
TFIDF-LR	0.545	0.495	0.492	0.222	0.0243	0.044	0.569	**0.514**	**0.540**	0.844	0.947	**0.893**
BERT-Talklife-focal	0.522	0.386	0.380	0.090	0.012	0.022	**0.723**	0.163	0.266	0.753	**0.983**	0.853
LSTM-SpaCy	0.353	0.336	0.305	0.055	0.024	0.033	0.272	0.028	0.052	0.733	0.956	0.830
LSTM-GloVe	0.376	0.343	0.316	0.1	0.061	0.075	0.3	0.0288	0.052	0.729	0.939	0.821
BERT	**0.552**	**0.534**	**0.523**	**0.165**	**0.353**	**0.225**	0.609	0.389	0.475	**0.881**	0.860	0.871
SubTask B												
	Macro-Average			Low			Moderate			Severe		
Majority	0.156	0.333	0.213	nan	0	0	nan	0	0	0.469	**1**	**0.638**
TFIDF-LR	0.303	0.338	0.295	0	0	0	**0.428**	**0.214**	**0.286**	0.48	0.8	0.6
RF3	**0.305**	**0.423**	**0.297**	**0.166**	**0.666**	**0.266**	0.25	0.071	0.111	**0.5**	0.533	0.516

References

Francisca Adoma Acheampong, Henry Nunoo-Mensah, and Wenyu Chen. 2021. Transformer models for text-based emotion detection: a review of BERT-based approaches. *Artificial Intelligence Review*.

Mohammed Ali Al-Garadi, Yuan-Chi Yang, Haitao Cai, Yucheng Ruan, Karen O'Connor, Gonzalez-Hernandez Graciela, Jeanmarie Perrone, and Abeed Sarker. 2021. Text classification models for the automatic detection of nonmedical prescription medication use from social media. *BMC Medical Informatics and Decision Making*, 21(1):27.

Falwah AlHamed and Aljohara AlGwaiz. 2020. A Hybrid Social Mining Approach for Companies Current Reputation Analysis. In *Recent Advances on Soft Computing and Data Mining*, pages 429–438, Cham. Springer International Publishing.

Gérard Biau and Erwan Scornet. 2016. A random forest guided tour. *TEST*, 25(2):197–227.

Jonnathan Carvalho and Alexandre Plastino. 2021. On the evaluation and combination of state-of-the-art features in Twitter sentiment analysis. *Artificial Intelligence Review*, 54(3):1887–1936.

Chun Yueh Chiu, Hsien Yuan Lane, Jia Ling Koh, and Arbee L.P. Chen. 2021. Multimodal depression detection on instagram considering time interval of posts. *Journal of Intelligent Information Systems*, 56(1):25–47.

Jacob Devlin, Ming Wei Chang, Kenton Lee, and Kristina Toutanova. 2019. BERT: Pre-training of deep bidirectional transformers for language understanding. *NAACL HLT 2019 - 2019 Conference of the North American Chapter of the Association for Computational Linguistics: Human Language Technologies - Proceedings of the Conference*, 1(Mlm):4171–4186.

David E. Losada and Fabio Crestani. 2016. A test collection for research on depression and language use. In *Experimental IR Meets Multilinguality, Multimodality, and Interaction - 7th International Conference of the CLEF Association, CLEF 2016, Évora, Portugal, September 5-8, 2016, Proceedings*, volume 9822 of *Lecture Notes in Computer Science*, pages 28–39. Springer.

David E. Losada, Fabio Crestani, and Javier Parapar. 2020. Overview of erisk 2020: Early risk prediction on the internet. In *Experimental IR Meets Multilinguality, Multimodality, and Interaction - 11th International Conference of the CLEF Association, CLEF 2020, Thessaloniki, Greece, September 22-25, 2020, Proceedings*, volume 12260 of *Lecture Notes in Computer Science*, pages 272–287. Springer.

Bahman Mirheidari and Supervisor Heidi Christensen. 2019. Detecting early signs of dementia in conversation. (March).

Tomas Pranckevivius and Virginijus Marcinkevicius. 2017. Comparison of naive bayes, random forest, decision tree, support vector machines, and logistic regression classifiers for text reviews classification. *Balt. J. Mod. Comput.*, 5.

Yada Pruksachatkun, Sachin R. Pendse, and Amit Sharma. 2019. Moments of change: Analyzing peer-based cognitive support in online mental health forums. In *Proceedings of the 2019 CHI Conference on Human Factors in Computing Systems*, CHI '19, page 1–13, New York, NY, USA. Association for Computing Machinery.

Ramit Sawhney, Harshit Joshi, Saumya Gandhi, and Rajiv Ratn Shah. 2020. A Time-Aware Transformer Based Model for Suicide Ideation Detection on Social Media. pages 7685–7697.

Han-Chin Shing, Suraj Nair, Ayah Zirikly, Meir Friedenberg, Hal Daumé III, and Philip Resnik. 2018. Expert, crowdsourced, and machine assessment of suicide risk via online postings. In *Proceedings of the Fifth Workshop on Computational Linguistics and Clinical Psychology: From Keyboard to Clinic*, pages 25–36, New Orleans, LA. Association for Computational Linguistics.

Adam Tsakalidis, Jenny Chim, Iman Munire Bilal, Ayah Zirikly, Dana Atzil-Slonim, Federico Nanni, Philip Resnik, Manas Gaur, Kaushik Roy, Becky Inkster, Jeff Leintz, and Maria Liakata. 2022a. Overview of the CLPsych 2022 shared task: Capturing moments of change in longitudinal user posts. In *Proceedings of the Eighth Workshop on Computational Linguistics and Clinical Psychology: Mental Health in the Face of Change*.

Adam Tsakalidis, Federico Nanni, Anthony Hills, Jenny Chim, Jiayu Song, and Maria Liakata. 2022b. Identifying moments of change from longitudinal user text. In *Proceedings of the 60th Annual Meeting of the Association for Computational Linguistics (Volume 1: Long Papers)*, pages 4647–4660, Dublin, Ireland. Association for Computational Linguistics.

Bing Xiang Yang, Pan Chen, Xin Yi Li, Fang Yang, Zhisheng Huang, Guanghui Fu, Dan Luo, Xiao Qin Wang, Wentian Li, Li Wen, et al. 2022. Characteristics of high suicide risk messages from users of a social network—sina weibo "tree hole". *Frontiers in psychiatry*, 13.

Ayah Zirikly, Philip Resnik, Özlem Uzuner, and Kristy Hollingshead. 2019. CLPsych 2019 shared task: Predicting the degree of suicide risk in Reddit posts. In *Proceedings of the Sixth Workshop on Computational Linguistics and Clinical Psychology*, pages 24–33, Minneapolis, Minnesota. Association for Computational Linguistics.

Towards Capturing Changes in Mood and Identifying Suicidality Risk

Sravani Boinepelli, Shivansh Subramanian,
Abhijeeth Singam, Tathagata Raha, and **Vasudeva Varma**
{sravani.boinepelli, shivansh.s,
tathagata.raha}@research.iiit.ac.in
abhijeeth.singam@students.iiit.ac.in
vv@iiit.ac.in
Information Retrieval and Extraction Lab,
IIIT-Hyderabad, Gachibowli, Hyderabad, Telangana, India

Abstract

This paper describes our systems for CLPsych's 2022 Shared Task[1]. Subtask A involves capturing moments of change in an individual's mood over time, while Subtask B asked us to identify the suicidality risk of a user. We explore multiple machine learning and deep learning methods for the same, taking real-life applicability into account while considering the design of the architecture. Our team, IIITH, achieved top results in different categories for both subtasks. Task A was evaluated on a post-level (using macro averaged F1) and on a window-based timeline level (using macro-averaged precision and recall). We scored a post-level F1 of 0.520 and ranked second with a timeline-level recall of 0.646. Task B was a user-level task where we also came in second with a micro F1 of 0.520 and scored third place on the leaderboard with a macro F1 of 0.380.

1 Introduction

Globally, close to 800,000 people die by suicide each year (WHO, 2014). Suicide is the fourth leading cause of death among 15-19 year-olds (WHO, 2021). Though suicide is such a dire issue, a myriad of obstacles such as social stigma, apprehensions about privacy, financial concerns, etc., prevent many from seeking professional help. Over the last couple of years, there has been an influx of suicide and mental health posts on social media, especially from the young users - social media's primary consumers. Anonymous social media platforms such as mental health blogs or Reddit forums have become increasingly popular as they can share their personal stories without judgment. People who face similar issues can share their experiences, give advice, motivate and persuade them to seek counsel from professionals. Therefore, social media has become a valuable source of linguistic cues for work to identify mental health problems from textual data (Cao et al., 2019; Masuda et al., 2013; Choudhury et al., 2016; Pruksachatkun et al., 2019). A challenge in the area of mental illness detection and suicide risk identification on social media is the importance of focusing on the individual and detecting the critical point where intervention is necessary from a batch of posts. This shared task (Tsakalidis et al., 2022a) breaks up the problem into two problem statements:

Subtask A: Given a user's posts over a certain period in time, this task aims to capture those sub-periods during which a user's mood deviates from their baseline mood. This is defined as a post-level sequential classification task (Tsakalidis et al., 2022b). It encourages us to identify moments of change in the individual's mood over a timeline of about two months. A moment of change (MOC) is defined as a post/sequence of posts in a timeline indicating that the user's behavior or mental health status is shifted. This is represented in the form of the following labels: IS (indicating a switch in the user's mood), IE (indicating an escalation of the user's mood) and, O (refers to all other cases).

Subtask B: This is a user-level classification problem to predict the degree of suicide risk on Reddit. A user is considered to belong to one of four categories: No Risk (or "None"), Low, Moderate, and Severe Risk based on their posts on r/SuicideWatch (Shing et al., 2018; Zirikly et al., 2019). We present several approaches to tackle both subtasks, keeping in mind real-life application and the temporal aspect of this problem.

In the first subtask, we observed that the post-level classification would be influenced by its context. Hence, due to the longitudinal nature of the task, we use a transformer-based LSTM architecture. Post-level representations are generated using sentence transformer models and passed through an LSTM layer to consider historical context before developing the final output label.

[1]https://clpsych.org/sharedtask2022/

Proceedings of the Eighth Workshop on Computational Linguistics and Clinical Psychology, pages 245 - 250
July 15, 2022

Our second task considers the need for detection mechanisms to continuously monitor suicide risk with the introduction of new posts to a user's history. We, therefore, first evaluate on a post-level using finetuned transformer models. We then adopt a majority voting strategy to assign the final label to the user. Our models outperform the baseline and rank in the top 3 submitted models for both subtasks across various categories.

2 Data

The dataset contains 255 timelines taken from users who have posted on mental health-related subreddits and /r/SuicideWatch. Each timeline consists of 10 to 122 posts each. The data given for this task is taken from 3 separate datasets. The E-Risk dataset (Losada et al., 2020; Losada and Crestani, 2016) is primarily used for Subtask A, while the Reddit datasets, such as the UMD dataset, are used for both subtasks (Shing et al., 2018; Zirikly et al., 2019). The dataset was split into a train and test dataset, with the training dataset having 149 users, 204 timelines, and 5143 posts, and the test dataset having 36 users, 51 timelines, and 1052 posts. Of the 149 users in the training dataset, 61 were labeled as 'Severe', 55 as 'Moderate', 11 as 'Low', and 22 remained unlabelled.

3 Baseline Experiments

We experiment with various popular machine learning, text classification algorithms on a post-level. Majority Voting is then applied for the task B experiments to generate the final label. Count Vectors and tf-idf vectors for different levels of input tokens (words, n-grams, etc.) served as the primary features for most of our baseline models. We used Scikit-learn (Pedregosa et al., 2011) and Keras (Chollet et al., 2015) libraries to develop and evaluate the models.

Logistic regression(LR): The logistic regression model uses tf-idf and n-grams as features for our baseline. Hyperparameter tuning proved the model to work best for ranges of unigrams and bigrams.

Random Forest Model(RF): Decision trees tend to overfit on the training set. Random decision forests with bagging help correct this behavior. We test their performance against tf-idf word-level vector and count vector features.

Xtreme Gradient Boosting Model(Xgb): The boosting algorithm is popularly used to optimize the performance of decision trees by reducing bias and variance. We test the performance of this model against other baselines with count vectors and word-level tf-idf.

4 Final architecture

4.1 Experimental Settings

The HuggingFace transformers library's RoBERTa model was used for finetuning, and all our architectures were implemented in Pytorch (Paszke et al., 2019). Our MOC-LSTM model was run with a learning rate of 2e-06 and a batch size of 8, while our finetuned RoBERTa model was run with batch size 16. The model uses the AdamW optimizer with an initial learning rate of 2e-5 and a linear warm-up schedule.

4.2 Detecting MOCs from a User timeline

To detect switches in a user's mood, our model must retain the knowledge of user history to assign the present post label. Therefore, our model uses an LSTM-based (Gers et al., 2000) architecture to capture the essence of the previous context and generate the labels for the latest posts. We initially convert the content and titles of each Reddit post within a timeline into 384-dimensional embeddings using paraphrase-MiniLM-L6-v2 (Reimers and Gurevych, 2019).

Based on the dataset, the maximum number of posts by a user was 125, so we created a hard limit of 128 posts per user. This required us to pad posts to users who had posted less than 128 posts to make them equal in length but allowed us to do batch-wise computation. We also replace specific posts with no content or title with pad tokens.

Once this preprocessing is complete, we use the sentence transformer to get text embeddings for each post's content and title. We took the truncated text as required by the transformer model. We concatenated the content and title representation to get a single post representation and used that as input to the LSTM layer. A window_size amount of posts is sent at a time to limit the previous posts' influence on the current output. The output of this LSTM layer was then used for the post-wise classification. We experimented with different window_sizes to see the effect of previous information on the quality of predictions. When comparing sizes 4 and 8, we found that window_size = 4 gives us the best results. We added a linear layer and SoftMax activation to get

probability distribution over the three classes. The class with the highest probability was considered the model's prediction. The final loss is computed based on WeightedCrossEntropyLoss to reflect the bias in the dataset. This gave superior results to regular cross entropy loss.
We then changed the embedding model from paraphrase-MiniLM-L6-v2 to robertaSTSB. The embedding dimension for RoBERTa was different since each text gave an output of 768 dimensions. We observed that RoBERTa embeddings performed equal or better in almost all parameters compared to paraphrase-MiniLM. But due to technical issues on our side, we could not use RoBERTa in our final submission. This experiment was performed on a validation set (80-20 split).

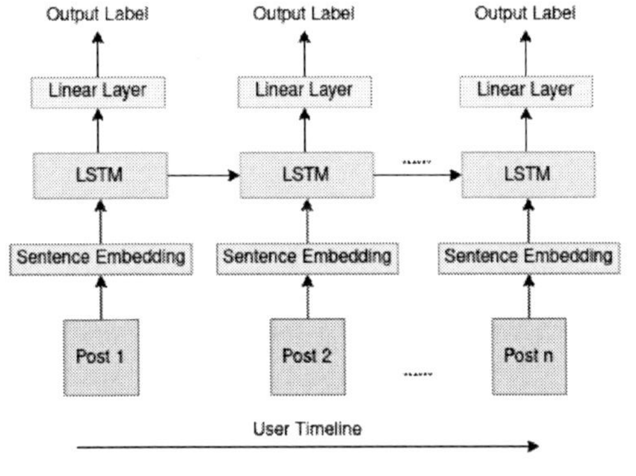

Figure 1: Framework to detect MOCs from a user timeline

4.3 Finetuned RoBERTa model for Assessing Suicidality Risk

Transformers and transfer learning architectures have previously achieved SOTA results in multiple datasets. They combine positional encoding and non-sequential single input processing to achieve better long-range dependencies than LSTMs and RNNs. The attention mechanisms are targeted toward such sequential data. Therefore, our final architecture uses RoBERTa for predicting suicide risk for a given text by fine-tuning it for classification on the given dataset. We also consider using sampling techniques to correct the imbalance in the dataset. However, models like RoBERTa are known to perform better on imbalanced data sets rather than on oversampled augmented datasets (Tayyar Madabushi et al., 2019). Weighted Random Sampling was preferred over other under-sampling techniques based on the results of our experiments.
The cleaned dataset is passed to the RoBERTa tokenizer. The length of each input is fixed to be 256 tokens. Only the first embedding produced by the RoBERTa for Sequence Classification Model is used for classification. This embedding is then passed to a linear layer and produces logits used to predict the post-level labels. Backward propagation of the loss is performed, and the weights of the linear layer and the model are updated. This fine-tuning process helps the model learn the unique domain related to the problem.
We now have the labels for each post made by the user. However, it is difficult to determine the number of posts (with a certain level of severity) required to assign a final label to the user. Because the span of each timeline is about two months, the most straightforward approach would be to simply take the most occurring label as the final assigned user label. This is called majority voting. Given that the posts we are considering represent the user's 'n' most relevant posts, we postulate that the 'degree' of suicide risk of the user can be ascertained by simply taking the mode of the outputted 'n' post labels. This approach performed well on the leaderboard for this task. Our team came in second with a micro F1 of 0.520 and scored third place with a macro F1 of 0.380.

4.4 Finetuning MOC-LSTM

We also tried to leverage our results from Task A to improve performance on Task B. As a preprocessing step, we assumed that the user's risk level is similarly reflected in their timeline's risk level. Once we do that, the task becomes a timeline-level classification task, and we determine the user's final label based on majority voting.
We used transfer learning to classify the entire timeline and give us better results. Our initial model was trained on task A (as described in MOC-LSTM) for the post-level classification task. We had to take special care of the window_size in this approach since our primary goal is a timeline-level classification. We used a window_size = 128, implying that all the posts are considered simultaneously. Hence, the output of the pre-trained model was a probability distribution over the three classes: IS, IE, and O, for 128 posts. We then utilize the pre-trained model, learned on task A, to finetune the task B

dataset. We added another linear layer followed by SoftMax to the output of the task A model to combine the post-level classification into a final timeline-level classification. Hence, we combined the post-level probability distribution to get the timeline-level probability distribution. Once we got out probability distribution, we used CrossEntropyLoss to train our model and Adam optimizer. Though the model performs well, we were unable to officially submit it to the shared task due to technical issues from our side.

4.5 Evaluation metrics

For Subtask A, the post-level results are calculated using macro F1 scores (represented as 'M-F1' in Table 2. The coverage and window-based results are evaluated using Precision('P') and Recall('R') oriented scores as specified by the shared task organizers. The details for calculating these evaluation metrics may be found in their overview of the shared task (Tsakalidis et al., 2022a).

Subtask B was evaluated using Macro and Micro averaged F1 scores. We look at the range per timeline and the distribution of labels. Since the range for each timeline in the dataset is about two months, we propose a majority voting approach. This performed well, and our model ranked in the top three with both macro F1 and micro F1 scores. However, this may fall short for more extended time periods, at which point it becomes increasingly imperative to adopt a more longitudinal and temporal approach to calibrate the level of a user's suicide risk.

Task	Model	Macro-F1
Task A	RF, Count	0.48
	RF, tf-idf, Word lvl	0.48
	Xgb, Count	0.50
	Xgb, tf-idf, Word lvl	0.49
	LR, Count	**0.54**
	LR, tf-idf, Word lvl	0.47
	LR, tf-idf, N-Gram	0.47
	MOC-LSTM	**0.59**
Task B	RF, Count	0.43
	RF, tf-idf, Word lvl	0.42
	Xgb, Count	0.40
	Xgb, tf-idf, Word lvl	0.40
	LR, Count	**0.46**
	LR, tf-idf, Word lvl	0.45
	LR, tf-idf, N-Gram	0.44
	Finetuned MOC-LSTM	**0.51**
	Finetuned RoBERTa	**0.54**

Table 1: Comparing baseline results with the final models for Task A and B on an 80/20 split of the training data.

Eval_type	Model	P	R	M-F1
Task A, Window	Baseline	0.496	0.539	-
	IIITH	0.530	**0.646**	-
Task A, Coverage	Baseline	0.377	0.424	-
	IIITH	0.346	0.405	-
Task A Post-level	Baseline	0.545	0.495	0.492
	IIITH	0.520	0.600	0.520
Task B, Macro-avg	Baseline	0.302	0.338	0.295
	IIITH	**0.396**	0.407	**0.380**
Task B, Micro-avg	Baseline	0.412	0.468	0.406
	IIITH	**0.538**	**0.562**	**0.520**

Table 2: CLPsych 2022 Official Results on the test set.

5 Results and Analysis

The comparison between the baseline models and our main models can be found in Table 1. LogisticRegression with CountVectorizer was the best baseline for both tasks with respect to Macro-F1. Our model MOC-LSTM beats the baselines of Task A comfortably with a 0.05 increase in Macro-F1. For task B, the finetuned MOC-LSTM has a slight edge over the baselines, whereas the finetuned RoBERTa model scores significantly better with a 0.08 F1 over the baseline models.

Results on the unseen test set for our submitted models can be found in Table 2 as provided to us by the workshop organizers. The 'Baseline' results belong to the Logistic Regression model trained on tf-idf features supplied by the organizers. Values in bold are amongst the top 3 ranked by the shared task. Our final submission included the 'MOC-LSTM', and Finetuned RoBERTa models. Our MOC-LSTM model scores a post-level F1 of 0.520 and ranks second with a timeline-level recall of 0.646. Our Finetuned-RoBERTa model ranks second with a micro F1 of 0.520 and scored third place with a macro F1 of 0.380.

6 Conclusion

In this shared task, we have worked towards detecting moments of change and the suicidality risk of a user based on their post history. Our MOC-LSTM model allows us to determine post-level information and timeline level classification, enabling us to better understand mental health by identifying the specific moments of change in emotions. The finetuned RoBERTa model works to identify at-risk users based on their post timeline to better care for them. In the future, we plan to consider multi-modal approaches for different social media platforms. This helps give a better picture of the user's mental health since people use different media for different purposes. We also plan to build more efficient models that work on longer timelines to provide these warnings in a real-time platform-agnostic manner and help identify at-risk users.

7 Ethical Statement

Secure access to the shared task dataset was provided with IRB approval under University of Maryland, College Park protocol 1642625 and approval by the Biomedical and Scientific Research Ethics Committee (BSREC) at the University of Warwick (ethical application reference BSREC 40/19-20).

8 Acknowledgements

The authors are particularly grateful to the anonymous users of Reddit whose data feature in this year's shared task dataset, to the annotators of the data for Task A, to the clinical experts from Bar-Ilan University who annotated the data for Task B, the American Association of Suicidology, to NORC for creating and administering the secure infrastructure and providing researcher support and to UKRI for providing funding to the CLPsych 2022 shared task organisers.

References

Lei Cao, Huijun Zhang, Ling Feng, Zihan Wei, Xin Wang, Ningyun Li, and Xiaohao He. 2019. Latent suicide risk detection on microblog via suicide-oriented word embeddings and layered attention. In *Proceedings of the 2019 Conference on Empirical Methods in Natural Language Processing and the 9th International Joint Conference on Natural Language Processing (EMNLP-IJCNLP)*, pages 1718–1728, Hong Kong, China. Association for Computational Linguistics.

François Chollet et al. 2015. Keras. `https://keras.io`.

Munmun De Choudhury, Emre Kıcıman, Mark Dredze, Glen A. Coppersmith, and Mrinal Kumar. 2016. Discovering shifts to suicidal ideation from mental health content in social media. *Proceedings of the 2016 CHI Conference on Human Factors in Computing Systems*.

Felix A. Gers, Jürgen Schmidhuber, and Fred Cummins. 2000. Learning to forget: Continual prediction with lstm. *Neural Computation*, 12:2451–2471.

David E. Losada and Fabio Crestani. 2016. A test collection for research on depression and language use. In *Experimental IR Meets Multilinguality, Multimodality, and Interaction - 7th International Conference of the CLEF Association, CLEF 2016, Évora, Portugal, September 5-8, 2016, Proceedings*, volume 9822 of *Lecture Notes in Computer Science*, pages 28–39. Springer.

David E. Losada, Fabio Crestani, and Javier Parapar. 2020. Overview of erisk 2020: Early risk prediction on the internet. In *Experimental IR Meets Multilinguality, Multimodality, and Interaction - 11th International Conference of the CLEF Association, CLEF 2020, Thessaloniki, Greece, September 22-25, 2020, Proceedings*, volume 12260 of *Lecture Notes in Computer Science*, pages 272–287. Springer.

Naoki Masuda, Issei Kurahashi, and Hiroko Onari. 2013. Suicide ideation of individuals in online social networks. *PLoS One*, 8(4):e62262.

Adam Paszke, Sam Gross, Francisco Massa, Adam Lerer, James Bradbury, Gregory Chanan, Trevor Killeen, Zeming Lin, Natalia Gimelshein, Luca Antiga, Alban Desmaison, Andreas Köpf, Edward Z. Yang, Zach DeVito, Martin Raison, Alykhan Tejani, Sasank Chilamkurthy, Benoit Steiner, Lu Fang, Junjie Bai, and Soumith Chintala. 2019. Pytorch: An imperative style, high-performance deep learning library. *CoRR*, abs/1912.01703.

Fabian Pedregosa, Gaël Varoquaux, Alexandre Gramfort, Vincent Michel, Bertrand Thirion, Olivier Grisel, Mathieu Blondel, Peter Prettenhofer, Ron Weiss, Vincent Dubourg, et al. 2011. Scikit-learn: Machine learning in python. *Journal of machine learning research*, 12(Oct):2825–2830.

Yada Pruksachatkun, Sachin R. Pendse, and Amit Sharma. 2019. Moments of change: Analyzing peer-based cognitive support in online mental health forums. In *Proceedings of the 2019 CHI Conference on Human Factors in Computing Systems*, CHI '19, page 1–13, New York, NY, USA. Association for Computing Machinery.

Nils Reimers and Iryna Gurevych. 2019. Sentence-bert: Sentence embeddings using siamese bert-networks. In *Proceedings of the 2019 Conference on Empirical Methods in Natural Language Processing*. Association for Computational Linguistics.

Han-Chin Shing, Suraj Nair, Ayah Zirikly, Meir Friedenberg, Hal Daumé III, and Philip Resnik. 2018. Expert, crowdsourced, and machine assessment of suicide risk via online postings. In *Proceedings of the Fifth Workshop on Computational Linguistics and Clinical Psychology: From Keyboard to Clinic*, pages 25–36, New Orleans, LA. Association for Computational Linguistics.

Harish Tayyar Madabushi, Elena Kochkina, and Michael Castelle. 2019. Cost-sensitive BERT for generalisable sentence classification on imbalanced data. In *Proceedings of the Second Workshop on Natural Language Processing for Internet Freedom: Censorship, Disinformation, and Propaganda*, pages 125–134, Hong Kong, China. Association for Computational Linguistics.

Adam Tsakalidis, Jenny Chim, Iman Munire Bilal, Ayah Zirikly, Dana Atzil-Slonim, Federico Nanni, Philip Resnik, Manas Gaur, Kaushik Roy, Becky Inkster, Jeff Leintz, and Maria Liakata. 2022a. Overview of the CLPsych 2022 shared task: Capturing moments of change in longitudinal user posts. In *Proceedings of the Eighth Workshop on Computational Linguistics and Clinical Psychology: Mental Health in the Face of Change*.

Adam Tsakalidis, Federico Nanni, Anthony Hills, Jenny Chim, Jiayu Song, and Maria Liakata. 2022b. Identifying moments of change from longitudinal user text. In *Proceedings of the 60th Annual Meeting of the Association for Computational Linguistics (Volume 1: Long Papers)*, pages 4647–4660, Dublin, Ireland. Association for Computational Linguistics.

WHO. 2014. Preventing suicide: A global imperative.

WHO. 2021. Suicide.

Ayah Zirikly, Philip Resnik, Özlem Uzuner, and Kristy Hollingshead. 2019. CLPsych 2019 shared task: Predicting the degree of suicide risk in Reddit posts. In *Proceedings of the Sixth Workshop on Computational Linguistics and Clinical Psychology*, pages 24–33, Minneapolis, Minnesota. Association for Computational Linguistics.

WWBP-SQT-lite: Difference Embeddings and Multi-level Models for Moments of Change Identification in Mental Health Forums

Adithya V Ganesan[1], **Vasudha Varadarajan**[1], **Juhi Mittal**[2], **Shashanka Subrahamanya**[3], **Matthew Matero**[1], **Nikita Soni**[1], **Sharath Chandra Guntuku**[2], **Johannes C. Eichstaedt**[3] and **H. Andrew Schwartz**[1]
[1]Stony Brook University, [2]University of Pennsylvania, [3]Stanford University
{avirinchipur, vvaradarajan}@cs.stonybrook.edu

Abstract

Psychological states unfold dynamically; to understand and measure mental health at scale we need to detect and measure these changes from sequences of online posts. We evaluate two approaches to capturing psychological changes in text: the first relies on computing the difference between the embedding of a message with the one that precedes it, the second relies on a "human-aware" multi-level recurrent transformer (HaRT). The mood changes of timeline posts of users were annotated into three classes, 'ordinary,' 'switching' (positive to negative or vice versa) and 'escalations' (increasing in intensity). For classifying these mood changes, the difference-between-embeddings technique – applied to RoBERTa embeddings – showed the highest overall F1 score (0.61) across the three different classes on the test set. The technique particularly outperformed the HaRT transformer (and other baselines) in the detection of switches (F1 = .33) and escalations (F1 = .61). Consistent with the literature, the language use patterns associated with mental-health related constructs in prior work (including depression, stress, anger and anxiety) predicted both mood switches and escalations.

1 Introduction

Detecting shifts in mental health from language use could assist in identifying episodes of mental ill health and providing in-time treatment for conditions such as depression or anxiety. The accessibility and abundant usage of social media (Coppersmith, 2022) in comparison to traditional healthcare data (e.g. hospital visits) is enabling first steps toward unprecedented assessment and understanding of mental health, including detection of elevated risks (Choudhury et al., 2016; Zirikly et al., 2019; Guntuku et al., 2021). However, most language datasets for mental health classification are annotated statically such that a person has just one label across all of their language (Coppersmith et al., 2014; Lynn et al., 2018). Longitudinal language datasets can help analyze the mental state of a person over time (Halder et al., 2017; Matero and Schwartz, 2020; Son et al., 2021), but also open the door for many sequential, differencing, and time-series modeling techniques.

Here, we explore two types of modeling techniques that can capture changes over time: Human-aware Recurrent Transformers (Soni et al., 2022) and difference embeddings. These techniques were used as part of the WWBP-SQT-lite[1] system for the CLPsych 2022 shared tasks (Tsakalidis et al., 2022a): (Task A) modeling user state changes over time (Tsakalidis et al., 2022b), and (Task B) the suicide risk associated with the user (Shing et al., 2018), our **contributions** are as follows: (a) evaluation of Human-aware Recurrent Transformers (HaRT) and difference embeddings for Task A (b) exploring SoTA methods for predicting state escalations and switches, and (c) exploring theoretically related linguistic assessments.

2 Data

2.1 Task A

Task Data. The training data for task A contained 5,143 Reddit posts comprising of titles and contents from 149 users spanning over 204 timelines. As described in Tsakalidis et al. 2022b, posts from each timeline were annotated with the Moment of Change (MoC) of the user's mood into three classes, namely, "Ordinary" (O), "In Switch" (IS) when the mood changes from positive to negative or vice versa, and "In Escalation" (IE) signifying mood progressions, i.e., changes from neutral or positive to more positive or negative to more negative. The posts were annotated in the context

[1]**SQT**: **S**eawolf, **Q**uaker, and **T**ree (the mascots of the schools composing our team); **lite**: due to constraints out of our control, we were restricted to just 4 days working with the data, covering only a portion of our planned human-level and temporal approaches.

Proceedings of the Eighth Workshop on Computational Linguistics and Clinical Psychology, pages 251 - 258
July 15, 2022

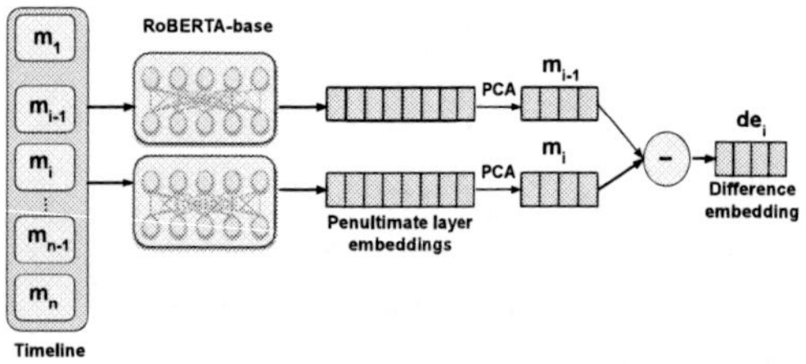

Figure 1: Difference embedding for the current message is obtained using a point-wise subtraction on the the current embedding and previous message's embedding.

of other posts from timelines as carried out in the CLEF eRisk 2020 dataset (Losada and Crestani, 2016; Losada et al., 2020). A small number of timelines in the CLPsych 2022 data was extracted from the CLEF eRisk 2020 dataset.

Internal Train & Validation sets. 119 (80%) randomly chosen users were sampled to form an internal train set, and the remaining 30 users were used for validation set. This resulted in 3,974 posts (78%) over 164 timelines in train set and 1,169 posts over 40 timelines in the validation set.

2.2 Task B

Task Data The goal of task B is to predict the Suicide risk level associated with the Reddit users into Low, Moderate or Severe. It utilizes the same data as Task A to the exclusion of 22 users which were annotated as "N/A". Thus a total of 127 users were present in task B, who collectively posted a total of 4,507 Reddit posts, averaging at around 35 posts per user. The risk level of the user was assigned as the maximum risk level annotation across all their posts. The suicide risk annotation followed the procedure described in Shing et al. 2018 and Zirikly et al. 2019.

Internal Train & Validation sets A random sample of 101 users (79.5%) were chosen for the internal train set – a total of 3,761 posts for training and the remaining 26 (20.5%) users for the validation set, resulting in 746 posts in the validation set.

2.3 Evaluation

For Task A, macro F1 and coverage-based (Arbeláez et al., 2011; Tsakalidis et al., 2022b) precision and recall scores were used to measure the performance of the models. The coverage based metrics are aimed at evaluating the model's ability to capture the regions of change. However, for task B, only macro F1 is used to evaluate model performance.

Dimension	β_O	β_{IE}	β_{IS}
Big 5 Traits			
Emotional Stability	.57‡	-.14‡	-.38‡
Extraversion	.18‡	-.05	-.12‡
Conscientiousness	.13‡	-.03	-.12†
Agreeableness	.13‡	-.01	-.12‡
Openness to Experience	-.04	.01	.03†
Anger	-.35‡	.09†	.24‡
Anxiety	-.48‡	.13‡	.31‡
Stress	-.58‡	.14‡	.39‡
Depression	-.59‡	.14‡	.39‡
Loneliness	-.82‡	.20‡	.56‡

Table 1: Association (standardized logistic regression coefficients, β) of theoretical features measures in language with the three classes of task A (ordinary, in-escalation, or in-switch). †: $p < .05$; ‡:$p < .001$

3 Methods

3.1 Task A

Beyond utilizing the best transformer based approaches, we also explore relevant theoretical features to understand the relationship between moment of change and psychological/demographic constructs. Furthermore, recent works (Sawhney et al., 2020, 2021) have shown the importance of joint modelling of such theoretical dimensions with the present-day neural approaches.

HypLex. To quantify the association of psychological and demographic constructs with the moments of change, 12 models trained on larger datasets were used to derive theoretical features which we call HypLex (short for Hypothesis-driven Lexica). These models include Cohen's stress (Guntuku et al., 2019a), depression, anger and anxiety (Schwartz et al., 2014; Son et al., 2021; Guntuku et al., 2019b), age and gender (Sap et al., 2014), loneliness expressions (Guntuku et al., 2019c), and the big 5 personality traits (Park et al., 2015). All these features were on a continuous scale.

HaRT. Recent works (Lynn et al., 2020; Matero et al., 2021b; Soni et al., 2022) have highlighted the importance of incorporating author context into the message representations through the use of history and multi-level modeling. We use the Human aware Recurrent Transformer model (Soni et al., 2022) which is built on GPT2 (Radford et al., 2019), to produce message representations that encode the latent representation of the author as well.

Method	IS			IE	Post-level Evaluation		O			macro avg			Coverage-based macro avg	
	P	R	F1	P	R	F1	P	R	F1	P	R	F1	P	R
Internal Validation														
HypLex	.00	.00	.00	.52	.34	.41	.84	.95	.89	.46	.43	.44	.27	.30
HaRT														
CLS	.16	.13	.14	.55	**.67**	**.60**	**.92**	.88	.90	.54	.56	.55	.39	.43
CLS+Last layer	.16	.15	.15	.56	.63	.59	.91	.89	.90	.55	.56	.56	.38	.44
PCA-Roba														
Curr	.09	.09	.09	.56	.46	.50	.90	.93	**.92**	.52	.49	.50	.36	.38
Diff	**.44**	**.34**	**.38**	.33	.03	.05	.81	**.98**	.88	.53	.45	.44	.39	.29
Curr+Prev	.42	.25	.31	.64	.53	.58	.89	.95	**.92**	**.65**	.57	.60	.43	.43
Curr+Diff	.38	.26	.31	.63	.46	.53	.88	.95	.91	.63	.56	.59	.42	.43
Curr+Prev+Diff	.42	**.34**	.37	**.65**	.52	.58	.89	.94	.91	**.65**	**.60**	**.62**	**.44**	**.46**
Test Set														
HaRT														
CLS	.23	.22	.22	.44	.46	.45	.85	.55	.85	.51	.51	.51	.34	.38
CLS+Last layer	.23	.20	.21	.43	.48	.45	.86	.84	.85	.50	.50	.50	.33	.37
PCA-Roba														
Curr+Prev+Diff*	**.42**	**.27**	**.33**	**.67**	**.56**	**.61**	**.86**	**.94**	**.90**	**.65**	**.59**	**.61**	**.46**	**.47**

Table 2: Results on internal validation and the test set for task A. IS, IE, and O refer to Switch, Escalation, and Ordinary classes respectively, and P, R and F1 refer to precision, recall, and F1 score. **Best** scores are highlighted. The variants of HaRT (Soni et al., 2022) refer to the fine tuning of the classification layer (CLS) and the last transformer layer (last Layer). The variants of PCA-Roba refer to the Current (Curr), Previous (Prev), and Difference (Diff) between the two on Roberta embeddings of text reduced using PCA. *The PCA-Roba (Curr+Prev+Diff) was turned in late due to technical difficulties.

We adapted HaRT in two ways. First, we try a frozen approach where we train using the message representation output from HaRT but only update weights of the classification layer. We call this approach **HaRT CLS**. Second, we allow a single transformer layer (the topmost layer) to also update its weights during fine-tuning, this variant is called **HaRT CLS+Last Layer**.

RoBERTa. Previous works have shown that contextual embeddings from large pre-trained language models can help improve downstream task performance (Matero et al., 2021a; Bao and Qiao, 2019). However, these models often output embeddings with a large number of dimensions, typically 768 or 1024, which can cause problems when training on small datasets (Li and Eisner, 2019; Bao and Qiao, 2019). Here, we leverage the dimensionality reduction approach proposed by V Ganesan et al. 2021, which suggests using RoBERTa embeddings (Liu et al., 2019) with PCA (Martinsson et al., 2011) to achieve the best performance in low data regime. Further, we incorporate techniques proposed in previous works on suicide risk-level assessment, such as modeling the title and message body of a post separately and concatenating them for a single representation (Matero et al., 2019).

To build our text representations, we extract separate transformer representations for title and body, from the second to last layer of RoBERTa. This allows us to keep highly relevant features, the individual words in the title, from getting underrepresented in the longer text from the body content. We then run our PCA reduction on each representation individually, down to 16 dimensions for title and 128 for the body, then concatenate them into a single representation of 144 dimensions. The number of reduced dimensions for title and body were chosen based on cross validation performance using 16, 64 and 128 dimensions. We observed no improvement in performance when increasing the dimensions for title from 16, but observed degradation in performance when decreasing the dimensions from 128 for the body.

Using dimension reduced RoBERTa (PCA-Roba) embeddings as a base, we build 5 separate models that each use different combinations of feature representations. (1) **Curr** uses only the current message as input features, (2) **Curr+Prev** uses both current and previous message representations concatenated, (3) **Diff** uses the difference in representation between the current and the previous messages as shown in figure 1, (4) **Curr+Diff** uses *diff* concatenated with only the *current*, and (5) **Curr+Prev+Diff** uses the *current*, *previous*, and *difference* representations all concatenated.

All feature representations are fit using a logistic regression model. To the exception of HaRT, experiments were performed using an open

Features	Cross Val F1(macro)	Internal Val F1(macro)
1gram	0.37	0.29
Roba	0.34	0.34
OpenVocab	0.39	**0.60** †
OpenVocab, HypLex	0.40	0.37
Roba, HypLex	0.36	0.35
OpenVocab+Roba, HypLex	**0.42**	0.37

Table 3: Results on the cross validation and internal validation set for task B. **Best scores** are highlighted. All the features were extracted for titles and message body separately. The OpenVocab consists of PCA reductions of the LDA Topics and 1-grams to 32 dimensions each. For Roba, we reduce 768 dimensions to 64 in case of contents and 16 in case of titles. HypLex is a set of 12 theoretical features as explained in §3.1.
† : the drastically high F-1 score is likely from chance due to the very low sample sizes afforded for the user-level task.

source python library for language analysis at scale, DLATK (Schwartz et al., 2017). The design of the library to support multiple levels of analysis for both linguistic and extra-linguistic features facilitated using it for both task A and task B, although the former maps an outcome to each message while the latter maps multiple messages to an outcome.

3.2 Task B

Open-Vocab Features. We explore three representations of a user's language for this task. First, **N grams** are extracted and normalized to obtain the frequencies, from the title and content for each user. The outliers are removed by retaining only the N grams that occur in at least 5% of users' posts. Next, we use the N grams to build **LDA Topics** which are generated using open-source data-driven word clusters Schwartz et al. (2013). These provide 2,000 topics trained on a corpus of 18 million Facebook posts. Each user is represented by the probability of usage for each topic across these 2,000 dimensions. The topic dimensions are then reduced down to 32 using PCA.

Additionally, we again used **PCA-Roba** as described in task A with the same dimension sizes, title/body split, and extraction layer. However, for this task we process all individual messages uttered by a user and average the message representations to build a user representation.

HypLex The HypLex (§3.1) models were run on the N gram counts of the user to obtain the theoretical HypLex features for task B.

We use both Open-Vocab and HypLex features as inputs for a logistic regression model. Internally we tested various combinations of features for this task, but only a single model was selected to be evaluated on the test set.

4 Results

4.1 Task A

As can be seen in Table 1, the mental-health-related hypothesis-driven lexica (HypLex)–including depression, anxiety, anger and loneliness–show high β associations (standardized logistic regression coefficients) with the outcome variables of task A. The 12 HypLex features alone produce a macro F1 of .44 on the internal validation set (Table 2) which demonstrates the power of these machine-learning-based language models learned on person-level survey responses. Throughout, mood 'switches' (from positive to negative and vice versa) where more easily predicted than mood 'escalations.' The absence of language signal related to negative affect (anger, anxiety, stress, depression, loneliness) predicted 'ordinary' mood states, as did the presence of language signal of the three personality traits typically associated with positive affect: extraversion, agreeableness and conscientiousness. Perhaps surprisingly, the language model for the low-arousal negative affect state of loneliness proved to be more predictive of both mood switches and escalations than the language models for high-arousal negative affect states (such as anger, stress, and anxiety).

Generally, the performance of auto-regressive transformer models are poorer than auto-encoder transformer models in classification tasks (V Ganesan et al., 2021; Zhou et al., 2020). However, the results on the internal validation set in Table 2 suggest that HaRT (CLS) performs better than RoBERTa embeddings (PCA-Roba Curr), primarily accounting for the importance of encoding history into text representations, especially for tasks spanning the temporal dimension. However, HaRT CLS+Last layer doesn't seem provide much improvement showing that fine tuning is not of much help. We would like to note that the hyperparameter values were chosen based on values reported in the paper due to the limited availability of time and computational resources.

It is evident from table 2 that the differencing approach of the PCA-Roba embeddings between the current and previous texts (PCA-Roba Diff) gives the best performance in capturing Switches on the internal validation set. However, the difference feature is very poor at capturing the gradual mood change (IE). It was found that 93% of the IE class was predicted as ordinary when using only the difference feature. This could potentially be because the difference in post embeddings for gradual mood changes are much more gradual and smaller, and may be mistaken for no mood changes - whereas the difference in switches have much more obvious and large differences in post embeddings.

The previous text representation in context to the current (PCA-Roba Curr+Prev) vastly improves the model to identify escalations besides improving the detection of switches. Overall the concatenation of Curr, Prev and Diff performs the best by bringing the best out of these individual features.

The strongest baseline from (Tsakalidis et al., 2022a) for task A utilizing tf-idf features trained with a logistic regression scores macro F1 of 0.49 on the test set. All our models perform significantly better than tf-idf features, particularly in capturing the switches by using transformer based embeddings and factoring previous message's representation for modelling the mood change.

4.2 Task B

The results on Table 3 suggests that using dimension-reduced RoBERTa (Roba) does not offer much advantage over dimension-reduced 1-gram features. This is likely due to the availability of small number of training samples where language models have shown to overfit (Bao and Qiao, 2019). The addition LDA Topics improves the performance of both 1gram model (OpenVocab) and the Roba HypLex model (OpenVocab+Roba, HypLex) showing the robustness of the topics trained on large external dataset in such low data regime.

The HypLex features too slightly improve the performance in cross validation. We get the best result in cross validation when we combine all the three – PCA-reduced RoBERTa embeddings + 1grams, PCA-reduced LDA Topics and HypLex features (OpenVocab+Roba, HypLex).

However, in the internal validation, we find that the best performing model was PCA-reduced OpenVocab. Since the performance in the cross validation was similar for all the listed models, we chose OpenVocab for the final predictions for test set on Task B, which scored an F1(macro) of 0.35.

5 Conclusion

We presented two approaches to detecting mental state changes in users through (a) a recurrent transformer model (HaRT) that encodes messages within context of previous ones and (b) a logistic regression model that relies on RoBERTA difference embeddings along with previous and current text representations to capture change in language over time. Compared to using other representation types, such as theoretically motivated (HypLex) or traditional open vocabulary features (N grams, Topics), both approaches saw improved model performance when predicting changes over time. Further, we found that theoretically relevant lexical scores had large associations with the change patterns. It showed emotional stability correlating with no change, and loneliness, depression, stress, anxiety and anger being associated with the mood change.

6 Ethical Consideration

We used publicly available data stripped of identifiable information which was collected in a non-intrusive manner for mental health research. Secure access to the shared task dataset was provided with IRB approval under University of Maryland, College Park protocol 1642625 and approval by the Biomedical and Scientific Research Ethics Committee (BSREC) at the University of Warwick (ethical application reference BSREC 40/19-20). Individuals of the study team who ran the analyses for this work are certified to conduct Human Subject Research and complied with the non-disclosure agreement signed with the dataset providers.

The findings of this work are intended for fellow researchers in Computational Linguistics and Psychology to improve technology for mental health assessments. Around 14 million adults in the United States face severe mental health issues (NIMH, 2022) and a very large part of this is marginalized communities that are underserved (Saraceno et al., 2007). However, given the prevalence of these communities in social media (Center, 2021), technology-enabled solutions can assist in detecting and providing assistance in a timely manner to a more diverse group of individuals. This work is a part of the growing body of mental health research aimed at applications for improving well-being. However, this shouldn't be deployed to use without collaboration of clinical practitioners.

Acknowledgements

The authors are grateful to the anonymous users of Reddit whose data feature in this year's shared task dataset, to the annotators of the data for Task A, to the clinical experts from Bar-Ilan University who annotated the data for Task B, the American Association of Suicidology, to NORC for creating and administering the secure infrastructure and providing researcher support and to UKRI for providing funding to the CLPsych 2022 shared task organisers.

References

Pablo Arbeláez, Michael Maire, Charlotte Fowlkes, and Julien Malik. 2011. Contour detection and hierarchical image segmentation. *IEEE Transactions on Pattern Analysis and Machine Intelligence*, 33:898–916.

Xingce Bao and Qianqian Qiao. 2019. Transfer learning from pre-trained BERT for pronoun resolution. In *Proceedings of the First Workshop on Gender Bias in Natural Language Processing*, pages 82–88, Florence, Italy. Association for Computational Linguistics.

Pew Research Center. 2021. Social media fact sheet.

Munmun De Choudhury, Emre Kıcıman, Mark Dredze, Glen A. Coppersmith, and Mrinal Kumar. 2016. Discovering shifts to suicidal ideation from mental health content in social media. *Proceedings of the 2016 CHI Conference on Human Factors in Computing Systems*.

Glen Coppersmith, Mark Dredze, and Craig Harman. 2014. Quantifying mental health signals in Twitter. In *Proceedings of the Workshop on Computational Linguistics and Clinical Psychology: From Linguistic Signal to Clinical Reality*, pages 51–60, Baltimore, Maryland, USA. Association for Computational Linguistics.

Glen A. Coppersmith. 2022. Digital life data in the clinical whitespace. *Current Directions in Psychological Science*, 31:34 - 40.

Sharath Chandra Guntuku, Anneke Buffone, Kokil Jaidka, Johannes C. Eichstaedt, and Lyle H. Ungar. 2019a. Understanding and measuring psychological stress using social media. *Proceedings of the International AAAI Conference on Web and Social Media*, 13(01):214–225.

Sharath Chandra Guntuku, Elissa V Klinger, Haley J McCalpin, Lyle H Ungar, David A Asch, and Raina M Merchant. 2021. Social media language of healthcare super-utilizers. *NPJ digital medicine*, 4(1):1–6.

Sharath Chandra Guntuku, Daniel Preotiuc-Pietro, Johannes C Eichstaedt, and Lyle H Ungar. 2019b. What twitter profile and posted images reveal about depression and anxiety. In *Proceedings of the international AAAI conference on web and social media*, volume 13, pages 236–246.

Sharath Chandra Guntuku, Rachelle Schneider, Arthur Pelullo, Jami Young, Vivien Wong, Lyle Ungar, Daniel Polsky, Kevin G Volpp, and Raina Merchant. 2019c. Studying expressions of loneliness in individuals using twitter: an observational study. *BMJ Open*, 9(11).

Kishaloy Halder, Lahari Poddar, and Min-Yen Kan. 2017. Modeling temporal progression of emotional status in mental health forum: A recurrent neural net approach. In *Proceedings of the 8th Workshop on Computational Approaches to Subjectivity, Sentiment and Social Media Analysis*, pages 127–135, Copenhagen, Denmark. Association for Computational Linguistics.

Xiang Lisa Li and Jason Eisner. 2019. Specializing word embeddings (for parsing) by information bottleneck. In *Proceedings of the 2019 Conference on Empirical Methods in Natural Language Processing and the 9th International Joint Conference on Natural Language Processing (EMNLP-IJCNLP)*, pages 2744–2754, Hong Kong, China. Association for Computational Linguistics.

Yinhan Liu, Myle Ott, Naman Goyal, Jingfei Du, Mandar Joshi, Danqi Chen, Omer Levy, Mike Lewis, Luke Zettlemoyer, and Veselin Stoyanov. 2019. Roberta: A robustly optimized bert pretraining approach.

D. Losada and F. Crestani. 2016. A test collection for research on depression and language use. In *Proc. of Experimental IR Meets Multilinguality, Multimodality, and Interaction, 7th International Conference of the CLEF Association, CLEF 2016*, pages 28–39, Evora, Portugal.

David E. Losada, Fabio Crestani, and Javier Parapar. 2020. Overview of eRisk 2020: Early Risk Prediction on the Internet. In *Experimental IR Meets Multilinguality, Multimodality, and Interaction Proceedings of the Eleventh International Conference of the CLEF Association (CLEF 2020)*. Springer International Publishing.

Veronica Lynn, Niranjan Balasubramanian, and H. Andrew Schwartz. 2020. Hierarchical modeling for user personality prediction: The role of message-level attention. In *Proceedings of the 58th Annual Meeting of the Association for Computational Linguistics*, pages 5306–5316, Online. Association for Computational Linguistics.

Veronica Lynn, Alissa Goodman, Kate Niederhoffer, Kate Loveys, Philip Resnik, and H. Andrew Schwartz. 2018. CLPsych 2018 shared task: Predicting current and future psychological health from childhood essays. In *Proceedings of the Fifth Workshop on Computational Linguistics and Clinical Psychology: From Keyboard to Clinic*, pages 37–46, New Orleans, LA. Association for Computational Linguistics.

Per-Gunnar Martinsson, Vladimir Rokhlin, and Mark Tygert. 2011. A randomized algorithm for the decomposition of matrices. *Applied and Computational Harmonic Analysis*, 30:47–68.

Matthew Matero, Albert Hung, and H. Andrew Schwartz. 2021a. Evaluating contextual embeddings and their extraction layers for depression assessment.

Matthew Matero, Akash Idnani, Youngseo Son, Salvatore Giorgi, Huy Vu, Mohammad Zamani, Parth Limbachiya, Sharath Chandra Guntuku, and H. Andrew Schwartz. 2019. Suicide risk assessment with multi-level dual-context language and BERT. In *Proceedings of the Sixth Workshop on Computational Linguistics and Clinical Psychology*, pages 39–44, Minneapolis, Minnesota. Association for Computational Linguistics.

Matthew Matero and H. Andrew Schwartz. 2020. Autoregressive affective language forecasting: A self-supervised task. In *Proceedings of the 28th International Conference on Computational Linguistics*, pages 2913–2923, Barcelona, Spain (Online). International Committee on Computational Linguistics.

Matthew Matero, Nikita Soni, Niranjan Balasubramanian, and H. Andrew Schwartz. 2021b. MeLT: Message-level transformer with masked document representations as pre-training for stance detection. In *Findings of the Association for Computational Linguistics: EMNLP 2021*, pages 2959–2966, Punta Cana, Dominican Republic. Association for Computational Linguistics.

NIMH. 2022. Mental Illness Statistics.

Gregory Park, H Andrew Schwartz, Johannes C Eichstaedt, Margaret L Kern, Michal Kosinski, David J Stillwell, Lyle H Ungar, and Martin EP Seligman. 2015. Automatic personality assessment through social media language. *Journal of personality and social psychology*, 108(6):934.

Alec Radford, Jeff Wu, Rewon Child, David Luan, Dario Amodei, and Ilya Sutskever. 2019. Language models are unsupervised multitask learners.

Maarten Sap, Gregory Park, Johannes Eichstaedt, Margaret Kern, David Stillwell, Michal Kosinski, Lyle Ungar, and Hansen Andrew Schwartz. 2014. Developing age and gender predictive lexica over social media. In *Proceedings of the 2014 Conference on Empirical Methods in Natural Language Processing (EMNLP)*, pages 1146–1151, Doha, Qatar. Association for Computational Linguistics.

Benedetto Saraceno, Mark van Ommeren, Rajaie Batniji, Alex Cohen, Oye Gureje, John Mahoney, Devi Sridhar, and Chris Underhill. 2007. Barriers to improvement of mental health services in low-income and middle-income countries. *The Lancet*, 370(9593):1164–1174.

Ramit Sawhney, Harshit Joshi, Lucie Flek, and Rajiv Ratn Shah. 2021. PHASE: Learning emotional phase-aware representations for suicide ideation detection on social media. In *Proceedings of the 16th Conference of the European Chapter of the Association for Computational Linguistics: Main Volume*, pages 2415–2428, Online. Association for Computational Linguistics.

Ramit Sawhney, Harshit Joshi, Saumya Gandhi, and Rajiv Ratn Shah. 2020. A time-aware transformer based model for suicide ideation detection on social media. In *Proceedings of the 2020 Conference on Empirical Methods in Natural Language Processing (EMNLP)*, pages 7685–7697, Online. Association for Computational Linguistics.

H. Andrew Schwartz, Johannes Eichstaedt, Margaret L. Kern, Gregory Park, Maarten Sap, David Stillwell, Michal Kosinski, and Lyle Ungar. 2014. Towards assessing changes in degree of depression through Facebook. In *Proceedings of the Workshop on Computational Linguistics and Clinical Psychology: From Linguistic Signal to Clinical Reality*, pages 118–125, Baltimore, Maryland, USA. Association for Computational Linguistics.

H. Andrew Schwartz, Johannes C. Eichstaedt, Margaret L. Kern, Lukasz Dziurzynski, Stephanie M. Ramones, Megha Agrawal, Achal Shah, Michal Kosinski, David Stillwell, Martin E. P. Seligman, and Lyle H. Ungar. 2013. Personality, gender, and age in the language of social media: The open-vocabulary approach. *PLOS ONE*, 8(9):1–16.

H. Andrew Schwartz, Salvatore Giorgi, Maarten Sap, Patrick Crutchley, Johannes Eichstaedt, and Lyle Ungar. 2017. Dlatk: Differential language analysis toolkit. In *Proceedings of the 2017 Conference on Empirical Methods in Natural Language Processing: System Demonstrations*, pages 55–60. Association for Computational Linguistics.

Han-Chin Shing, Suraj Nair, Ayah Zirikly, Meir Friedenberg, Hal Daumé III, and Philip Resnik. 2018. Expert, crowdsourced, and machine assessment of suicide risk via online postings. In *Proceedings of the Fifth Workshop on Computational Linguistics and Clinical Psychology: From Keyboard to Clinic*, pages 25–36, New Orleans, LA. Association for Computational Linguistics.

Youngseo Son, Sean A. P. Clouston, Roman Kotov, Johannes C. Eichstaedt, Evelyn J. Bromet, Benjamin J. Luft, and H. Andrew Schwartz. 2021. World trade center responders in their own words: predicting ptsd symptom trajectories with ai-based language analyses of interviews. *Psychological Medicine*, page 1-9.

Nikita Soni, Matthew Matero, Niranjan Balasubramanian, and H. Schwartz. 2022. Human language modeling. In *Findings of the Association for Computational Linguistics: ACL 2022*, pages 622–636, Dublin, Ireland. Association for Computational Linguistics.

Adam Tsakalidis, Jenny Chim, Iman Munire Bilal, Ayah Zirikly, Dana Atzil-Slonim, Federico Nanni, Philip Resnik, Manas Gaur, Kaushik Roy, Becky Inkster, Jeff Leintz, and Maria Liakata. 2022a. Overview of the CLPsych 2022 shared task: Capturing moments of change in longitudinal user posts. In *Proceedings of the Eighth Workshop on Computational Linguistics and Clinical Psychology: Mental Health in the Face of Change*.

Adam Tsakalidis, Federico Nanni, Anthony Hills, Jenny Chim, Jiayu Song, and Maria Liakata. 2022b. Identifying moments of change from longitudinal user text. In *Proceedings of the 60th Annual Meeting of the Association for Computational Linguistics (Volume 1: Long Papers)*, pages 4647–4660, Dublin, Ireland. Association for Computational Linguistics.

Adithya V Ganesan, Matthew Matero, Aravind Reddy Ravula, Huy Vu, and H. Andrew Schwartz. 2021. Empirical evaluation of pre-trained transformers for human-level NLP: The role of sample size and dimensionality. In *Proceedings of the 2021 Conference of the North American Chapter of the Association for Computational Linguistics: Human Language Technologies*, pages 4515–4532, Online. Association for Computational Linguistics.

Xuhui Zhou, Yue Zhang, Leyang Cui, and Dandan Huang. 2020. Evaluating commonsense in pre-trained language models. In *Proceedings of the AAAI Conference on Artificial Intelligence*, volume 34, pages 9733–9740.

Ayah Zirikly, Philip Resnik, Özlem Uzuner, and Kristy Hollingshead. 2019. CLPsych 2019 shared task: Predicting the degree of suicide risk in Reddit posts. In *Proceedings of the Sixth Workshop on Computational Linguistics and Clinical Psychology*, pages 24–33, Minneapolis, Minnesota. Association for Computational Linguistics.

Author Index